Lada Owners Workshop Manual

J H Haynes
Member of the Guild of Motoring Writers

Marcus S Daniels and Peter G Strasman

Models covered
All Lada Saloon, Estate and Van models, including Riva
1198 cc, 1294 cc, 1452 cc & 1570 cc

Does not cover Niva or Samara hatchback variants

(413-2T7)   ABC

Haynes Publishing Group
Sparkford Nr Yeovil
Somerset BA22 7JJ England

Haynes Publications, Inc
861 Lawrence Drive
Newbury Park
California 91320 USA

Acknowledgements

Thanks are due to the Champion Sparking Plug Company Limited, who supplied the illustrations showing spark plug conditions, to Holt Lloyd Limited who supplied the illustrations showing bodywork repair, and to Duckhams Oils, who provided lubrication data. Thanks are also due to John Clarke of Lada Cars (UK) Limited for the supply of technical information, Sykes-Pickavant Limited who provided some of the workshop tools, and to all those people at Sparkford who assisted in the production of this manual.

© **Haynes Publishing Group 1991**

A book in the **Haynes Owners Workshop Manual Series**

Printed by J. H. Haynes & Co. Ltd., Sparkford, Nr Yeovil, Somerset BA22 7JJ, England

All rights reserved. No part of this book may be reproduced or transmitted in any form or by any means, electronic or mechanical, including photocopying, recording or by any information storage or retrieval system, without permission in writing from the copyright holder.

ISBN 1 85010 756 4

British Library Cataloguing in Publication Data
Daniels, Marcus S. (Marcus Stewart) 1942-
Lada 1240, 1300, 1500 & 1600 owners workshop manual.-5th ed.
 1. Cars. Maintenance
 I. Title II. Strasman, Peter G. 1923- III. Haynes, J. H. (John Harold). Lada owners workshop manual IV. Series
629.28722

ISBN 1-85010-756-4

Whilst every care is taken to ensure that the information in this manual is correct, no liability can be accepted by the authors or publishers for loss, damage or injury caused by any errors in, or omissions from, the information given.

Restoring and Preserving our Motoring Heritage

Few people can have had the luck to realise their dreams to quite the same extent and in such a remarkable fashion as John Haynes, Founder and Chairman of the Haynes Publishing Group.

Since 1965 his unique approach to workshop manual publishing has proved so successful that millions of Haynes Manuals are now sold every year throughout the world, covering literally thousands of different makes and models of cars, vans and motorcycles.

A continuing passion for cars and motoring led to the founding in 1985 of a Charitable Trust dedicated to the restoration and preservation of our motoring heritage. To inaugurate the new Museum, John Haynes donated virtually his entire private collection of 52 cars.

Now with an unrivalled international collection of over 210 veteran, vintage and classic cars and motorcycles, the Haynes Motor Museum in Somerset is well on the way to becoming one of the most interesting Motor Museums in the world.

A 70 seat video cinema, a cafe and an extensive motoring bookshop, together with a specially constructed one kilometre motor circuit, make a visit to the Haynes Motor Museum a truly unforgettable experience.

Every vehicle in the museum is preserved in as near as possible mint condition and each car is run every six months on the motor circuit.

Enjoy the picnic area set amongst the rolling Somerset hills. Peer through the William Morris workshop windows at cars being restored, and browse through the extensive displays of fascinating motoring memorabilia.

From the 1903 Oldsmobile through such classics as an MG Midget to the mighty 'E' Type Jaguar, Lamborghini, Ferrari Berlinetta Boxer, and Graham Hill's Lola Cosworth, there is something for everyone, young and old alike, at this Somerset Museum.

Haynes Motor Museum

Situated mid-way between London and Penzance, the Haynes Motor Museum is located just off the A303 at Sparkford, Somerset (home of the Haynes Manual) and is open to the public 7 days a week all year round, except Christmas Day and Boxing Day.

Contents

	Page
Acknowledgements	2
About this manual	5
Introduction to the Lada	5
Buying spare parts and vehicle identification numbers	6
Tools and working facilities	7
Jacking and towing	9
Recommended lubricants and fluids	10
Routine maintenance *(also see Chapter 13, page 159)*	11
Chapter 1 Engine *(also see Chapter 13, page 159)*	13
Chapter 2 Cooling system *(also see Chapter 13, page 159)*	38
Chapter 3 Fuel and exhaust systems *(also see Chapter 13, page 159)*	44
Chapter 4 Ignition system *(also see Chapter 13, page 159)*	55
Chapter 5 Clutch *(also see Chapter 13, page 159)*	61
Chapter 6 Gearbox *(also see Chapter 13, page 159)*	67
Chapter 7 Propeller shaft *(also see Chapter 13, page 159)*	81
Chapter 8 Rear axle *(also see Chapter 13, page 159)*	85
Chapter 9 Braking system *(also see Chapter 13, page 159)*	93
Chapter 10 Electrical system *(also see Chapter 13, page 159)*	105
Chapter 11 Suspension and steering *(also see Chapter 13, page 159)*	134
Chapter 12 Bodywork and underframe *(also see Chapter 13, page 159)*	149
Chapter 13 Supplement: Revisions and information on later models	159
Fault diagnosis	214
General dimensions, weights and capacities	218
General repair procedure	219
Safety first!	220
Conversion factors	221
Index	222

Spark plug condition and bodywork repair colour pages between pages 32 and 33

Lada 1200 Saloon

Lada 1500 ES Estate

About this manual

Its aim

The aim of this manual is to help you get the best value from your car. It can do so in several ways. It can help you decide what work must be done (even should you choose to get it done by a garage), provide information on routine maintenance and servicing, and give a logical course of action and diagnosis when random faults occur. However, it is hoped that you will use the manual by tackling the work yourself. On simpler jobs it may even be quicker than booking the car into a garage and going there twice to leave and collect it. Perhaps most important, a lot of money can be saved by avoiding the costs the garage must charge to cover its labour and overheads.

The manual has drawings and descriptions to show the function of the various components so that their layout can be understood. Then the tasks are described and photographed in a step-by-step sequence so that even a novice can do the work.

Its arrangement

The manual is divided into thirteen Chapters, each covering a logical sub-division of the vehicle. The Chapters are each divided into Sections, numbered with single figures, eg 5; and the Sections into paragraphs (or sub-sections), with decimal numbers following on from the Section they are in, eg 5.1, 5.2, 5.3 etc.

It is freely illustrated, especially in those parts where there is a detailed sequence of operations to be carried out. There are two forms of illustration: figures and photographs. The figures are numbered in sequence with decimal numbers, according to their position in the Chapter -- eg Fig. 6.4 is the fourth drawing/illustration in Chapter 6. Photographs carry the same number (either individually or in related groups) as the Section or sub-section to which they relate.

There is an alphabetical index at the back of the manual as well as a contents list at the front. Each Chapter is also preceded by its own individual contents list.

References to the 'left' or 'right' of the vehicle are in the sense of a person in the driver's seat facing forwards.

Unless otherwise stated, nuts and bolts are removed by turning anti-clockwise, and tightened by turning clockwise.

Vehicle manufacturers continually make changes to specifications and recommendations, and these, when notified, are incorporated into our manuals at the earliest opportunity.

Whilst every care is taken to ensure that the information in this manual is correct, no liability can be accepted by the authors or publishers for loss, damage or injury caused by any errors in, or omissions from, the information given.

Introduction to the Lada

The Lada 1200 Saloon and Estate models were introduced in the UK in May 1974 and caused quite a few raised eyebrows. Not because they possessed any startling design innovations but because of the extremely low price. This price included a great number of standard features that on most cars would be optional extras and therefore offered extremely good value for money.

The bodyshell of the Lada is based on the Fiat 124 as are some of the mechanical components. The engine differs in design however, by having a five-bearing crankshaft and an overhead camshaft that operates the valves via adjustable rockers.

The mechanical layout is fairly conventional, comprising a front mounted engine driving the rear wheels through a manual gearbox, propeller shaft and live rear axle. The front suspension is of the independent coil spring and wishbone type, whilst the rear axle is supported on coil springs and located by four suspension arms. Telescopic shock absorbers are used all round. An anti-roll bar is fitted at the front whilst a Panhard rod is located transversely at the rear.

An ES version of the 1200 was offered in September 1976 and featured a vinyl roof, body coach lines, heated rear window and carpets, all as standard equipment. In May 1976 the 1500 saloon and estate models became available; the 1500 saloon being the first model to incorporate twin headlamps. The 1300 ES was introduced in November 1977 as a more refined and powerful version of the 1200. In the autumn of 1978 the 1600 and 1600 ES models were launched. Apart from having a larger engine, the cars were refinished internally and externally by adding more chrome trim and vinyl roof covering. Mechanically, the gear ratios were modified and certain other minor changes in component design were incorporated.

All saloon models were redesignated Riva during 1983 and 1984, and the 1300 engine was modified to incorporate a timing belt instead of a timing chain.

There is no doubt that the complete Lada range offers rugged vehicles at extremely competitive prices. The design retains many attractive features for the enthusiastic home mechanic including, on early models, a starting handle, screw-type valve clearance adjusters, and gearbox and rear axle oil drain plugs. Altogether the Lada is a reliable vehicle of heavy duty construction from which a few early teething troubles have been eradicated.

Buying spare parts and vehicle identification numbers

Buying spare parts

Spare parts are available from many sources, for example: Lada dealers, other garages and accessory shops, and motor factors. Our advice regarding spare part sources is as follows:

Officially appointed Lada garages – This is the best source of parts which are peculiar to your car and are otherwise not generally available (eg; complete cylinder heads, internal gearbox components, badges, interior trim etc). It is also the only place at which you should have repairs carried out if your car is still under warranty – non-Lada components may invalidate the warranty. To be sure of obtaining the correct parts it will always be necessary to give the storeman your car's vehicle identification number, and if possible, to take the old part along for positive identification. It obviously makes good sense to go straight to the specialists on your car for this type of part for they are best equipped to supply you.

Other garages and accessory shops – These are often very good places to buy materials and components needed for the maintenance of your car (eg; spark plugs, bulbs, fan belts, oils and greases, filler paste etc.). They also sell general accessories, usually have convenient opening hours, charge reasonable prices and can often be found not far from home.

Motor factors – Good factors will stock all the more important components which wear out relatively quickly (eg; clutch components, pistons, valves, exhaust systems, brake cylinders/pipes/hoses/seals/shoes and pads etc). Motor factors will often provide new or reconditioned components on a part exchange basis – this can save a considerable amount of money.

Vehicle identification numbers

Although many individual parts, and in some cases sub-assemblies, fit a number of different models it is dangerous to assume that just because they look the same, they are the same. Differences are not always easy to detect except by serial numbers. Make sure therefore, that the appropriate identity number for the model or sub-assembly is known and quoted when a spare part is ordered.

The vehicle identification plate is located on the forward engine compartment crossmember adjacent to the right-hand bonnet hinge. The plate gives the car model number and type of engine. On later models it is located next to the body serial number.

The body serial number is stamped on the right-hand side of the engine compartment rear bulkhead. This number should be quoted when ordering body parts.

The engine number is stamped on the left-hand side of the cylinder block above the oil filter and should be quoted when obtaining new engine parts.

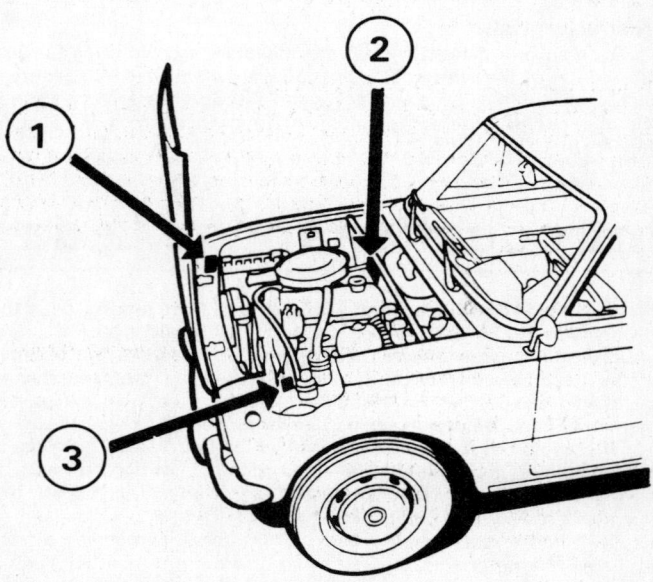

Location of identification numbers

1 Vehicle identification plate
2 Body serial number
3 Engine number

Tools and working facilities

Introduction

A selection of good tools is a fundamental requirement for anyone contemplating the maintenance and repair of a motor vehicle. For the owner who does not possess any, their purchase will prove a considerable expense, offsetting some of the savings made by doing-it-yourself. However, provided that the tools purchased meet the relevant national safety standards and are of good quality, they will last for many years and prove an extremely worthwhile investment.

To help the average owner to decide which tools are needed to carry out the various tasks detailed in this manual, we have compiled three lists of tools under the following headings: *Maintenance and minor repair*, *Repair and overhaul*, and *Special*. The newcomer to practical mechanics should start off with the *Maintenance and minor repair* tool kit and confine himself to the simpler jobs around the vehicle. Then, as his confidence and experience grows, he can undertake more difficult tasks, buying extra tools as, and when, they are needed. In this way, a *Maintenance and minor repair* tool kit can be built-up into a *Repair and overhaul* tool kit over a considerable period of time without any major cash outlays. The experienced do-it-yourselfer will have a tool kit good enough for most repair and overhaul procedures and will add tools from the *Special* category when he feels the expense is justified by the amount of use to which these tools will be put.

It is obviously not possible to cover the subject of tools fully here. For those who wish to learn more about tools and their use there is a book entitled *How to Choose and Use Car Tools* available from the publishers of this manual.

Maintenance and minor repair tool kit

The tools given in this list should be considered as a minimum requirement if routine maintenance, servicing and minor repair operations are to be undertaken. We recommend the purchase of combination spanners (ring one end, open-ended the other); although more expensive than open-ended ones, they do give the advantages of both types of spanner. All nuts, bolts, screws and threads on the Lada are to metric standards.

Combination spanners - 10, 11, 13, 14, 17 mm
Adjustable spanner - 9 inch
Engine sump/gearbox/rear axle drain plug key (where applicable)
Spark plug spanner (with rubber insert)
Spark plug gap adjustment tool
Set of feeler gauges
Brake adjuster spanner (where applicable)
Brake bleed nipple spanner
Screwdriver - 4 in long x $\frac{1}{4}$ in dia (flat blade)
Screwdriver - 4 in long x $\frac{1}{4}$ in dia (cross blade)
Combination pliers - 6 inch
Hacksaw, junior
Tyre pump
Tyre pressure gauge
Grease gun (where applicable)
Oil can
Fine emery cloth (1 sheet)
Wire brush (small)
Funnel (medium size)

Repair and overhaul tool kit

These tools are virtually essential for anyone undertaking any major repairs to a motor vehicle, and are additional to those given in the *Maintenance and minor repair* list. Included in this list is a comprehensive set of sockets. Although these are expensive they will be found invaluable as they are so versatile - particularly if various drives are included in the set. We recommend the $\frac{1}{2}$ in square-drive type, as this can be used with most proprietary torque wrenches. If you cannot afford a socket set, even bought piecemeal, then inexpensive tubular box spanners are a useful alternative.

The tools in this list will occasionally need to be supplemented by tools from the *Special* list.

Sockets (or box spanners) to cover range in previous list
Reversible ratchet drive (for use with sockets)
Extension piece, 10 inch (for use with sockets)
Universal joint (for use with sockets)
Torque wrench (for use with sockets)
'Mole' wrench - 8 inch
Ball pein hammer
Soft-faced hammer, plastic or rubber
Screwdriver - 6 in long x $\frac{5}{16}$ in dia (flat blade)
Screwdriver - 2 in long x $\frac{5}{16}$ in square (flat blade)
Screwdriver - 1$\frac{1}{2}$ in long x $\frac{1}{4}$ in dia (cross blade)
Screwdriver - 3 in long x $\frac{1}{8}$ in dia (electricians)
Pliers - electricians side cutters
Pliers - needle nosed
Pliers - circlip (internal and external)
Cold chisel - $\frac{1}{2}$ inch
Scriber (this can be made by grinding the end of a broken hacksaw blade)
Scraper (this can be made by flattening and sharpening one end of a piece of copper pipe)
Centre punch
Pin punch
Hacksaw
Valve grinding tool
Steel rule/straight edge
Allen keys
Selection of files
Wire brush (large)
Axle-stands
Jack (strong scissor or hydraulic type)

Special tools

The tools in this list are those which are not used regularly, are expensive to buy, or which need to be used in accordance with their manufacturers' instructions. Unless relatively difficult mechanical jobs are undertaken frequently, it will not be economic to buy many of these tools. Where this is the case, you could consider clubbing together with friends (or a motorists' club) to make a joint purchase, or borrowing the tools against a deposit from a local garage or tool hire specialist.

The following list contains only those tools and instruments freely available to the public, and not those special tools produced by the vehicle manufacturer specifically for its dealer network. You will find occasional references to these manufacturers' special tools in the text of this manual. Generally, an alternative method of doing the job without the vehicle manufacturer's special tool is given. However, sometimes, there is no alternative to using them. Where this is the case and the relevant tool cannot be bought or borrowed you will have to entrust the work to a franchised garage.

Valve spring compressor
Piston ring compressor
Balljoint separator
Universal hub/bearing puller
Impact screwdriver
Micrometer and/or vernier gauge
Dial gauge
Stroboscopic timing light
Dwell angle meter/tachometer
Universal electrical multi-meter
Cylinder compression gauge
Lifting tackle
Trolley jack
Light with extension lead

Buying tools

For practically all tools, a tool factor is the best source since he will have a very comprehensive range compared with the average garage or accessory shop. Having said that, accessory shops often offer excellent quality tools at discount prices, so it pays to shop around.

There are plenty of good tools around at reasonable prices, but always aim to purchase items which meet the relevant national safety standards. If in doubt, ask the proprietor or manager of the shop for advice before making a purchase.

Care and maintenance of tools

Having purchased a reasonable tool kit, it is necessary to keep the tools in a clean serviceable condition. After use, always wipe off any dirt, grease and metal particles using a clean, dry cloth, before putting the tools away. Never leave them lying around after they have been used. A simple tool rack on the garage or workshop wall, for items such as screwdrivers and pliers is a good idea. Store all normal spanners and sockets in a metal box. Any measuring instruments, gauges, meters, etc, must be carefully stored where they cannot be damaged or become rusty.

Take a little care when tools are used. Hammer heads inevitably become marked and screwdrivers lose the keen edge on their blades from time to time. A little timely attention with emery cloth or a file will soon restore items like this to a good serviceable finish.

Working facilities

Not to be forgotten when discussing tools, is the workshop itself. If anything more than routine maintenance is to be carried out, some form of suitable working area becomes essential.

It is appreciated that many an owner mechanic is forced by circumstances to remove an engine or similar item, without the benefit of a garage or workshop. Having done this, any repairs should always be done under the cover of a roof.

Wherever possible, any dismantling should be done on a clean flat workbench or table at a suitable working height.

Any workbench needs a vice: one with a jaw opening of 4 in (100 mm) is suitable for most jobs. As mentioned previously, some clean dry storage space is also required for tools, as well as the lubricants, cleaning fluids, touch-up paints and so on which become necessary.

Another item which may be required, and which has a much more general usage, is an electric drill with a chuck capacity of at least $\frac{5}{16}$ in (8 mm). This, together with a good range of twist drills, is virtually essential for fitting accessories such as mirrors and reversing lights.

Last, but not least, always keep a supply of old newspapers and clean, lint-free rags available, and try to keep any working area as clean as possible.

Spanner jaw gap comparison table

Jaw gap (in)	Spanner size
0.250	$\frac{1}{4}$ in AF
0.275	7 mm
0.312	$\frac{5}{16}$ in AF
0.315	8 mm
0.344	$\frac{11}{32}$ in AF; $\frac{1}{8}$ in Whitworth
0.354	9 mm
0.375	$\frac{3}{8}$ in AF
0.393	10 mm
0.433	11 mm
0.437	$\frac{7}{16}$ in AF
0.445	$\frac{3}{16}$ in Whitworth; $\frac{1}{4}$ in BSF
0.472	12 mm
0.500	$\frac{1}{2}$ in AF
0.512	13 mm
0.525	$\frac{1}{4}$ in Whitworth; $\frac{5}{16}$ in BSF
0.551	14 mm
0.562	$\frac{9}{16}$ in AF
0.590	15 mm
0.600	$\frac{5}{16}$ in Whitworth; $\frac{3}{8}$ in BSF
0.625	$\frac{5}{8}$ in AF
0.629	16 mm
0.669	17 mm
0.687	$\frac{11}{16}$ in AF
0.708	18 mm
0.710	$\frac{3}{8}$ in Whitworth; $\frac{7}{16}$ in BSF
0.748	19 mm
0.750	$\frac{3}{4}$ in AF
0.812	$\frac{13}{16}$ in AF
0.820	$\frac{7}{16}$ in Whitworth; $\frac{1}{2}$ in BSF
0.866	22 mm
0.875	$\frac{7}{8}$ in AF
0.920	$\frac{1}{2}$ in Whitworth; $\frac{9}{16}$ in BSF
0.937	$\frac{15}{16}$ in AF
0.944	24 mm
1.000	1 in AF
1.010	$\frac{9}{16}$ in Whitworth; $\frac{5}{8}$ in BSF
1.023	26 mm
1.062	$1\frac{1}{16}$ in AF; 27 mm
1.100	$\frac{5}{8}$ in Whitworth; $\frac{11}{16}$ in BSF
1.125	$1\frac{1}{8}$ in AF
1.181	30 mm
1.200	$\frac{11}{16}$ in Whitworth; $\frac{3}{4}$ in BSF
1.250	$1\frac{1}{4}$ in AF
1.259	32 mm
1.300	$\frac{3}{4}$ in Whitworth; $\frac{7}{8}$ in BSF
1.312	$1\frac{5}{16}$ in AF
1.390	$\frac{13}{16}$ in Whitworth; $\frac{15}{16}$ in BSF
1.417	36 mm
1.437	$1\frac{7}{16}$ in AF
1.480	$\frac{7}{8}$ in Whitworth; 1 in BSF
1.500	$1\frac{1}{2}$ in AF
1.574	40 mm; $\frac{15}{16}$ in Whitworth
1.614	41 mm
1.625	$1\frac{5}{8}$ in AF
1.670	1 in Whitworth; $1\frac{1}{8}$ in BSF
1.687	$1\frac{11}{16}$ in AF
1.811	46 mm
1.812	$1\frac{13}{16}$ in AF
1.860	$1\frac{1}{8}$ in Whitworth; $1\frac{1}{4}$ in BSF
1.875	$1\frac{7}{8}$ in AF
1.968	50 mm
2.000	2 in AF
2.050	$1\frac{1}{4}$ in Whitworth; $1\frac{3}{8}$ in BSF
2.165	55 mm
2.362	60 mm

Jacking and towing

If it is necessary to tow the car, the tow-rope should be attached to the front axle. If towing another car, attach the rope to the rear axle.

For wheel changing or servicing purposes, the car can be raised using the jack supplied with the car or preferably an hydraulic trolley jack. When using the latter type, place the jack beneath the rear axle casing when raising the rear end of the car and beneath the front suspension crossmember if raising the front end. In all cases the car must be supported on axle stands or concrete blocks before starting to work beneath it. **Note**: *Never rely on the jack alone for support.*

Sill jacking points

Raising front of vehicle under crossmember

Raising rear of vehicle under rear axle casing

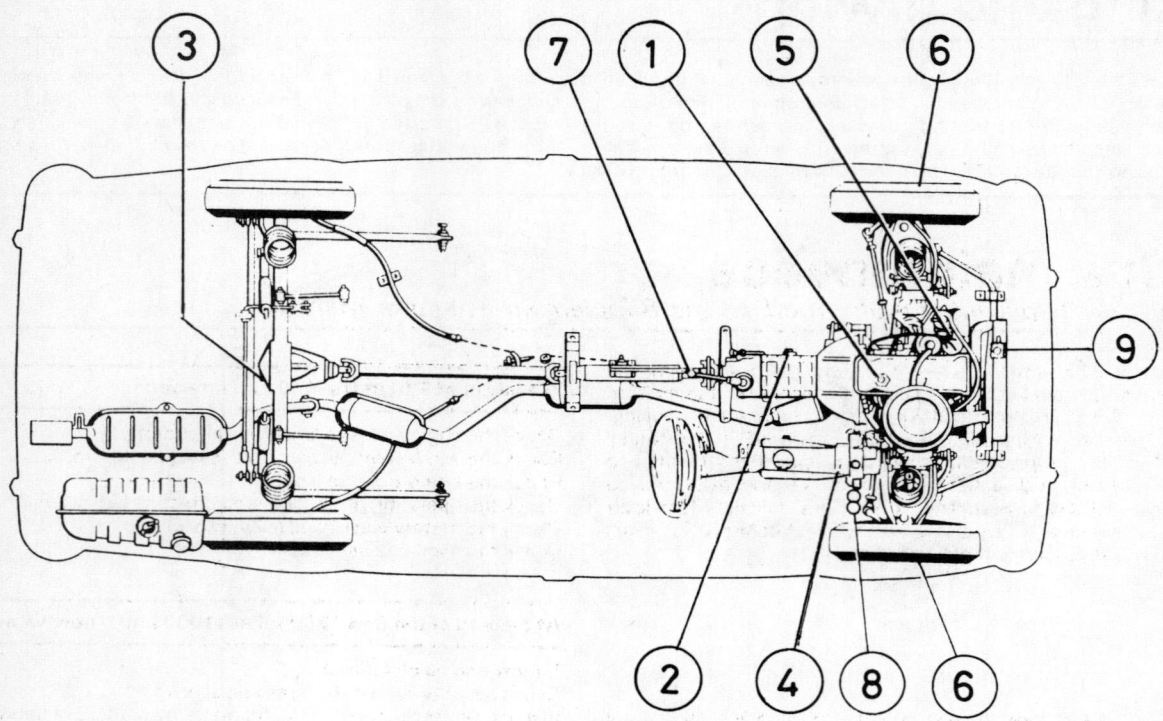

Recommended lubricants and fluids

Component or system	Lubricant type/specification	Duckhams recommendation
1 Engine	Multigrade engine oil, viscosity range SAE 10W/30 to 20W/50, to API SE or SF	Duckhams QXR, Hypergrade, or 10W/40 Motor Oil
2 Gearbox	Hypoid gear oil, viscosity SAE 90 or 75W/90 to API GL5	Duckhams Hypoid 75W/90S
3 Rear axle	Hypoid gear oil, viscosity SAE 90 or 75W/90 to API GL5	Duckhams Hypoid 75W/90S
4 Steering box	Hypoid gear oil, viscosity SAE 90 or 75W/90 to API GL5	Duckhams Hypoid 75W/90S
5 Distributor	Multigrade engine oil, viscosity range SAE 10W/30 to 20W/50, to API SE or SF	Duckhams QXR, Hypergrade, or 10W/40 Motor Oil
6 Front wheel bearings	Multi-purpose lithium-based grease, to NLGI No 3	Duckhams LB 10
7 Propeller shaft splines	Multi-purpose lithium-based grease with molybdenum disulphide, to NLGI No 2	Duckhams LBM 10
8 Brake and clutch reservoirs	Hydraulic fluid to SAE J1703, or FMVSS 116 DOT 3 or DOT 4	Duckhams Universal Brake and Clutch Fluid
9 Cooling system	Ethylene glycol based antifreeze, to BS3151, 3152 or 6580	Duckhams Universal Antifreeze and Summer Coolant

Routine maintenance
For information applicable to later models, see Supplement at end of manual

Maintenance is essential for ensuring safety and desirable for the purpose of getting the best in terms of performance and economy from your car. Over the years the need for periodic lubrication – oiling, greasing and so on – has been drastically reduced if not totally eliminated. This has unfortunately tended to lead some owners to think that because no such action is required, components either no longer exist, or will last forever. This is a serious delusion. It follows therefore that the largest initial element of maintenance is visual examination. This may lead to repairs or renewals.

Note: *The right to withdraw the warranty on the vehicle is reserved if the services, as specified in the vehicle service book, have not been carried out at the specified intervals and/or mileages by a Satra authorised Lada dealer/distributor and the appropriate service counterfoil dated, stamped and the mileage entered accordingly.*

Weekly or at 250 miles (400 km) intervals

Check the engine oil level and top-up (photo).
Check the windscreen washer fluid level and top-up.
Check the tyre pressures (cold).
Check the operation of all lights and direction indicators.
Check the battery electrolyte level (photo).
Check the coolant level.

At the end of the first 1200 miles (1900 km) – new vehicles

Renew engine oil (photo).
Check security of all nuts, bolts and screws.
Inspect all mechanical units and brake hydraulic assemblies and pipe lines for leaks.
Check oil level in gearbox (photo).
Renew oil in rear axle (photo).
Check fluid level in brake and clutch hydraulic reservoirs (photo).

Topping-up the engine oil

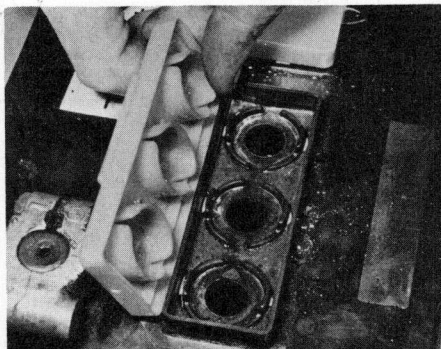
Checking battery electrolyte level

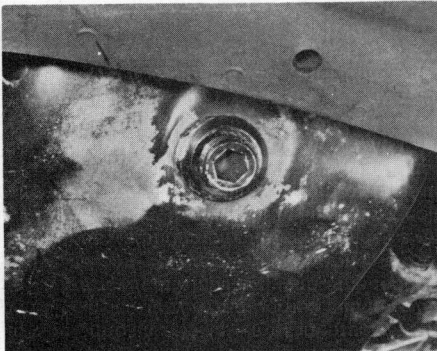
Sump drain plug

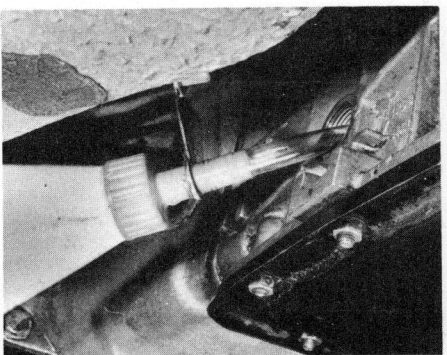
Topping up the gearbox oil level

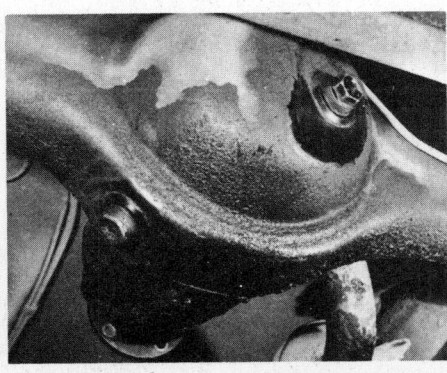
Rear axle oil drain and filler/level plugs

Topping-up brake fluid reservoir level

Routine maintenance

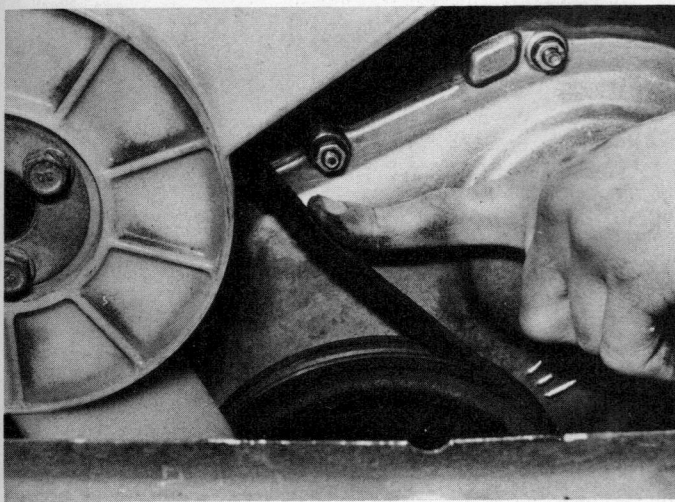
Checking alternator drive belt tension

Gearbox drain plug

Filling the radiator

Topping-up the expansion tank

At 6000 mile (9600 km) intervals or every six months

Renew engine oil and filter.
Check and adjust valve clearances.
Adjust camshaft timing chain tension.
Adjust alternator drive belt tension (photo).
Renew air cleaner element.
Clean fuel pump filter.
Clean carburettor fuel filter.
Clean distributor points and adjust gap.
Adjust ignition timing.
Clean spark plugs and adjust gap.
Lubricate all controls, door locks and hinges.
Check gearbox and rear axle oil levels.
Check and top-up brake and clutch fluid reservoirs.
Clean battery terminals and check security.
Check all suspension and steering components for wear or damaged dust excluders.
Inspect front disc pads for wear.
Move position of roadwheels to even out tread wear on tyres and re-balance if necessary.
Check front wheel alignment.
Adjust clutch pedal free travel.

At 12 000 mile (19 300 km) intervals or every twelve months

Clean out carburettor fuel bowl.
Adjust slow idle and mixture.
Clean crankcase breather.
Check rear brake shoe linings for wear.
Adjust rear brakes (where applicable).
Check and adjust handbrake travel.
Adjust camshaft timing chain tension.
Lubricate steering column upper bush.
Renew brake hydraulic fluid by bleeding system.
Check headlamp beam alignment.
Adjust front hub bearings.
Check underbody corrosion and exhaust system.
Renew spark plugs.

At 24 000 mile (38 000 km) intervals or every two years

Adjust camshaft timing chain tension.
Examine condition of brake flexible hoses.
Lubricate propeller shaft front sliding joint.
Drain cooling system, flush and refill with antifreeze mixture (photos).
Renew gearbox and rear axle oils (photo).

Chapter 1 Engine

For modifications, and information applicable to later models, see Supplement at end of manual

Contents

Auxiliary driveshaft – removal and refitting ... 13	Engine – reassembly (general) ... 33
Big-end bearings – removal ... 15	Engine – removal ... 5
Big-end and main bearing shells – examination and renovation ... 23	Engine – removing ancillary components ... 7
	Engine – refitting ... 48
Camshaft and camshaft housing – examination and renovation ... 27	Fault diagnosis – engine ... 50
Camshaft – refitting ... 43	Final assembly – general ... 47
Connecting rods/big-end bearings – refitting to crankshaft ... 38	Flywheel and starter ring gear – examination and renovation ... 32
Connecting rods – examination and renovation ... 26	Flywheel – refitting ... 41
Connecting rods/pistons and piston rings – removal and dismantling ... 17	General description ... 1
	Major operations possible with engine in car ... 2
Crankshaft – examination and renovation ... 22	Major operations requiring engine removal ... 3
Crankshaft and main bearings – removal (including rear crankshaft oil seal) ... 16	Methods of engine removal ... 4
	Oil filter cartridge – renewal ... 20
Crankshaft – refitting ... 34	Oil pump – removal, dismantling and inspection ... 19
Cylinder block dismantling – flywheel removal ... 14	Oil pump and sump – refitting ... 39
Cylinder bores – examination and renovation ... 24	Pistons and connecting rods – reassembly ... 35
Cylinder head – decarbonisation ... 30	Pistons and connecting rods – refitting ... 37
Cylinder head dismantling – camshaft and rocker removal ... 10	Pistons and piston rings – examination and renovation ... 25
Cylinder head – refitting ... 44	Piston rings – refitting ... 36
Cylinder head – removal (engine in car) ... 8	The engine lubrication system – general description ... 18
Cylinder head – removal (engine out of car) ... 9	Timing chain – removal and refitting ... 12
Distributor, oil pump and fuel pump drive – examination and renovation ... 31	Timing chain and sprockets – refitting ... 40
	Valves and valve seats – examination and renovation ... 28
Distributor – refitting ... 46	Valve assemblies – refitting to cylinder head ... 42
Engine – dismantling (general) ... 6	Valve clearances – adjustment ... 45
Engine examination and renovation – general ... 21	Valve guides – examination and renovation ... 29
Engine – initial start-up after overhaul ... 49	Valves – removal ... 11

Specifications

Engine (general)

Type ... Four cylinder in-line, chain driven single overhead camshaft

	1200	1300	1500	1600
Identification number	2101 or 2102	2105 or 21011	2103	2106
Displacement	73·1 cu in (1198 cc)	79·0 cu in (1294 cc)	88·6 cu in (1452 cc)	95·8 cu in (1570 cc)
Bore	2·99 in (76 mm)	3·11 in (79 mm)	2·99 in (76 mm)	3·11 in (79 mm)
Stroke	2·59 in (66 mm)	2·59 in (66 mm)	3·15 in (80 mm)	3·15 in (80 mm)
BHP at 5600 rpm	62	67	75	80
Maximum torque (lbf ft at 3000 rpm)	64	69	78	88

Valve clearances (all engines):
  Hot ... 0·008 in (0·2 mm)
  Cold ... 0·006 in (0·15 mm)
Firing order (all models) ... 1 – 3 – 4 – 2 (No 1 cylinder nearest radiator)

Cylinder block
Type ... Cast iron cylinders, cast integrally
Reboring limit ... 0·6 mm above standard (in steps of 0·1 mm)

Pistons and connecting rods
Oversize pistons available ... mm
  0·1, 0·2, 0·4 and 0·6
Gudgeon pin hole diameter:
  Category 1 ... 21·982 to 21·986
  Category 2 ... Over 21·986 to 21·990
  Category 3 ... Over 21·990 to 21·994
Ring groove width: ... mm
  Top groove ... 1·535 to 1·555
  2nd grove ... 2·015 to 2·035
  3rd groove ... 3·957 to 3·977
Standard gudgeon pin diameter:
  Category 1 ... 21·970 to 21·974

Category 2	Over 21·974 to 21·978
Category 3	Over 21·978 to 21·982
Gudgeon pin oversizes	0·2 and 0·5
Piston ring thickness:	
Top ring, compression	1·478 to 1·490
2nd ring, compression, scraper type	1·978 to 1·990
3rd ring, oil control, expander type	3·925 to 3·937
Piston to cylinder clearance (measured square to pin 52·40 mm from crown):	
Assembly limits	0·050 to 0·070
Wear limit	0·15
Gudgeon pin to hole clearance:	
Assembly limits	0·008 to 0·016
Wear limit	0·05
Piston ring side clearance:	
Top compression ring:	
Assembly limits	0·045 to 0·077
Wear limit	0·15
2nd compression ring:	
Assembly limits	0·025 to 0·057
Wear limit	0·15
Oil control ring:	
Assembly limits	0·020 to 0·052
Wear limit	0·15
Piston ring endgap (assembly limits)	0·20 to 0·35
Ring oversizes	0·1, 0·2, 0·4, 0·6
Connecting rod big-end bore	51·330 to 51·346
Connecting rod small-end bore	21·940 to 21·960
Standard crankpin shell thickness	1·723 to 1·730
Crankpin shell thickness oversizes	0·125, 0·25, 0·5, 0·75, 1·00
Interference between gudgeon pin and connecting rod bore	0·010 to 0·042
Crankpin to shell clearance:	
Assembly limits	0·036 to 0·086
Wear limit	0·10
Connecting rod big and small-end bores aligned within (measured 125 mm from shank)	± 0·10
Width of connecting rod big-end and cap	26·90 to 26·98

Crankshaft and main bearings

	mm
Standard main journal diameter	50·775 to 50·795
Standard main bearing shell thickness	1·824 to 1·831
Main bearing shell undersize	0·25, 0·50, 0·75, 1·0
Standard crankpin diameter	47·814 to 47·834
Main journal to shell clearance:	
Assembly limits	0·050 to 0·095
Wear limit	0·15
Length of rear main journal between thrust surfaces	27·975 to 28·025
Standard thrust half-ring thickness	2·310 to 2·360
Oversize thrust half-ring thickness	2·437 to 2·487
Crankshaft endfloat (at rear main bearing):	
Assembly limits	0·055 to 0·265
Wear limit	0·35

Camshaft and bearings

	mm
Diameter of camshaft bearing housing bores:	
Front	46·000 to 46·025
Front intermediate	45·700 to 45·725
Centre	45·400 to 45·425
Rear intermediate	45·100 to 45·125
Rear	43·500 to 43·525
Diameter of camshaft journals:	
Front	45·915 to 45·931
Front intermediate	45·615 to 45·631
Centre	45·315 to 45·331
Rear intermediate	45·015 to 45·031
Rear	45·013 to 45·031
Camshaft journal clearance in bore	0·069 to 0·110
Cam height	36·36
Camshaft centre journal radial run-out (camshaft supported on end journals)	0·02

Cylinder head and valves

Cylinder head material	Light alloy
	mm
Diameter of cylinder head bores for valve guides	13·950 to 13·977
Valve guide outside, diameter	14·040 to 14·058
Valve guide outside diameter oversize	0·25

Valve guide inside diameter (guides fitted in cylinder head):
    Intake valve guide ... 8·022 to 8·040
    Exhaust valve guide ... 8·029 to 8·047
Interference between valve guides and bores in cylinder head, assembly limits ... 0·063 to 0·108
Valve stem diameter ... 7·985 to 8·000
Valve stem to guide clearance:
    Intake valve ... 0·022 to 0·055
    Exhaust valve ... 0·029 to 0·062
    Wear limit ... 0·15
Valve head diameter:
    Intake ... 37 ± 0·15
    Exhaust ... 31·5 ± 0·15
Valve face angle ... 45° 30' ± 5'
Valve face concentric with stem (measured at face centre) within ... 0·03
Diameter of cylinder head bores for valve seats:
    Intake ... 37·489 to 37·514
    Exhaust ... 32·489 to 32·514
Valve seat outside diameter:
    Intake ... 37·595 to 37·610
    Exhaust ... 32·585 to 32·600
Interference between valve seats and bores in cylinder head, assembly limits:
    Intake ... 0·081 to 0·121
    Exhaust ... 0·071 to 0·111
Valve seat inside diameter:
    Intake ... 32·5 to 32·7
    Exhaust ... 27·5 to 27·7
Valve seat angle ... 45° ± 5'
Valve seat face concentric with valve guide within ... 0·05
Intake valve seat 30° chamfer concentric with valve guide within ... 0·1

Valve springs:	Inner	Outer
No of effective coils	5	4·5
Total No of coils	6·5	6
Inside diameter	17·6	25·5
Wire diameter	2·7	3·6
Free length	39·2	50
Length under load:		
13·9 ± 0·78 kg	29·7	
28·9 ± 1·5 kg		33·7
28·1 ± 1·4 kg	20	
46·1 ± 2·3 kg		24

Rated valve lift:
    Intake ... 9·728
    Exhaust ... 9·728
Rocker spring:
    Inside diameter ... 20
    Free length ... 35

Lubrication
Oil type/specification ... Multigrade engine oil viscosity range SAE 10W/30 to 20W/50, to API SE or SF (Duckhams QXR, Hypergrade, or 10W/40 Motor Oil)
Oil filter ... Champion C117

Oil pump
Oil pressure ... 50 to 65 lbf in² (3·5 to 4·5 kgf cm²)

mm

Diameter of cylinder block holes for oil pump driveshaft bushes:
    Front hole ... 51·120 to 51·150
    Rear hole ... 25·036 to 25·066
Outside diameter of oil pump driveshaft bushes:
    Front bush ... 51·230 to 51·271
    Rear bush ... 25·130 to 25·170
Oil pump driveshaft bush interference:
    Front bush ... 0·080 to 0·151
    Rear bush ... 0·064 to 0·134
Inside diameter of oil pump driveshaft bushes (fitted and finished):
    Front bush ... 48·084 to 48·104
    Rear bush ... 22·000 to 22·020
Oil pump driveshaft journal diameter:
    Front journal ... 48·013 to 48·038
    Rear journal ... 21·940 to 21·960

Clearance between oil pump driveshaft journals and bushes:
  Front journal .... 0·046 to 0·091
  Rear journal .... 0·040 to 0·080
  Wear limit .... 0·15
Oil pump gear backlash:
  Assembly dimension .... 0·15
  Wear limit .... 0·25
Oil pump gear tip clearance in pump body:
  Assembly limits .... 0·11 to 0·18
  Wear limit .... 0·25
Oil pump gear face clearance in pump body:
  Assembly limits .... 0·031 to 0·116
  Wear limit .... 0·15
Oil pump driven gear to shaft clearance:
  Assembly limits .... 0·017 to 0·057
  Wear limit .... 0·1
Oil pump shaft clearance in body:
  Assembly limits .... 0·016 to 0·055
  Wear limit .... 0·1
Oil pump relief valve spring:
  No of effective coils .... 7·75
  Total No of coils .... 9·25
  Inside diameter .... 13·3
  Wire diameter .... 1·7
  Free length 38
Engine oil capacity (including new filter) .... 6·6 pints (3·75 litres)

Torque wrench settings

	lbf ft	kgf m
Cylinder head bolts (except bolt No 11)	82	11·5
Bolt No 11 (see Fig. 1.4)	28	3·9
Main bearing cap bolts	60	8·2
Connecting rod bolts	37	5·2
Flywheel bolts	60	8·5
Camshaft housing nuts	16	2·2
Oil sump bolts	6	0·8
Intake and exhaust manifold nuts	18	2·5

1 General description

The engines covered in this Chapter comprise the 1200, 1300, 1500 and 1600 cc types. The mechanical design of these engines is basically the same, the main differences being the bore diameter and the stroke.

On all engines the crankshaft rotates in five main bearings of the replaceable shell type. The connecting-rod (big-end) bearings utilize the same type of bearing. The forward end of the crankshaft protrudes through the front timing cover and drives the camshaft via a sprocket and chain and the oil pump, fuel pump and distributor via a sprocket and auxiliary drive shaft.

The camshaft runs in the cylinder head and operates the valves via adjustable stud mounted rocker arms. This gives the advantages of an ohc design but enables the valve clearances to be adjusted without having to remove the camshaft. The double chain that drives the camshaft is fitted with a self-adjusting tensioner.

The valve gear layout coupled with a rigidly supported crankshaft and 'oversquare' bore/stroke configuration, has resulted in an extremely reliable engine with a very respectable power output for the cubic capacity.

The ancillary components (distributor, carburettors, fuel pump, inlet and exhaust manifolds) are all conventional; and detailed descriptions of these items and the systems in which they serve may be found in the relevant Chapters of this manual. Section 18 of this Chapter describes the lubrication system and the overhaul of the major components of that system.

2 Major operations possible with engine in car

The following major operations can be carried out on the engine with it in place:

1 Removal and refitting of cylinder head
2 Removal and refitting of camshaft and timing chain
3 Removal and refitting of engine front mountings
4 Removal and refitting of the auxiliary drive shaft

Note: *Although it is possible to remove the pistons and connecting rods with the engine in the car it is necessary to raise the engine in order to remove the sump. Therefore if piston removal is necessary it is recommended that the engine is first removed from the car. This will also enable other major tasks to be carried out if required.*

3 Major operations requiring engine removal

The following major operations can be carried out with the engine out of the body frame on the bench or floor:

1 *Removal and refitting of the main bearings*
2 *Removal and refitting of the crankshaft*
3 *Removal and refitting of the flywheel*
4 *Removal and refitting of the crankshaft rear oil seal*
5 *Removal and refitting of the sump*
6 *Removal and refitting of the pistons, connecting rods and big-end bearings*

4 Methods of engine removal

As there is not a great deal of clearance between the engine sump and the front suspension crossmember, the best method is to remove the engine after first disconnecting it from the gearbox. As the gearbox is only secured to the engine by four large bolts this does not present any real problem.

5 Engine – removal

1 The do-it-yourself owner should be able to remove the engine fairly easily in about 3½ hours. It is essential to have a good hoist, and two strong axle stands if a pit is not available. Engine removal will be much easier if you have someone to help you. Before beginning work, it is worthwhile to get all the accumulated debris cleaned off the engine unit at a service station which is equipped with steam or high pressure air and water cleaning equipment. It helps to make the job

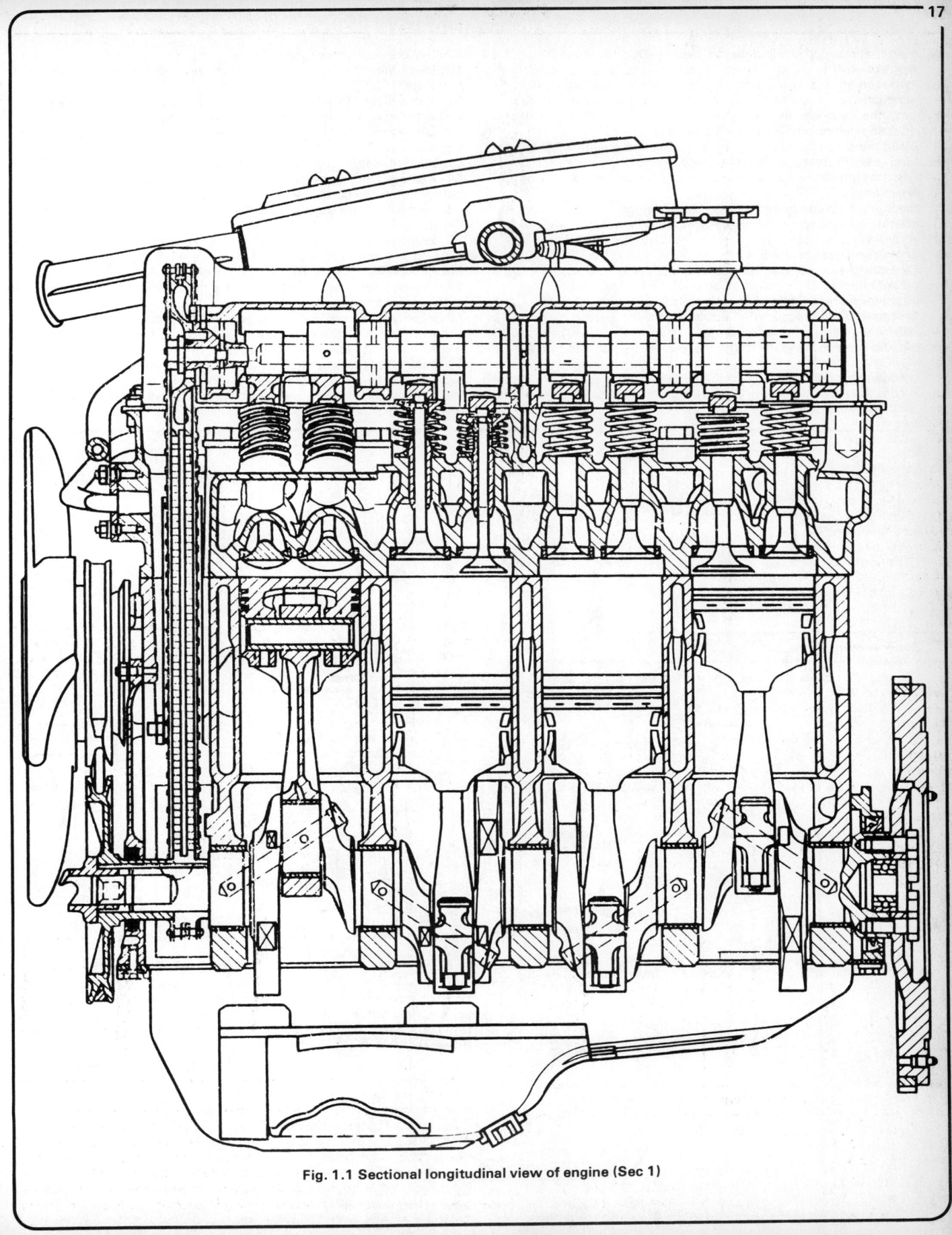

Fig. 1.1 Sectional longitudinal view of engine (Sec 1)

quicker, easier and, of course, much cleaner. Decide whether you are going to jack up the car and support it on axle stands or raise the front end on wheel ramps. If the latter, run the car up now (and chock the rear wheels) whilst you still have engine power available. Remember that with the front wheels supported on ramps the working height and engine lifting height is going to be increased.

2 Open the bonnet and prop it up to expose the engine and ancillary components. Disconnect the battery leads and lift the battery out of the car. This prevents accidental short circuits while working on the engine.

3 Undo the nuts and bolts from the bonnet hinges and lift the bonnet off (photo). Place it somewhere safe where it will not be knocked over or bumped into. Remove the undertrays beneath the engine.

4 Drain the cooling system.

5 Remove the sump drain plug and drain the oil out of the engine into a container (an old 1 gallon oil tin with the side cut out).

6 Disconnect the HT and LT leads from the coil to the distributor.

7 Disconnect the leads to the alternator, starter motor, oil pressure indicator and water temperature sender units. On 1500 and 1600 models, disconnect the leads to the thermal fan switch and the electric fan.

8 Disconnect the accelerator rod from the relay lever and the choke cable from the carburettor. On 1500 and 1600 models, disconnect the lead from the solenoid anti run-on valve.

9 Disconnect the fuel line from the pump.

10 Disconnect the exhaust pipe from the manifold by removing the four nuts.

11 Disconnect the water hoses connecting the radiator to the engine and also those connecting the heater to the engine.

12 Disconnect the radiator hose to the auxiliary tank.

13 Remove the securing bolts and lift off the radiator cowl (photo). On 1600 models, unbolt the electric fan assembly.

14 Remove the two top radiator retaining bolts and lift out the radiator from the bottom support bracket (photo).

15 Remove the air cleaner from the carburettor, referring to Chapter 3 if necessary.

16 Remove the three bolts that secure the starter motor to the clutch housing and pull it forward until it is free of the housing. Disconnect the electrical leads from the starter motor.

17 Remove the four bolts that secure the flywheel cover to the front of the transmission.

18 Support the front end of the gearbox on a jack and remove the four bolts holding it to the crankcase. An articulated socket adaptor piece will be needed to reach two of these.

19 Now sling the engine to whatever lifting device you are using and support the weight. Undo the engine mounting nuts, the upper one on the right and the lower one on the left. This will enable the studs to disengage easily from the body mountings. On no account unscrew

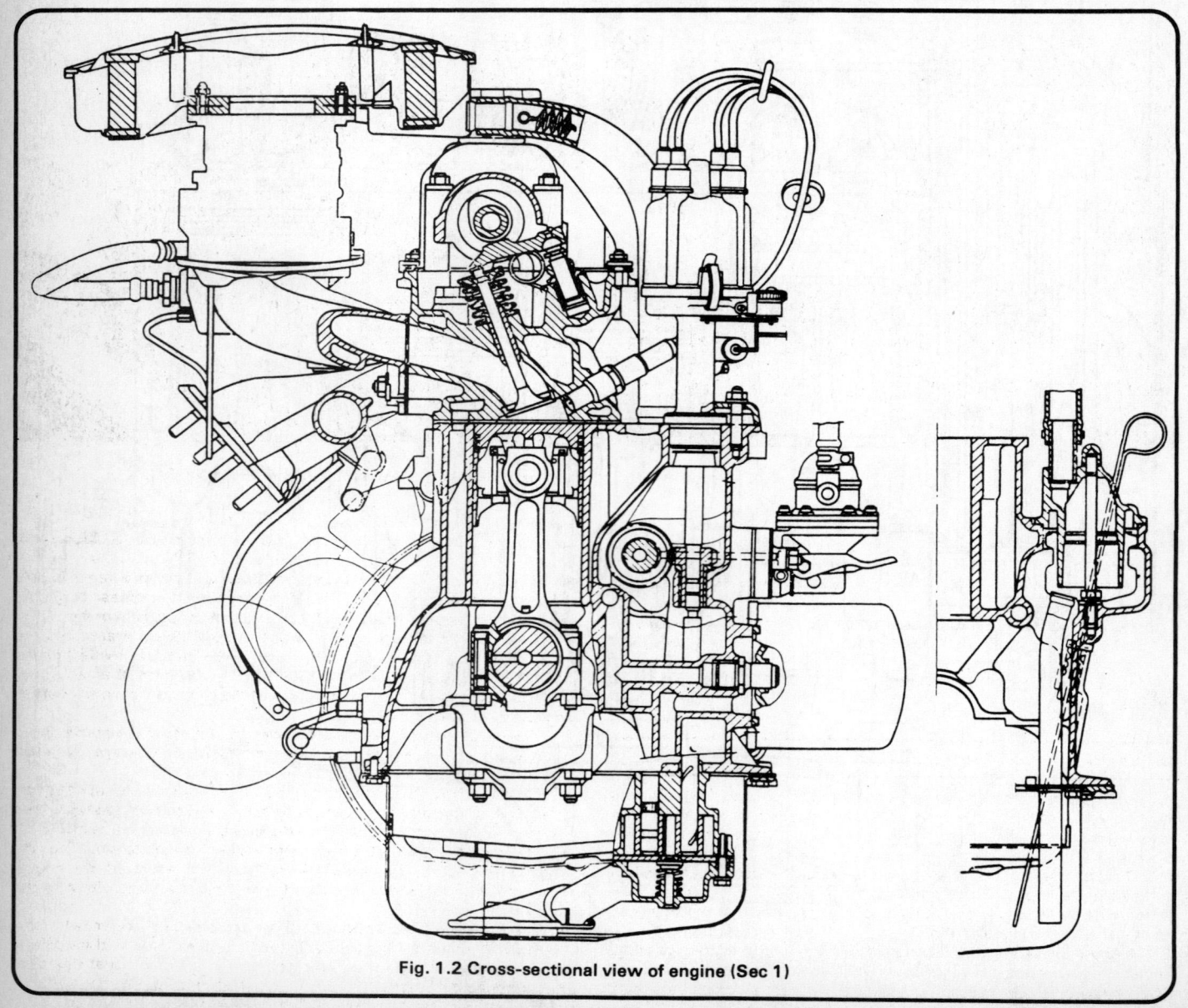

Fig. 1.2 Cross-sectional view of engine (Sec 1)

Chapter 1 Engine

5.3 Bonnet hinge retaining bolts

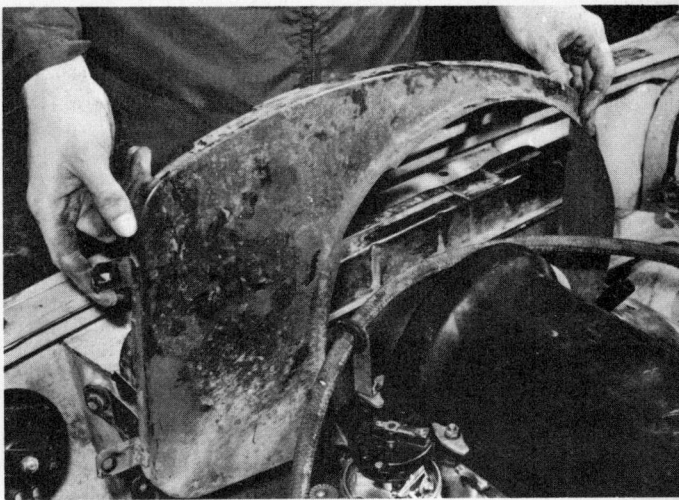
5.13 Removing the radiator cowl

5.14 Lifting out the radiator

5.20 Hoisting out the engine
The rocker cover is removed for fitting lifting hooks

the two small nuts which secure the plate to the flexible mounting unless the mounting has been slightly compressed in the jaws of a vice. The plate retains a heavy coil spring which could fly out and cause injury.
20 Pull the engine forward, supported by the sling, until it disengages from the gearbox input shaft and then lift it up clear and out (photo). If some difficulty is experienced in drawing the engine forward do not impose any lifting or lowering strains which could cause damage to the gearbox input shaft.

6 Engine – dismantling (general)

1 It is best to mount the engine on a dismantling stand, but if this is not available, stand the engine on a strong bench at a comfortable working height. Failing this, it will have to be stripped down on the floor.
2 During the dismantling process, the greatest care should be taken to keep the exposed parts free from dirt. As an aid to achieving this, thoroughly clean down the outside of the engine, first removing all traces of oil and congealed dirt.
3 A good grease solvent will make the job much easier, for, after the solvent has been applied and allowed to stand for a time, a vigorous jet of water will wash off the solvent and grease with it. If the dirt is thick and deeply embedded, work the solvent into it with a strong stiff brush.
4 Finally, wipe down the exterior of the engine with a rag and only then, when it is quite clean, should the dismantling process begin. As the engine is stripped, clean each part in a bath of paraffin or petrol.
5 Never immerse parts with oilways in paraffin (eg; crankshaft and camshaft). To clean these parts, wipe down carefully with a petrol dampened rag. Oilways can be cleaned out with wire. If an air line is available, all parts can be blown dry and the oilways blown through as an added precaution.
6 Re-use of old gaskets is false economy. To avoid the possibility of trouble after the engine has been reassembled **always** use new gaskets throughout.
7 Do not throw away the old gaskets, for sometimes it happens that an immediate replacement cannot be found and the old gasket is then very useful as a template. Hang up the gaskets as they are removed.
8 To strip the engine, it is best to work from the top down. When the stage is reached where the crankshaft must be removed, the engine can be turned on its side and all other work carried out with it in this position.
9 Wherever possible, refit nuts, bolts and washers finger tight from wherever they were removed. This helps to avoid loss and muddle. If they cannot be refitted then arrange them in a fashion that it is clear from whence they came.

7 Engine – removing ancillary components

Before basic engine dismantling begins, it is necessary to strip it of ancillary components.

(a) Fuel components
   Carburettor and manifold assembly
   Exhaust manifold
   Fuel pump
   Fuel line
(b) Ignition system components
   Spark plugs
   Distributor
(c) Electrical system components
   Alternator
   Starter motor
(d) Cooling system components
   Fan and hub (not 1600 models)
   Water pump
   Thermostat housing and thermostat
   Water temperature indicator sender unit
(e) Engine
   Oil filter
   Oil pressure sender unit
   Oil level dipstick
   Oil filler cap and top cover
   Engine mountings
   Crankcase ventilation valve and oil separator
(f) Clutch
   Clutch pressure plate assembly
   Clutch friction plate assembly
(g) All nuts and bolts associated with the foregoing

Some of these items have to be removed for individual servicing or renewal periodically and details can be found in the appropriate Chapter.

8 Cylinder head – removal (engine in car)

1 Open the bonnet and using a soft pencil mark the outline of both the hinges at the bonnet to act as a datum for refitting.
2 With the help of a second person to take the weight of the bonnet, undo and remove the hinge to bonnet securing bolts, plain and spring washers. There are two bolts to each hinge.
3 Lift away the bonnet and put in a safe place where it will not be scratched.
4 Refer to Chapter 10, and remove the battery.
5 Refer to Chapter 3, and remove the air cleaner assembly from the top of the carburettor. Also remove the rocker cover.
6 Mark the HT leads so that they may be refitted in their original positions and detach them from the spark plugs.
7 Release the HT lead rubber moulding from the clip on the top of the cover.
8 Spring back the clips securing the distributor cap to the distributor body. Lift off the distributor cap.
9 Detach the HT lead from the centre of the ignition coil. Remove the distributor cap from the engine compartment.
10 Refer to Chapter 2, and drain the cooling system.
11 Disconnect the throttle linkage and choke control cable from the carburettor. On 1500 and 1600 models, disconnect the solenoid anti run-on valve electric lead.
12 Disconnect the fuel pipe, water inlet pipe and vacuum pipes from the carburettor.
13 Detach the front exhaust pipe bracket from the gearbox, remove the four securing nuts and pull the exhaust pipe flange away from the manifold. **Note:** *If difficulty is experienced in clearing the flange from the manifold studs it will be necessary to remove the heat shield, undo the manifold nuts and pull the exhaust and inlet manifolds away from the cylinder head.*
14 Remove the top radiator hose and heater hoses from the cylinder head.
15 Disconnect the temperature sensor lead and remove the camshaft cover.
16 Now remove the spark plugs and rotate the engine by means of the starting handle or a spanner on the crankshaft pulley dog nut until the detent in the camshaft sprocket is in line with the pointer on the camshaft housing (see Fig. 1.3). Place the car in gear and apply the handbrake to ensure the engine cannot be rotated when the cylinder head is removed; otherwise it will be necessary to reset the timing.
17 Bend back the lockwasher from the camshaft sprocket retaining bolt and remove the bolt, taking care not to rotate the engine.
18 Unscrew the cap, then unscrew the nuts and remove the timing chain tensioner from the side of the cylinder head (photo).
19 Ease the sprocket from the end of the camshaft and remove the sprocket from the timing chain. Attach some wire to the timing chain so that it can be guided through the cylinder head when the latter is removed from the block.
20 Unscrew the nuts evenly and lift off the camshaft housing, complete with camshaft.
21 Unscrew the cylinder head bolts evenly in the order shown in Fig. 1.4.
22 Lift the cylinder head from the block, at the same time guiding the timing chain through the timing chest aperture. If necessary, rock the head to release it from the gasket, but **do not** attempt to prise it from the block with a lever. Take care not to disturb the position of the auxiliary drive sprocket, otherwise the ignition timing will require resetting. If the cylinder head is being removed for decarbonising it is recommended that the timing chain is completely removed, as described in Section 12, so that the engine can be easily turned for cleaning the piston crowns.
23 Remove the gasket from the cylinder block.

9 Cylinder head – removal ((engine out of car)

The procedure for removing the cylinder head with the engine on the bench is similar to that for removal when the engine is in the car; with the exception of disconnecting the controls and services. Refer to Section 8, and follow the sequence given in paragraphs 16 to 21 inclusive, after first removing the rocker cover and gasket.

10 Cylinder head dismantling – camshaft and rocker removal

1 The cylinder head is made of aluminium, and the camshaft runs in a separate housing secured to the cylinder head by studs. The camshaft actuates the valves via rocker arms pivoting on adjustable studs.
2 The camshaft can be withdrawn from the housing by simply removing the thrust plate at the front of the housing and carefully sliding the camshaft out.
3 The rockers can be removed by prising off the spring retaining clips. Make a careful note of which way around the clips fit and keep the rockers and clips in the correct order to ensure that they are replaced in their original positions.
4 If necessary, the rocker pivot studs can be removed by slackening the locknut and unscrewing them from the head.

11 Valves – removal

1 The valves can be removed from the cylinder head by the following method. With a valve spring compressor, compress each spring in turn until the two halves of the collets can be removed. Release the compressor and remove the valve spring cap, springs (inner and outer), and lower spring seating. The inlet valve guides have oil seals fitted which should be removed together with the inlet and exhaust valves themselves.
2 If when the valve spring compressor is screwed down, the valve spring cap refuses to free and expose the split collets, do not continue to screw down on the compressor as there is a likelihood of damaging it.
3 Gently tap the top of the tool directly over the cap with a light hammer. This will free the cap. To avoid the compressor jumping off the spring cap when it is tapped, hold the compressor firmly in position with one hand.
4 It is essential that valves keep to their respective places in the head, unless they are so badly worn that they need to be renewed. If they are going to be kept and used again, place them in a sheet of card having two rows of four holes numbered 1 to 4 exhaust and 1 to 4 inlet, corresponding with the positions the valves occupied in the

Chapter 1 Engine

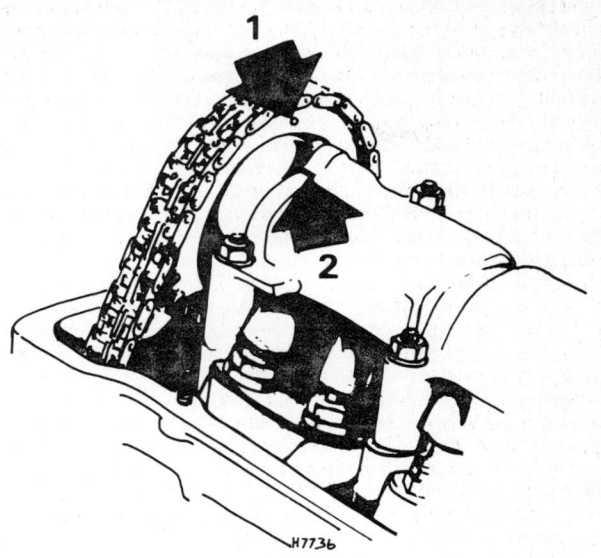

Fig. 1.3 Camshaft timing marks (Sec 8)

1 Mark on camshaft sprocket
2 Pointer on camshaft housing

8.18 Removing the timing chain tensioner

cylinder head. Keep the valve springs and caps in their correct order too.
5 Numbering from the front of the engine, exhaust valves are 1-4-5-8 and inlet valves 2-3-6-7.

12 Timing chain – removal and refitting

1 The Lada engines have an overhead camshaft which is chain driven from the crankshaft. The chain also drives an auxiliary shaft which in turn drives the oil pump, fuel pump and distributor.
2 To remove the timing chain it is first necessary to remove the

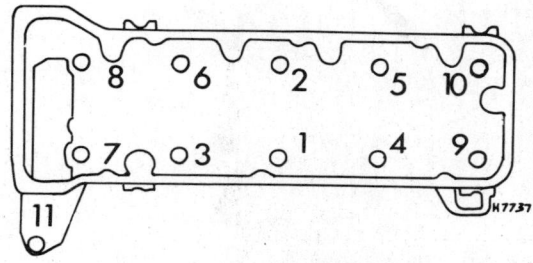

Fig. 1.4 Correct sequence for slackening and tightening the cylinder head bolts (Sec 8)

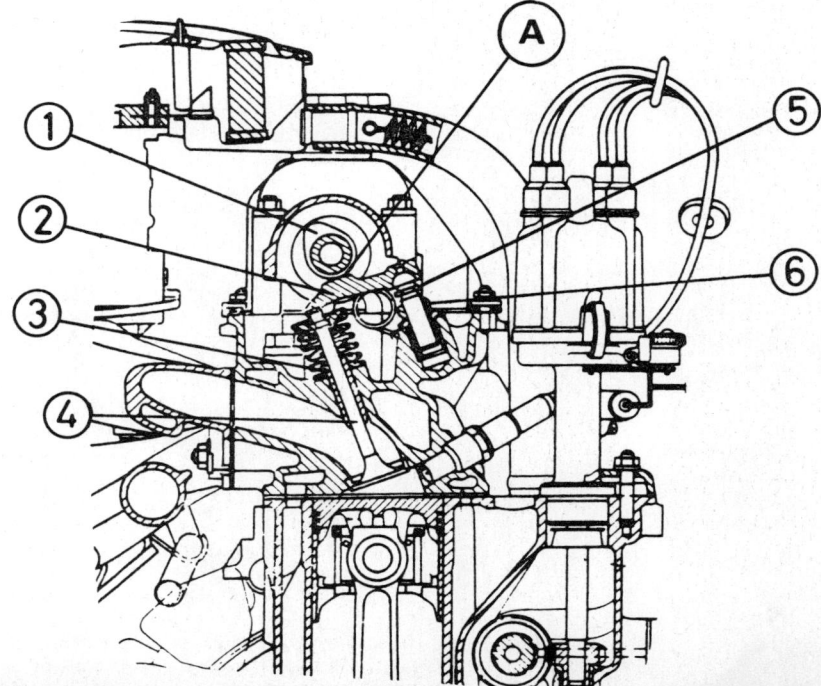

Fig. 1.5 Sectional view of camshaft and valve gear (Sec 10)

A Rocker to cam clearance
1 Camshaft cam
2 Valve rocker
3 Valve seal
4 Valve
5 Rocker adjusting screw
6 Adjusting screw locknut

camshaft sprocket (as described in Section 8), the cooling fan, crankshaft pulley and timing cover. If the task is being carried out with the engine still in the car it will be necessary to remove the radiator and front grille to gain access.

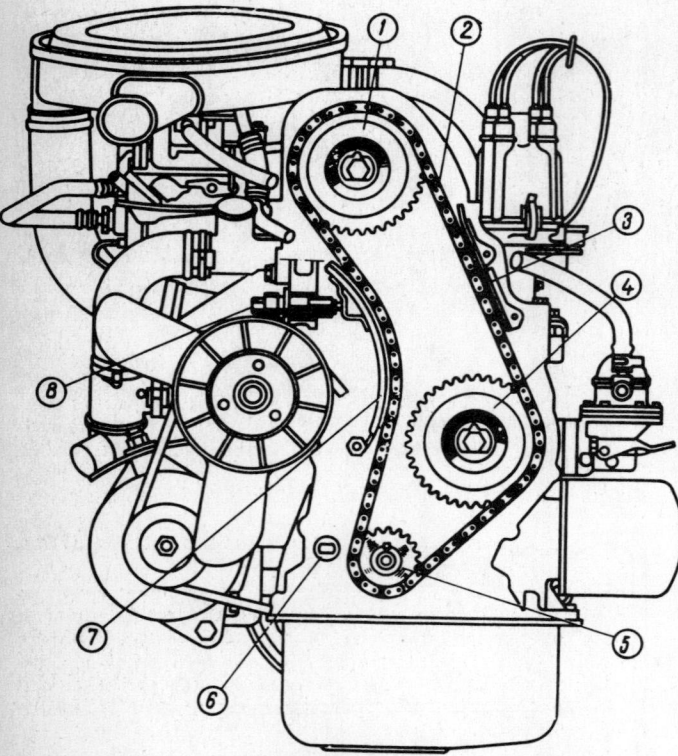

Fig. 1.6 Timing chain and sprockets (Sec 12)

1 Camshaft sprocket
2 Chain
3 Chain damper
4 Auxiliary drive sprocket
5 Crankshaft sprocket
6 Stop pin
7 Chain tensioner shoe
8 Chain tensioner

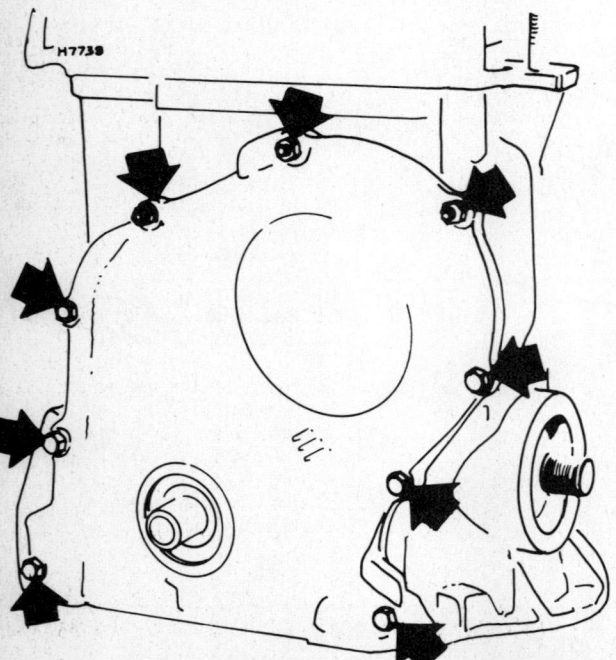

Fig. 1.7 Timing cover retaining nuts and bolts (Sec 12)

3 Remove the fan and pulley as described in Chapter 2 and then remove the large crankshaft pulley nut using a suitable size ring spanner or socket. If the engine is out of the car, it will be necessary to prevent the crankshaft rotating by jamming a tyre lever between the cylinder block and the teeth of the flywheel ring gear. If the engine is in the vehicle, it may be possible to unscrew the pulley nut if the crankshaft is prevented from rotating by engaging a gear and applying the handbrake fully. If this does not succeed, remove the starter motor (Chapter 10) and jam the ring gear as previously described.

4 Withdraw the pulley from the end of the crankshaft. If it is really tight a suitable puller will have to be used.

5 To remove the aluminium, timing cover it is first necessary to remove the sump front retaining bolts that screw into the bottom flange of the cover and then remove the timing cover retaining nuts (see Fig. 1.7).

6 Carefully ease the cover away from the block and remove the gasket.

7 With the camshaft sprocket removed from the timing chain, the chain can now be unhooked from the crankshaft sprocket and auxiliary drive sprocket and removed from the engine.

8 If required, the crankshaft sprocket can be pulled off the end of the crankshaft.

9 To refit the timing chain, slide the crankshaft sprocket onto the keyed end of the crankshaft. Temporarily screw in the starting handle dog and turn the crankshaft with the starting handle, until the index mark on the sprocket is aligned with the raised pointer cast in the block above the sprocket (photo). Remove the distributor cap and rotate the auxiliary shaft until the rotor arm points towards the No 4 spark plug HT lead segment.

10 Loop the chain over the crankshaft sprocket and auxiliary shaft sprocket and lift the end of the chain up through the cylinder head aperture.

11 Align the detent in the camshaft sprocket with the pointer on the camshaft housing and fit the chain around the sprocket (photo).

12 Rotate the camshaft so that the dowel on the front of the shaft lines up with the hole in the camshaft sprocket.

13 Fit the sprocket onto the end of the camshaft and screw in the retaining bolt finger tight (photo).

14 Fit the chain tensioner into the side of the cylinder head and check that the crankshaft sprocket and camshaft sprocket timing marks are still correctly aligned. If everything is correct tighten the camshaft retaining bolt and bend over the locking washer. Adjust the timing chain tension as described in Section 13, paragraph 9.

15 Refit the timing cover using a new gasket and replace the crankshaft pulley.

16 If the engine is still in the car, refit the radiator and refill the cooling system as described in Chapter 2.

17 Before replacing the camshaft cover, check the ignition timing as described in Chapter 4.

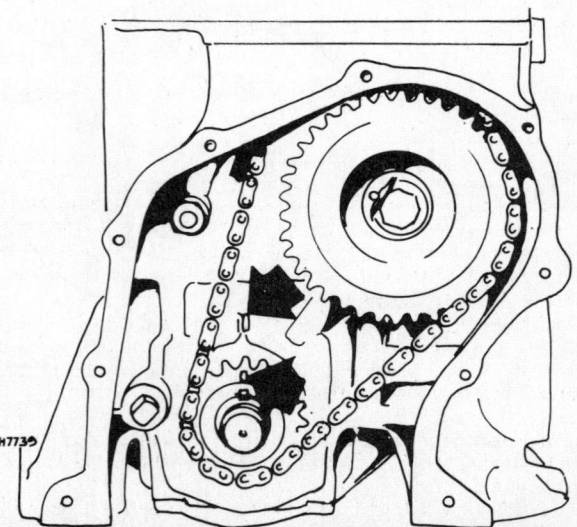

Fig. 1.8 Timing marks on crankshaft sprocket and cylinder block (Sec 12)

Chapter 1 Engine

12.9 Crankshaft sprocket timing marks

12.11 Camshaft sprocket timing mark and pointer

12.13 Camshaft sprocket retaining bolt and lockwasher

13 Auxiliary drive shaft – removal and refitting

1 To remove the auxiliary drive shaft, it is necessary to remove the timing cover as described in the previous section. The fuel pump must also be removed (Chapter 3).
2 If the engine is still in the car, align the crankshaft and camshaft timing marks and scribe a line across the auxiliary shaft sprocket and the front of the cylinder block. This will enable the shaft to be replaced in the correct position and avoid having to reset the ignition timing.
3 Bend back the locking washer and remove the sprocket retaining bolt.
4 Unscrew the timing chain tensioner from the side of the cylinder head and then draw the sprocket off the auxiliary shaft disengaging it from the timing chain.
5 Remove the thrust plate retaining bolts, lift off the plate and withdraw the shaft.
6 Examine the shaft skew gear and bearing surfaces. If signs of wear are evident a new shaft should be obtained.
7 The auxiliary shaft bearings require special tools to extract them. If they require replacement the task should be entrusted to your Lada dealer.
8 Refit the shaft and sprocket using the reverse procedure to removal. Make sure the mark made on the sprocket lines up with the mark on the front of the cylinder block.
9 It is most important that the timing chain is adjusted in the following way, both now and as a routine maintenance operation at 6000 mile (9600 km) intervals. Loosen the timing chain tensioner cap nut on the side of the cylinder head. Turn the crankshaft (using the engine starting handle) through 1½ turns. The tensioner will now have automatically taken up any slack in the chain. A check to verify this can be made if the rocker cover is removed and the chain tested with the fingers. Tighten the tensioner cap nut and remove the starting handle.
10 Check that the crankshaft and camshaft sprocket timing marks are correctly aligned as described in the previous Section and then refit the timing cover.
11 If the engine is in the car, check the ignition timing as described in Chapter 4.

14 Cylinder block dismantling – flywheel removal

1 Having removed all the ancillary items described in Section 7, the cylinder head as described in Section 8, and then removed the timing case and chain as described in Section 12, the cylinder block should now be ready for further dismantling.
2 Refer to Chapter 5, and remove the clutch from the flywheel. Mark the position of the clutch assembly on the flywheel so that it may be refitted into the exact position from which it was removed. Undo the fixing bolts and remove the clutch mechanism together with the friction disc.
3 It should now be possible to undo and remove the six bolts that secure the flywheel to the crankshaft. The flywheel can then be lifted from the crankshaft.
4 Remove the retaining screws and lift off the pressed steel plate from the rear of the cylinder block.
5 Roll the engine on its side, undo the retaining bolts and lift off the sump.
6 Remove the retaining bolts and prise off the crankshaft rear oil seal housing.
7 Unscrew the retaining bolts and withdraw the oil pump and shaft.
8 Finally, remove the centre stud from inside the crankcase ventilation baffle, detach the pipe bracket from the sump flange and remove the baffle and pipe.
9 The crankcase is now ready for the crankshaft to be removed.

15 Big-end bearings – removal

1 The big-end bearings are those items which may require attention more frequently than any other moving part on the engine. They are the usual shell bearings running on a hardened steel crankpin and the shell surface is very soft. If the shells are scored and worn, it will mean the crankshaft is also worn and it will have to be removed for regrinding (see Section 22)
2 Although it may be possible to raise the engine sufficiently to enable the sump to be removed with the engine still in the car, access is extremely limited and the best policy is to remove the engine completely and then remove the cylinder head as described in previous sections.
3 Work on a strong clean bench and when the engine has been thoroughly cleaned, turn it on its manifold side to give access to the sump.
4 Remove the sump and the oil pump.
5 Wipe the bearing caps clean and check that they have the correct identification markings. Normally the numbers 1 to 4 are stamped on the adjacent sides of the big-end caps and connecting rods, indicating which cap fits which rod and which way round the cap fits the rod; the number 1 should be found on the connecting rod operated by the No 1 piston at the front of the engine, 2 for the second piston/cylinder, and so on.
6 If for some reason no identification marks can be found, then centre-punch mating marks on the rods and caps. One dot for the rod and cap for the No 1 piston/cylinder, two for the No 2 and so on.
7 Undo and remove the big-end cap retaining bolts and keep them in their respective order for correct refitting.
8 Lift off the big-end bearing caps; if they are difficult to remove they may be gently tapped with a soft faced mallet to separate the cap and rod.
9 To remove the shell bearings, press the bearing end opposite the groove in both the connecting rod and bearing cap and the shells will slide out easily.
10 If the bearings are being attended to only, and it is not intended to remove the crankshaft, make sure that the pistons do not slide too far down the cylinder bores. The piston could reach a position where the lower piston ring passes out from the bottom of the cylinder bores. If this should happen, the crankshaft would have to be removed in order to compress the ring and refit the piston.

16 Crankshaft and main bearings – removal (including rear crankshaft oil seal)

1 The crankshaft is forged steel and is supported in five main bearings. The endfloat of the shaft is controlled by thrust bearings on the

rear main bearing. The rear crankshaft oil seal is mounted in a housing which is bolted to the rear of the cylinder block.

2   In order to be able to remove the crankshaft it will be necessary to have completed the following tasks:

(a) *Removal of the engine*
(b) *Removal of the flywheel*
(c) *Removal of the cylinder head*
(d) *Removal of the sump*
(e) *Removal of the timing case, timing chain*
(f) *Removal of the oil pump*
(g) *Removal of the big-end bearings*
(h) *Removal of the rear oil seal housing*

It is not essential to have extracted the pistons and connecting rods, but it makes for a less cluttered engine block during the crankshaft removal.

3   Check that the main bearing caps have identification grooves marked in them, No 1 should be at the front and No 5 should be the rear main cap. If for some reason identification cannot be found, use a centre-punch to identify the caps.

4   Before removing the crankshaft it is wise to check the existing endfloat. Use feeler gauges or a dial gauge to measure the gap between the rearmost main bearing journal wall on the crankshaft and the shoulder of the bearing shell in the bearing cap.

5   Move the crankshaft forwards as far as it will go with two tyre levers to obtain a maximum reading. The endfloat should be within the dimensions given in the Specifications. If the endfloat exceeds those figures, then oversize thrust bearings should be fitted. See the Specifications Section at the beginning of this Chapter for thrust and main bearing data.

6   Having ensured that the bearing caps have been marked in a manner which facilitates correct refitting, undo the bolts retaining the caps by one turn only to begin with. Once all have been loosened proceed to unscrew and remove them.

7   The bearing caps can now be lifted away together with the shells inside them. Finally the crankshaft can be removed and then the shells seated in the crankcase.

17 Connecting rods/pistons and piston rings – removal and dismantling

1   In order to remove the pistons and connecting rods it will have been necessary to have removed the engine from the car, removed the cylinder head, the sump and the big-end bearing caps.

2   With the engine stripped down as indicated, proceed as follows:

3   Extract the pistons and connecting rods out of the top of the cylinder bores; be careful not to allow the rough edges of the connecting rods to scratch and score the smooth surface of the cylinder bore.

4   The piston and connecting rods will not pass downwards out of the crankcase, without first removing the crankshaft.

5   Once the piston/connecting rod has been removed from the engine the piston rings may be slid off the piston.

6   Slide the piston rings carefully over the surface of the piston taking care not to scratch the aluminium alloy from which the piston is made. It is all too easy to break the cast iron piston ring if they are pulled off roughly, so this operation must be done with extreme care. It is helpful to make use of an old feeler gauge.

7   Lift one end of the piston ring to be removed out of its groove and insert the end of a feeler gauge under it.

8   Move the feeler gauge slowly around the piston and apply slight upward pressure to the ring as it comes out of the groove so that it rests on the land above. It can then be eased off the piston and the feeler gauge used to prevent it slipping into other grooves as necessary.

9   The piston and connecting rod are joined by a gudgeon pin. The pin is an interference fit in the connecting rod and a clearance fit in the piston.

10   Lada provide a special mandrel A60308 for the task of removing the gudgeon pin and a fixture A95605 to support the piston and conrod whilst the pin held in the special mandrel is being forced out by a hydraulic press.

11   The method to remove the gudgeon pin with the Lada tools is as follows:

12   Remove the knurled nut from the mandrel, leave the spacer on and insert the centre shaft through the centre of the gudgeon pin. Refit the knurled nut.

13   Heat the piston and connecting rod assembly in an oil bath to around 200°C and then swiftly transfer them to the support fixture so that the pin may be drifted out of the piston-rod assembly.

14   Providing one had the equipment, a wooden support cradle could be made for the piston and rod assembly and a conventional drift used to drive out the gudgeon pin from the connecting rod and piston once they have been heated.

15   It will be appreciated that the removal of the gudgeon pin requires quite a few items of specialised equipment and therefore in all probability it will be more practical, if the gudgeon pin really must be removed, to take the piston/rod assembly to your local Lada dealer.

18 The engine lubrication system – general description

1   The engine lubrication system is quite conventional. A gear type oil pump draws oil up from the sump, via the suction pipe and strainer, and pumps the oil under pressure into the cartridge oil filter. From the oil filter, the oil flows into galleries drilled in the engine block to feed the main bearing on the crankshaft and the moving components of the cylinder head. Oil is bled from the main bearing journals in the crankshaft to supply the big-end bearings.

2   Therefore, the bearings which receive pressure lubrication are the main crankshaft bearings, the big-end bearings, the camshaft bearings and the rockers.

3   The remaining moving parts receive oil by splash or drip-feed and these include the timing chain and associated items, the distributor and fuel pump drive, the rockers, the valve stems and to a certain extent the pistons.

4   The oil filter has a ball valve which, in the event of the filter becoming clogged with dirt, opens and allows the oil to bypass the filter.

5   A pressure relief valve in the oil pump allows oil to discharge directly into the sump if the oil pressure is excessive; usually during cold starting conditions.

6   An oil pressure transmitter is screwed into the left-hand side of the cylinder block, (photo). On 1500 and 1600 models, both a warning lamp switch and an electric pressure transmitter are screwed into a T-piece.

19 Oil pump – removal, dismantling and inspection

1   Remove the engine from the car (preferably – although one may detach the sump with the engine in place if the engine is disconnected from the front mountings and jacked up).

2   Remove the sump.

3   Undo the bolts securing the pump and suction pipe unit to the

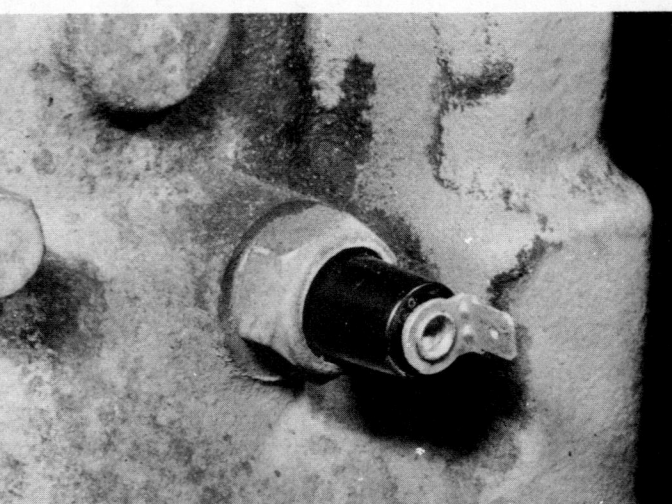

18.6 Oil pressure sensor unit

Chapter 1 Engine

crankcase and withdraw the pump.
4 If the oil pump is worn, it is best to purchase a replacement unit, as to rebuild the oil pump is a job that calls for engineering shop facilities.
5 To check if the pump is still serviceable, first check if there is any slackness in the spindle bushes, and then remove the bottom cover held by three bolts (photo).
6 Then check the two gears (the rotors) and the inside of the pump body for wear with the aid of a feeler gauge. Measure the gearwheels radial clearance (blades inserted between the end of the gearwheel teeth and the inside of the body) and the gearwheel end clearance (place a straight edge across the bottom flange of the pump body and measure with the feeler blades the gap between the straight edge and the sides of the gearwheel). The correct clearances are listed in the Specifications.
7 Fit a replacement pump if the clearances are incorrect.

20 Oil filter cartridge – renewal

1 When work is of a routine maintenance nature and the filter is due for renewal, a useful tool to loosen a disobliging filter cartridge is a small chain strap wrench. This type of wrench will grip the old filter firmly and enable the most obstinate of cartridges to be freed. Do not

Fig. 1.9 Oil pressure switch and transmitter assembly (1500 and 1600 models) (Sec 18)

1 Oil pressure warning lamp switch
2 Electric pressure transmitter
3 Coolant drain plug

19.5 Oil pump with bottom cover removed

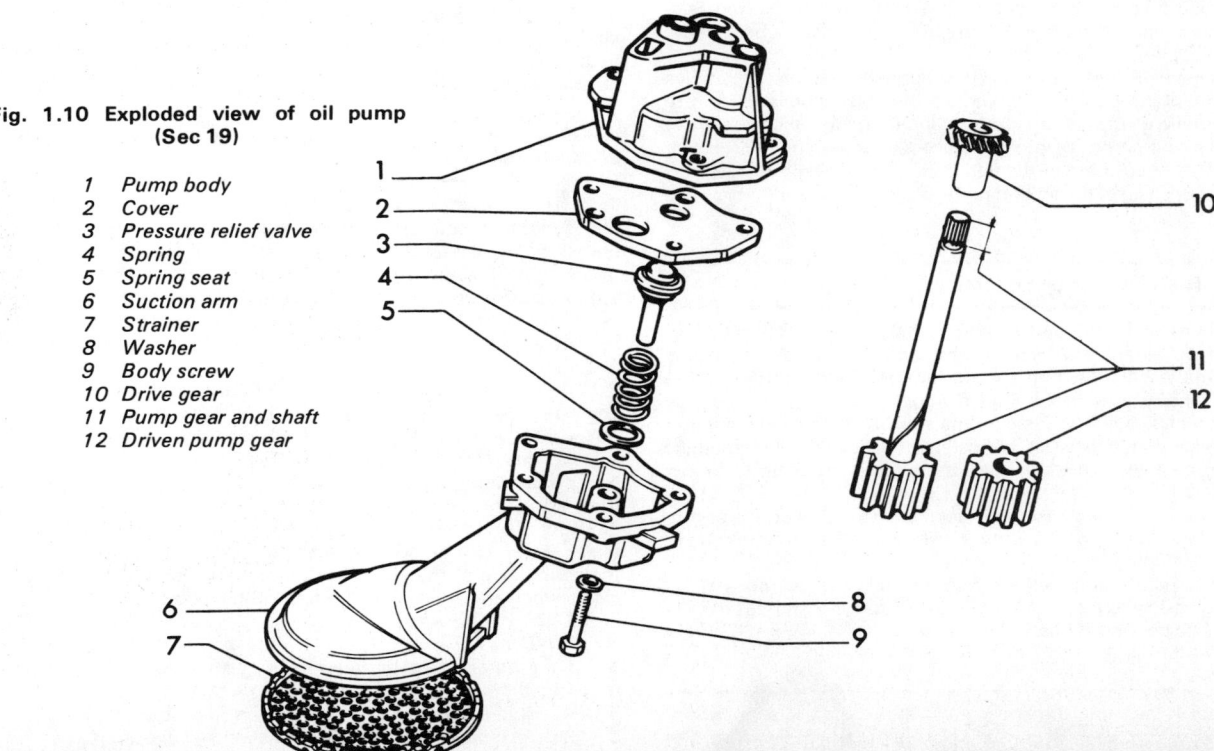

Fig. 1.10 Exploded view of oil pump (Sec 19)

1 Pump body
2 Cover
3 Pressure relief valve
4 Spring
5 Spring seat
6 Suction arm
7 Strainer
8 Washer
9 Body screw
10 Drive gear
11 Pump gear and shaft
12 Driven pump gear

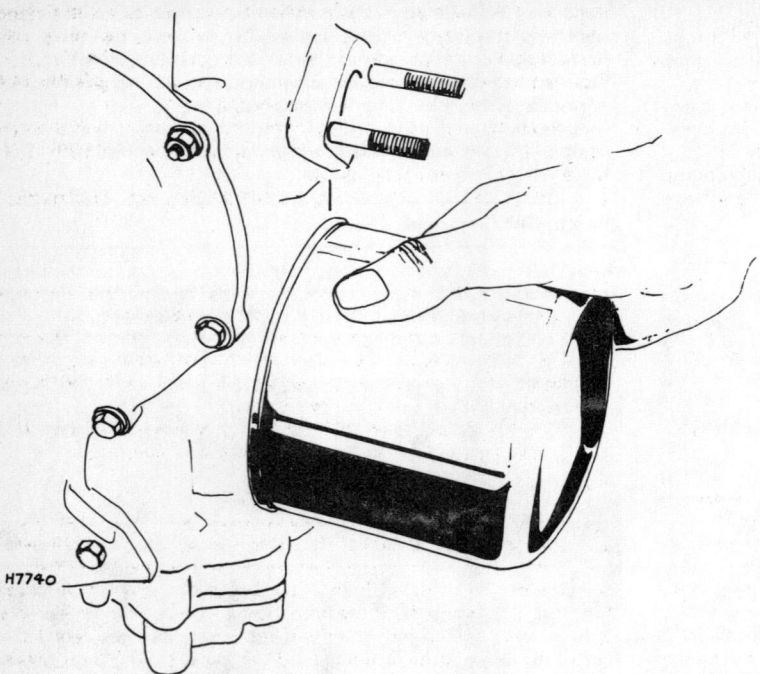

Fig. 1.11 Fitting the oil filter (Sec 20)

use a wrench to tighten the new filter, use hand pressure only.
2 Use genuine Lada oil filters only, as spurious types can allow oil to drain back to the sump when the engine is switched off. This will give rise to a knocking noise when the engine is re-started; also the oil pressure light will stay on for a short period. It is worth noting that low oil pressure can be due to loose plugs in the crankshaft webs or camshaft plugs.

21 Engine examination and renovation – general

1 With the engine stripped and all parts thoroughly cleaned, every component should be examined for wear. The items listed in the Sections following should receive particular attention and where necessary be renewed or renovated.
2 So many measurements of engine components require accuracies down to tenths of a thousandth of an inch. It is advisable therefore to either check your micrometer against a standard gauge occasionally to ensure that the instrument zero is set correctly, or use the micrometer as a comparative instrument. This last method however, necessitates that a comprehensive set of slip and bore gauges is available.

22 Crankshaft – examination and renovation

1 Examine the crankpin and main journal surfaces for signs of scoring or scratches and check the ovality and taper of the crankpins and main journals. If the bearing surface dimensions do not fall within the tolerance ranges given in the Specifications at the beginning of this Chapter, the crankpins and/or main journals will have to be reground.
2 Big-end and crankpin wear is accompanied by distinct metallic knocking, particularly noticeable when the engine is pulling from low revs.
3 Main bearings and main journal wear is accompanied by severe engine vibration rumble, getting progressively worse as engine revs increase.
4 If the crankshaft requires regrinding, take it to an engine reconditioning specialist, who will machine it for you and supply the correct undersize bearing shells.

23 Big-end and main bearing shells – examination and renovation

1 Big-end bearing failure is accompanied by a noisy knocking from the crankcase and a slight drop in oil pressure. Main bearing failure is accompanied by vibration which can be quite severe as the engine speed rises and falls, and a drop in oil pressure.
2 Bearings which have not broken up, but are badly worn will give rise to low oil pressure and some vibration. Inspect the big-ends, main bearings and thrust washers for signs of general wear, scoring, pitting and scratches. The bearings should be matt grey in colour. With lead-indium bearings, should a trace of copper colour be noticed, the bearings are badly worn as the lead bearing material has worn away to expose the indium underlay. Renew the bearings if they are in this condition or if there is any sign of scoring or pitting. **Note**: *You are strongly advised to renew the bearings regardless of their condition. Refitting used bearings is a false economy.*
3 The undersizes available are designed to correspond with crankshaft regrind sizes, ie –0.25 mm, –0.50 mm and –0.75 mm. The bearings are in fact, slightly more than the stated undersize as running clearances have been allowed for during their manufacture.

24 Cylinder bores – examination and renovation

1 The cylinder bores must be examined for taper, ovality, scoring and scratches. Start by carefully examining the top of the cylinder bores. If they are at all worn a very slight ridge will be found on the thrust side. This marks the top of the piston travel. The owner will have a good indication of the bore wear prior to dismantling the engine, or removing the cylinder head. Excessive oil consumption accompanied by blue smoke from the exhaust is a sure sign of worn cylinder bores and piston rings.
2 Measure the bore diameter just under the ridge with a micrometer and compare it with the diameter at the bottom of the bore, which is not subject to wear. If the difference between the two measurements is more than 0.006 in (0.15 mm) then it will be necessary to fit oversize pistons and rings. If no micrometer is available, remove the rings from a piston and place the piston in each bore in turn, about three quarters of an inch below the top of the bore. If an 0.010 in (0.25 mm) feeler gauge can be slid between the piston and the cylinder wall on the thrust side of the bore then remedial action must be taken. Oversize pistons are available in the following sizes:
0.020 in (0.5 mm)
0.040 in (1.0 mm)
3 If the bores are slightly worn but not so badly worn as to justify reboring them, special oil control rings can be fitted to the existing pistons which will restore compression and stop the engine burning oil. Several different types are available and the manufacturer's instructions concerning their fitting must be followed closely.

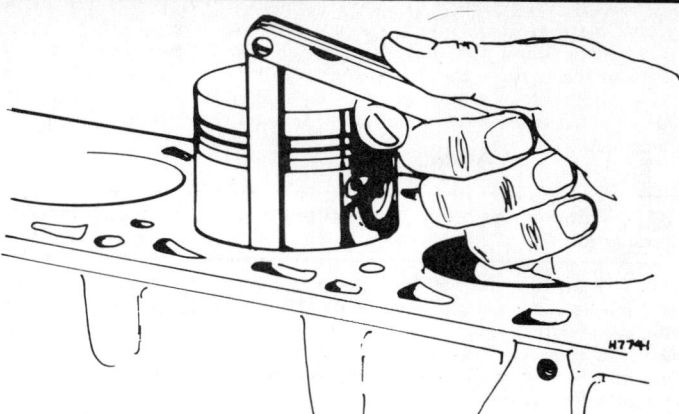

Fig. 1.12 Checking the piston clearance in the cylinder bore (Sec 24)

25 Pistons and piston rings – examination and renovation

1   If the old pistons are to be refitted, carefully remove the piston rings and thoroughly clean them. Take particular care to clean out the piston ring grooves. At the same time do not scratch the aluminium. If new rings are to be fitted to the old pistons, then the top ring should be stepped to clear the ridge left above the previous top ring. If a normal but oversize new ring is fitted, it will hit the ridge and break, because the new ring will not have worn in the same way as the old, which will have worn in unison with the ridge.
2   Before fitting the rings on the pistons each should be inserted approximately 3 in (75 mm) down the cylinder bore and the gap measured with a feeler gauge as shown in Fig. 1.13. This should be as detailed in the Specifications at the beginning of this Chapter. It is essential that the gap is measured at the bottom of the ring travel. If it is measured at the top of a worn bore and gives a perfect fit it could easily seize at the bottom. If the ring gap is too small rub down the ends of the ring with a fine file, until the gap, when fitted, is correct. To keep the rings square in the bore for measurement, line each up in turn with an old ring about 3 in down the bore. Remove the piston and measure the piston ring gap.
3   When fitting new pistons and rings to a rebored engine, the ring gap can be measured at the top of the bore as the bore will now not taper. It is not necessary to measure the side clearance in the piston ring groove with rings fitted, as the groove dimensions are accurately machined during manufacture. When fitting new oil control rings to old pistons it may be necessary to have the groove widened by machining to accept the new wider rings. In this instance the manufacturer's representative will make this quite clear and will supply the address to which the pistons must be sent for machining.
4   The face of the top ring is slightly convex and chrome plated. The middle ring has a recess on the face and must be fitted with the recess towards the bottom of the piston.
5   The bottom oil control ring has drain slots and is expanded by an internal spring.

26 Connecting rods – examination and renovation

1   The connecting rods are not subject to wear but can, in the case of engine seizure, become bent or twisted. If any distortion is visible or even suspected the rod must be renewed.
2   The rods should also be checked for hairline cracks or deep nicks and if in evidence the rod discarded and a new one fitted.

27 Camshaft and camshaft housing – examination and renovation

1   The camshaft runs direct in the bores of the aluminium camshaft housing. Check the fit of the camshaft in the housing bores, if wear is evident the housing must be renewed. It is recommended that when a camshaft and housing are replaced the full set of eight rockers are also replaced.
2   The camshaft itself should show no signs of wear, but, if very slight scoring marks on the cam are noticed, the score marks can be removed by very gentle rubbing down with a very fine emery cloth or an oil stone. The greatest care should be taken to keep the cam profiles smooth.
3   Examine the faces of the rockers; slight score marks can be removed with the careful use of an oil stone. However if the faces are badly worn the rocker(s) should be renewed.

28 Valves and valve seats – examination and renovation

1   Examine the heads of the valves for pitting and burning, especially the heads of the exhaust valves. The valve seatings should be examined at the same time. If the pitting on the valve and seats is very slight the marks can be removed by grinding the seats and valves together with coarse, and then fine, valve grinding paste. Where bad pitting has occurred to the valve seats it will be necessary to recut them to fit new valves. If the valve seats are so worn that they cannot be recut, then it will be necessary to fit valve seat inserts. These latter two jobs should be entrusted to the local Lada dealer or automobile engineering works. In practice it is very seldom that the seats are so badly worn that they require renewal. Normally, it is the valve that is too badly worn for replacement, and the owner can easily purchase a new set of valves and match them to the seats by valve grinding.
2   Valve grinding is carried out as follows: Place the cylinder head upside down on a bench with a block of wood at each end to give clearance for the valve stems. Alternatively, place the head at 45° to a wall with the combustion chambers away from the wall.
3   Smear a trace of coarse carborundum paste on the seat face and apply a suction grinder tool to the valve heads. With a semi-rotary action, grind the valve head to its seat, lifting the valve occasionally to redistribute the grinding paste. When a dull matt even surface finish is produced on both the valve seat and the valve, then wipe off the paste and repeat the process with fine carborundum paste, lifting and turning the valve to redistribute the paste as before. A light spring placed under the valve head will greatly ease this operation. When a smooth unbroken ring of light grey matt finish is produced, on both the valve and the valve seat faces, the grinding operation is complete.
4   Scrape away all carbon from the valve head and the valve stem. Carefully clean away every trace of grinding compound, taking great care to leave none in the ports or in the valve guides. Clean the valves and valve seats with a paraffin soaked rag, then with a clean rag, and finally if an airline is available, blow the valve, valve guides and valve ports clean.

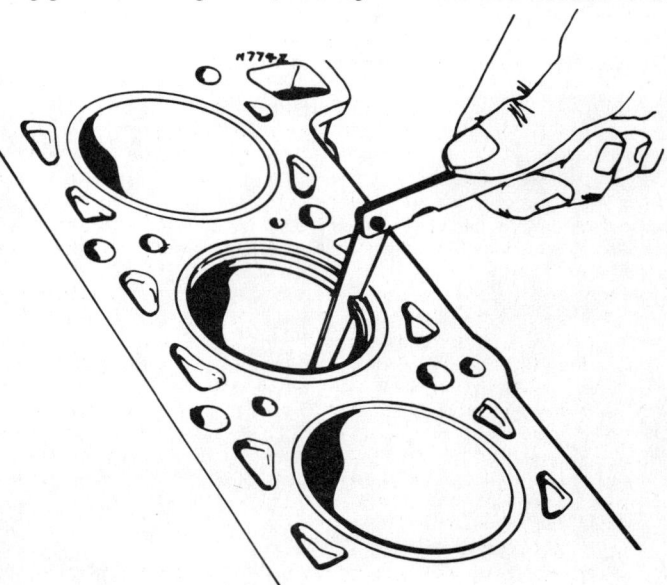

Fig. 1.13 Checking a piston ring gap in the cylinder bore (Sec 25)

29 Valve guides – examination and renovation

1 If the valves are a very slack fit in the guides calling for renewal of the guides, new guides may be fitted. The circlip is removed from the top of the guide and they are then drifted out from the valve port side with a stepped drift. New guides are driven in from the top of the cylinder head, also with a stepped drift. Great care must be taken to ensure that they are driven in 'clean' to start with and that the drift is so shaped that it cannot spread or split the guide. It is possible for the guides to distort slightly when being fitted and it may be necessary to ream them to prevent binding on the valve stem.

2 This job calls for proper tools and experience and is best entrusted to a competent fitter. Otherwise costly damage may be caused to the cylinder head.

30 Cylinder head – decarbonisation

1 It is very unlikely that with modern fuels and oils, decarbonisation will be necessary at anything shorter than 30 000 mile (48 000 km) intervals.

2 This operation can be carried out with the engine either in or out of the car. With the cylinder head off, carefully remove with a wire brush and blunt scraper, all traces of carbon deposits from the combustion spaces and ports. The valve stems and valve guides should be also freed from any carbon deposits. Wash the combustion spaces and ports down with petrol and scrape the cylinder head surface of any foreign matter with the side of a steel rule or a similar article. Take care not to scratch the surface.

3 Clean the pistons and top of the cylinder bores. If the pistons are still in the cylinder bores, then it is essential that great care is taken to ensure that no carbon gets into the cylinder bores as this could scratch the cylinder walls or cause damage to the piston and rings. To ensure that this does not happen, first turn the crankshaft so that two of the pistons are on top of the bores. Place clean non-fluffy rag into the other two bores or seal them off with paper and masking tape. The waterways and pushrod holes should also be covered with a small piece of masking tape to prevent particles or carbon entering the cooling system and damaging the water pump, or entering the lubrication system and damaging the oil pump or bearing surfaces.

4 Press a little grease into the gap between the cylinder walls and the two pistons which are to be worked on. With a blunt scraper carefully scrape away the carbon from the piston crown, taking care not to scratch the aluminium. Also scrape away the carbon from the surrounding lip of the cylinder wall. When all carbon has been removed, scrape away the grease which will now be contaminated with carbon particles, taking care not to press any into the bores. To assist prevention of carbon build up the piston crown can be polished with a metal polish. Remove the rags or masking tape from the other two cylinders and turn the crankshaft so that the two pistons which were at the bottom are now at the top. Place non-fluffy rag into the other two bores or seal them off with paper and masking tape. Do not forget the waterways and oilways as well. Proceed as previously described.

31 Distributor, oil pump and fuel pump drive – examination and renovation

1 The drive for the distributor, oil pump and fuel pump is taken from the auxiliary drive shaft (see Section 13) via a helical gear located in the cylinder block.

2 The gear can be withdrawn through the distributor mounting aperture in the cylinder using a suitable sized piece of wooden dowel forced into the internal splines of the gear, (photo).

3 The gear drives the splined ends of the distributor and oil pump drive shafts and the internal splines of the gear should be examined very carefully for wear.

4 If any doubts exist regarding the condition of gear it should be renewed as failure of the oil pump drive could be disastrous.

32 Flywheel and starter ring gear – examination and renovation

1 The flywheel assembly is a finely balanced item and really there are few jobs which can be undertaken by the home mechanic.

2 Having removed the flywheel from the engine examine the driving face of the flywheel for scores and roughness that will give excessive wear on the clutch disc and a rough 'pick-up' of drive. If the surface is rough and needs re-machining then this task should be entrusted to reputable motor repairers.

3 It will not be a matter of just skimming the surface of the flywheel but also rebalancing it.

4 Starter ring gear: The ring gear is an interference fit on the periphery of the flywheel. If on inspection several teeth are found to be missing and the state of the other teeth is also shabby, the gear should be renewed. As usual check that a spare gear is available before removing the old one.

5 The ring gear may be removed by drilling a 0.24 in (6 mm) diameter hole, some 0.32 in (8 mm) deep into the ring gear. This hole should weaken the gear sufficiently for a sharp blow with a cold chisel, right by the drilled hole, to break the gear. Once the gear is broken it can be lifted from the flywheel.

6 The new ring gear should be put into an oven (domestic electric) and heated to between 180° and 230°C (356° to 446°F). Maintain that temperature for five minutes, to ensure that the ring is fully expanded.

7 Quickly remove the ring from the oven, and place onto the

Fig. 1.14 Drifting out a valve guide (Sec 29)

1 Special tool 60153/R

31.2 Withdrawing the oil pump/distributor drive gear

Chapter 1 Engine

flywheel, with the inner chamfer towards the flywheel. Using a soft metal drift (eg. brass) tap the still hot ring gear evenly into position on the flywheel (Fig. 1.15).

8 The lateral run-out of the fitted gear must not exceed 0.020 in (0.508 mm) measured with a dial gauge. **Note:** *On some models a separate ring gear is not available and it is necessary to fit a complete new flywheel/ring gear assembly. Before attempting to remove the ring gear, check with your Lada dealer regarding the availability of parts.*

33 Engine – reassembly (general)

1 To ensure maximum life with minimum trouble from a rebuilt engine, not only must every part be correctly assembled, but everything must be spotlessly clean, all the oilways must be clear, locking washers and spring washers must always be fitted where indicated and all bearings and other working surfaces must be thoroughly lubricated during assembly. Before assembly begins renew any bolts or studs whose threads are in any way damaged; whenever possible use new spring washers.

2 Apart from your normal tools, a supply of non-fluffy rags, an oil can filled with engine oil (an empty washing-up fluid plastic bottle thoroughly clean and washed out will invariably do just as well), a supply of new spring washers, a set of new gaskets and a torque wrench should be collected together.

3 The order of assembly for the engine is as follows:

  (a) Assemble crankshaft into engine block
  (b) Assemble piston and conrods
  (c) Assemble piston/conrod into engine
  (d) Fit big-end bearings
  (e) Fit the sump
  (f) Assemble timing chain and associated components onto engine
  (g) Assemble the timing case onto the engine
  (h) Fit the flywheel onto the engine
  (i) Assemble the cylinder head
  (j) Fit the cylinder head to the engine
  (k) Fit oil pump, fuel pump, distributor, water pump and clutch

4 The engine should then be ready for refitment to the car.

5 Details of the engine assembly tasks are given in the following Sections.

34 Crankshaft – refitting

1 Ensure that the crankcase is thoroughly clean and that all oilways are clear. A thin twist drill is useful for clearing the oilways, or if possible they may be blown out with compressed air. Treat the crankshaft in the same fashion, and then inject engine oil into the oilways.

2 Never re-use old bearing shells; wipe the shell seats in the crankcase clean and then fit the upper halves of the main bearing shells into their seats (photo).

3 Note that there is a tab on the back of each bearing which engages with a groove in the shell seating (in both crankcase and

Fig. 1.15 Fitting a new starter ring gear (Sec 32)

bearing cap). Wipe away all traces of protective grease on the new shells.

4 Now fit the remaining shells into the bearing caps (photo).

5 It should be noted the shells for bearings Nos 1, 2, 4 and 5 have an oil groove and are identical in size while the centre, (No 3) shells are wider than the rest and have no oil groove.

6 Fit the thrust washer halves in the recess in the rear of the cylinder block and the rear bearing cap (photo).

7 Ensure that the holes in the shells fitted in the crankcase seats line up and do not diminish the oil supply hole in the crankcase. Inject oil into each supply oilway and coat the shells liberally with new engine oil (photo).

8 Lower the crankshaft into position and fit each cap in turn. The mating surfaces of the cap and crankcase must be spotlessly clean. Screw in the cap bolts and tighten each bolt a little. Check the crankshaft for ease of rotation (photos).

9 Should the crankshaft be stiff to turn or possess high spots, a most careful inspection should be made, preferably by a skilled mechanic, to trace the cause of the trouble; fortunately it is very seldom that any trouble of this nature will be experienced when fitting a crankshaft.

10 Tighten all the main bearing cap bolts, progressively to the specified torque wrench setting and recheck the crankshaft for freedom of rotation (photo).

11 Use feeler gauges or a clock gauge to check the crankshaft endfloat which should be within the dimensions given in the Specifications. Various thicknesses of rear bearing flange are available (photo).

12 Check that the square head engine plate retaining bolts are in place in the crankshaft rear seal housing (photo). Fit the seal and housing to the rear of the cylinder block and ensure the seal is central on the crankshaft hub before tightening the retaining bolts (photo).

13 The crankshaft is now ready to receive the piston and connecting rod assemblies.

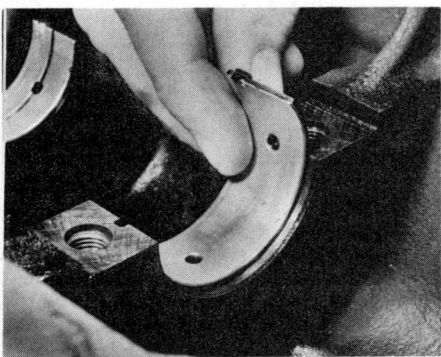

34.2 Inserting a main bearing shell into the cylinder block

34.4 Fitting a shell into a main bearing cap

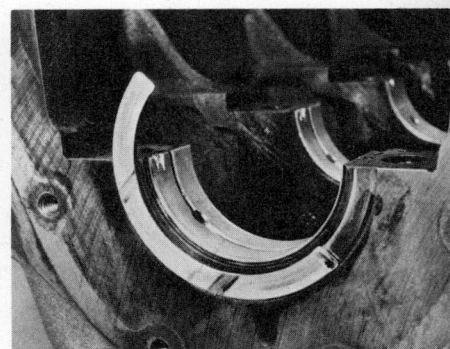

34.6 Fitting the upper thrust bearing

34.7 Lubricating the main bearings prior to fitting the crankshaft

34.8A Fitting the crankshaft

34.8B Fitting the main bearing caps

34.10 Tightening the main bearing cap bolts

34.11 Checking the crankshaft endfloat

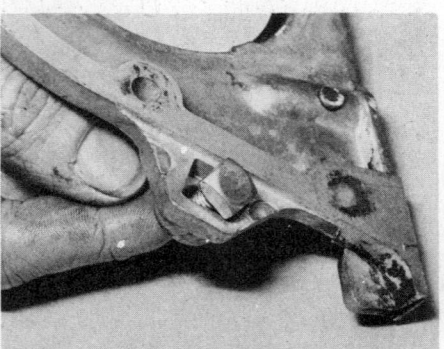
34.12A Checking that the engine plate retaining bolts are in place

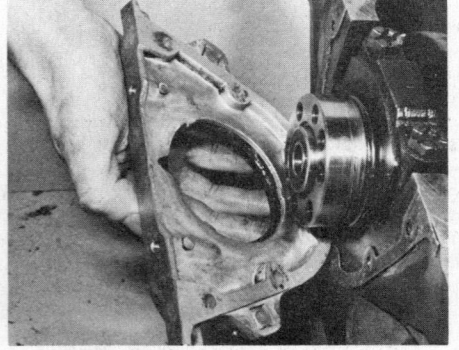
34.12B Fitting the crankshaft rear seal and housing

35 Pistons and connecting rods – reassembly

1 The pistons and connecting rods are supplied separately and whenever the engine block has been rebored the machine shop will supply the correct pistons and piston rings. **Note**: *The gudgeon pin holes in the pistons are offset and it is essential that the pistons are assembled to the connecting rods with the 'll' mark on the piston crown on the same side as the oil hole in the connecting rod.*

2 A special Lada tool No A60325 is available for inserting the gudgeon pins into the pistons but providing the connecting rods are heated to the correct temperature it should be possible to push the gudgeon pins into position using a drift.

3 The technique for assembling the piston and connecting rod is to heat the rod up to 240°C (528°F) and then place the rod and piston together and insert the gudgeon pin while the rod is still hot.

4 Place all four connecting rods in an oven (domestic) and heat to a temperature of around 240°C (528°F). Alternatively, use temperature indicator crayons and a hot plate to bring the small end of the connecting rod to the required temperature. A gas torch can be used directly on the connecting rod but only if you have had experience with such methods before. Care must be taken to 'brush' the small end of the rod evenly with the flame, so that hot spots are not created. It requires care to judge when the material has reached the required temperature, any overheating will seriously weaken the rod. One guide as to when the rod reaches the right temperature is when the clean metal surface is just changing colour to a darker shade of grey.

5 The importance of getting the connecting rod to the correct temperature and maintaining it whilst the gudgeon pin is fitted, cannot be overstressed. If the pin is caught by the contacting small end of the rod before the pin has been positioned properly, the piston will most probably be damaged if any attempt is made to adjust the pin's position.

6 Having heated the rod, place the rod into a vice; hold the piston over the small end. Make sure that you have the connecting rod and piston in the correct relative positions.

Chapter 1 Engine

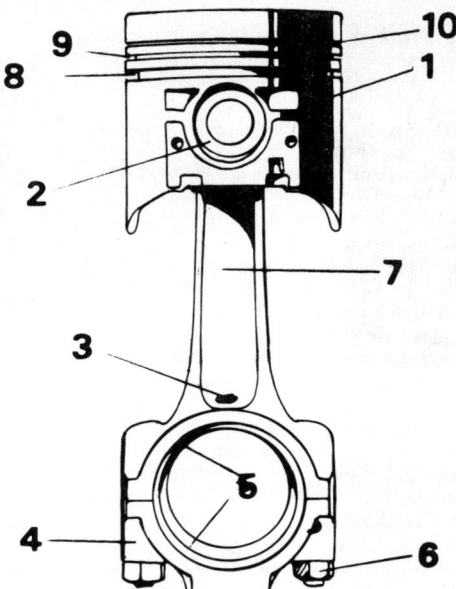

Fig. 1.16 Details of piston and connecting rod (Sec 35)

1 Piston
2 Gudgeon pin
3 Oil hole
4 Connecting rod cap
5 Shell
6 Connecting rod cap nut
7 Connecting rod
8 Oil control ring groove
9 Second compression ring groove
10 Top compression ring groove

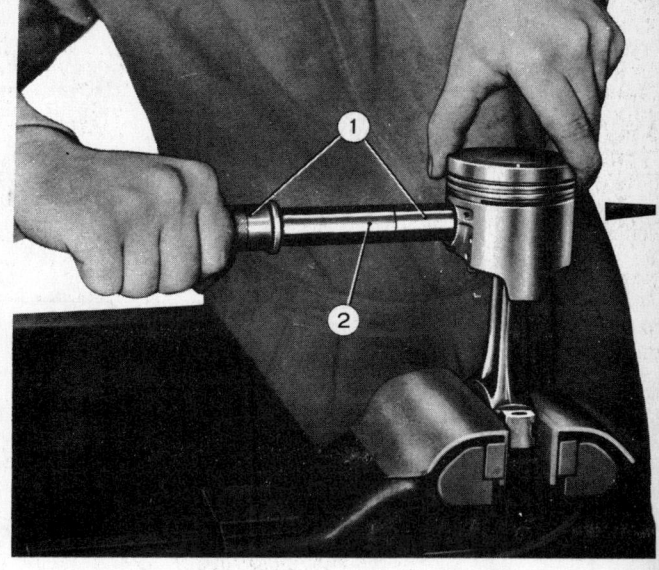

Fig. 1.17 Inserting gudgeon pin (Sec 35)

1 Special tool 60325
2 Gudgeon pin

7 Push the gudgeon pin into position with the tool chosen. When the connecting rod has cooled down, remove the assembly from the vice and check the gudgeon pin position in the assembly. The pin should be exactly central in the piston when the connecting rod is held centrally in the piston. The use of spacers will help in achieving this.
8 The assembly is now ready to accept the piston rings.

36 Piston rings – refitting

1 Check that the piston ring grooves and oilways are thoroughly clear. Always move the rings into position from the top of the piston.
2 The easiest method of fitting piston rings is to use a 0.015 in (0.38 mm) feeler gauge (or similar) around the top of the piston and move the rings into position over the feeler blade. This sequence is the reversal of the removal procedure detailed in Section 17 of this Chapter.
3 Note that as far as the oil control ring (the lowest), is concerned, it is necessary to fit the central spring ring first and then the two steel bands immediately above and below the spacer ring; then ensure that the steel band rings are free to move.
4 When all the rings are in position on the pistons move them around to bring each ring gap to be some 120° away from the adjacent ring. This rule applies to the individual rings that comprise the oil control ring.

37 Pistons and connecting rods – refitting

1 Lay the piston/connecting rod assemblies out in their correct order ready for refitting into their respective bores in the cylinder block. Remember that the connecting rods have been numbered to indicate to which cylinder they are to be fitted.
2 Note that the 'II' mark on the piston crown must face towards the front of the engine.
3 Clean the cylinder bores with a clean non-fluffy rag.
4 Apply some engine oil to the piston rings and then wrap the piston ring compressor around the first assembly to be fitted. A large diameter worm drive hose clip will serve as a ring compressor if a proper tool is not available.
5 Insert the connecting rod and piston into the top of the cylinder

37.5 Inserting a piston into the cylinder block

block and gently tap the piston through the ring compressor into the cylinder bore with a wooden or soft headed mallet. Guide the big-end of the connecting rod near to its position on the crankshaft (photo).
6 Repeat the sequence described for the remaining three piston-connecting rod assemblies.

38 Connecting rods/big-end bearings – refitting to crankshaft

1 Wipe the shell seat in the big-end of the connecting rod clean, and the underside of the new shell bearing. Fit the shell into position in the connecting rod with its locating tongue engaged with the appropriate groove in the big-end. Check that the oil squirt hole in the rod is aligned with the hole in the bearing shell.
2 Generously lubricate the crankpin journals with engine oil and turn

the crankshaft so that it is in its most advantageous position for the rod to be drawn onto it.
3   Wipe the bearing shell seat in the bearing cap clean, and then the underside of the new shell. Fit the shell into the cap, engaging the shell tongue with the groove in the cap.
4   Draw the big-end of the connecting rod onto the crankpin and then fit the cap into position. Make sure it is the correct way around (the numbers must be on the same side) and then insert the two retaining bolts.
5   Tighten the big-end nuts a little at first, and do not tighten fully to the specified torque until all the piston/connecting rod assemblies have been fitted and the rotational freedom of the crankshaft checked (photo).

39 Oil pump and sump – refitting

1   Refit the oil pump and drive shaft to the bottom of the cylinder block using a new gasket and tighten the two retaining bolts, (photo).
2   Refit the crankcase ventilation baffle into the cylinder block housing and bolt the pipe bracket onto the sump flange (photos).
3   Fit a new gasket to the bottom of the cylinder block, lower the sump into position and tighten the retaining bolts evenly and progressively (photo).
4   The engine can now be rolled over onto its sump in readiness for refitting the timing chain and cylinder head.

40 Timing chain and sprockets – refitting

1   Slide the small sprocket onto the keyed end of the crankshaft.
2   Insert the auxiliary drive shaft into the front of the cylinder block. Fit the shaft thrust plate and tighten the two retaining bolts (photos).
3   Fit the large sprocket to the dowelled end of the auxiliary drive shaft. Fit the lockwasher and retaining bolt, but do not attempt to tighten the bolt at this stage.
4   Loop the chain over both sprockets and lock the large auxiliary drive sprocket by inserting a screwdriver between the sprocket teeth and chain. Tighten the sprocket retaining bolt and bend over the lockwasher (photo).
5   Do not fit the timing cover at this stage; engine timing can only be carried out after the cylinder head and camshaft have been fitted.

41 Flywheel – refitting

1   Bolt the large pressed steel plate to the rear of the cylinder block (photo).
2   Offer up the flywheel to the crankshaft flange and line up the bolt holes. Note that the holes are not equally spaced and the flywheel will only fit in one position.
3   Insert the six retaining bolts and tighten them to the specified torque wrench setting (photo).

42 Valve assemblies – refitting to cylinder head

1   To refit the valves and valve springs to the cylinder head proceed as follows:
2   Lay the cylinder head on its side with the spark plug apertures facing downward.
3   Fit the valves into the same guides from which they were removed (photo).
4   Fit the inlet valve stem oil seals in position over the guide protrusion (photo).
5   Place the lower spring seat in position (photo).
6   Fit both inner and outer valve springs (photo).
7   Fit the springs cap onto the spring and position the spring compression tool onto the valve assembly.
8   Compress the springs sufficiently for the cotters to be slipped into place in the cotter groove machined into the top of the valve stem (photo).
9   Remove the valve spring compressor and repeat this procedure until all eight valves have been assembled into the cylinder head.
10 The rocker arms can now be refitted to the adjustable ball-studs and retained in place with the wire clips. Make sure the clips are correctly seated (photo).

43 Camshaft – refitting

1   Take the camshaft and wipe the lobes and journals clean with a non-fluffy rag, and then liberally coat with new engine oil.
2   Clean and then lubricate the bearing bores in the camshaft housing.
3   Carefully insert the camshaft into the housing, being careful not to scratch the surfaces of the bearings in the head. Support the shaft through the access aperture, whilst it is being passed down into position (photo).
4   Fit the thrust plate to the front end of the camshaft housing and tighten the two retaining bolts. Do not fit the camshaft sprocket at this stage.

44 Cylinder head – refitting

1   If the distributor/oil pump drive gear has been removed, carefully lower it into position in the cylinder block using a suitable sized piece of dowelling or tubing jammed in the internal splines.
2   Clean the faces of the cylinder block and head and place a new gasket in place on the block. Note that it can only go on one way.
3   Carefully lower the cylinder head into position on the block ensuring the gasket stays in the right position (photo).
4   Using a piece of wire, hook the timing chain up through the cylinder head aperture and temporarily tie it in place.
5   Now tighten the cylinder head bolts evenly and progressively in the sequence shown in Fig. 1.4 to the specified torque wrench setting (photo).
6   To avoid contact between the piston crowns and the valve heads, set the crankshaft and camshaft sprocket (temporarily refitted) timing

38.5 Tightening the big-end bearing nuts

39.1 Refitting the oil pump

39.2A Inserting the crankcase ventilation baffle into the cylinder block

Are your plugs trying to tell you something?

WHY DOUBLE COPPER IS BETTER FOR YOUR ENGINE.

Normal.
Grey-brown deposits, lightly coated core nose. Plugs ideally suited to engine, and engine in good condition.

Heavy Deposits.
A build up of crusty deposits, light-grey sandy colour in appearance.
Fault: Often caused by worn valve guides, excessive use of upper cylinder lubricant, or idling for long periods.

Lead Glazing.
Plug insulator firing tip appears yellow or green/yellow and shiny in appearance.
Fault: Often caused by incorrect carburation, excessive idling followed by sharp acceleration. Also check ignition timing.

Carbon fouling.
Dry, black, sooty deposits.
Fault: over-rich fuel mixture.
Check: carburettor mixture settings, float level, choke operation, air filter.

Oil fouling.
Wet, oily deposits. Fault: worn bores/piston rings or valve guides; sometimes occurs (temporarily) during running-in period.

Overheating.
Electrodes have glazed appearance, core nose very white – few deposits. Fault: plug overheating. Check: plug value, ignition timing, fuel octane rating (too low) and fuel mixture (too weak).

Electrode damage.
Electrodes burned away; core nose has burned, glazed appearance. Fault: pre-ignition. Check: for correct heat range and as for 'overheating'.

Split core nose.
(May appear initially as a crack). Fault: detonation or wrong gap-setting technique. Check: ignition timing, cooling system, fuel mixture (too weak).

Unique Trapezoidal Copper Cored Earth Electrode — 50% Larger Spark Area — Copper Cored Centre Electrode

Champion Double Copper plugs are the first in the world to have copper core in both centre and earth electrode. This innovative design means that they run cooler by up to 100°C – giving greater efficiency and longer life. These double copper cores transfer heat away from the tip of the plug faster and more efficiently. Therefore, Double Copper runs at cooler temperatures than conventional plugs giving improved acceleration response and high speed performance with no fear of pre-ignition.

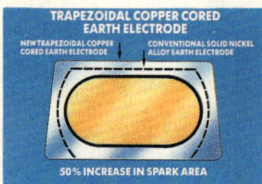

Champion Double Copper plugs also feature a unique trapezoidal earth electrode giving a 50% increase in spark area. This, together with the double copper cores, offers greatly reduced electrode wear, so the spark stays stronger for longer.

 FASTER COLD STARTING

 FOR UNLEADED OR LEADED FUEL

 ELECTRODES UP TO 100°C COOLER

 BETTER ACCELERATION RESPONSE

 LOWER EMISSIONS

 50% BIGGER SPARK AREA

 THE LONGER LIFE PLUG

Plug Tips/Hot and Cold.
Spark plugs must operate within well-defined temperature limits to avoid cold fouling at one extreme and overheating at the other.
Champion and the car manufacturers work out the best plugs for an engine to give optimum performance under all conditions, from freezing cold starts to sustained high speed motorway cruising.
Plugs are often referred to as hot or cold. With Champion, the higher the number on its body, the hotter the plug, and the lower the number the cooler the plug.

Plug Cleaning
Modern plug design and materials mean that Champion no longer recommends periodic plug cleaning. Certainly don't clean your plugs with a wire brush as this can cause metal conductive paths across the nose of the insulator so impairing its performance and resulting in loss of acceleration and reduced m.p.g.
However, if plugs are removed, always carefully clean the area where the plug seats in the cylinder head as grit and dirt can sometimes cause gas leakage.
Also wipe any traces of oil or grease from plug leads as this may lead to arcing.

This photographic sequence shows the steps taken to repair the dent and paintwork damage shown above. In general, the procedure for repairing a hole will be similar; where there are substantial differences, the procedure is clearly described and shown in a separate photograph.

First remove any trim around the dent, then hammer out the dent where access is possible. This will minimise filling. Here, after the large dent has been hammered out, the damaged area is being made slightly concave.

Next, remove all paint from the damaged area by rubbing with coarse abrasive paper or using a power drill fitted with a wire brush or abrasive pad. 'Feather' the edge of the boundary with good paintwork using a finer grade of abrasive paper.

Where there are holes or other damage, the sheet metal should be cut away before proceeding further. The damaged area and any signs of rust should be treated with Turtle Wax Hi-Tech Rust Eater, which will also inhibit further rust formation.

For a large dent or hole mix Holts Body Plus Resin and Hardener according to the manufacturer's instructions and apply around the edge of the repair. Press Glass Fibre Matting over the repair area and leave for 20-30 minutes to harden. Then ...

... brush more Holts Body Plus Resin and Hardener onto the matting and leave to harden. Repeat the sequence with two or three layers of matting, checking that the final layer is lower than the surrounding area. Apply Holts Body Plus Filler Paste as shown in Step 5B.

For a medium dent, mix Holts Body Plus Filler Paste and Hardener according to the manufacturer's instructions and apply it with a flexible applicator. Apply thin layers of filler at 20-minute intervals, until the filler surface is slightly proud of the surrounding bodywork.

For small dents and scratches use Holts No Mix Filler Paste straight from the tube. Apply it according to the instructions in thin layers, using the spatula provided. It will harden in minutes if applied outdoors and may then be used as its own knifing putty.

Use a plane or file for initial shaping. Then, using progressively finer grades of wet-and-dry paper, wrapped round a sanding block, and copious amounts of clean water, rub down the filler until glass smooth. 'Feather' the edges of adjoining paintwork.

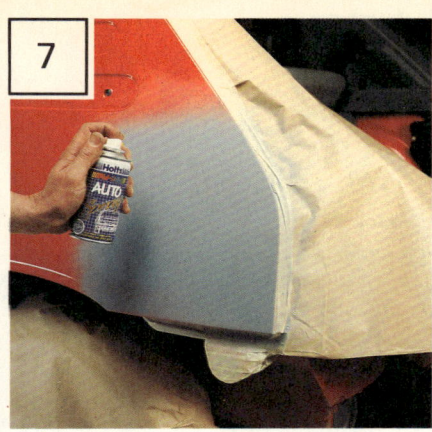

7 Protect adjoining areas before spraying the whole repair area and at least one inch of the surrounding sound paintwork with Holts Dupli-Color primer.

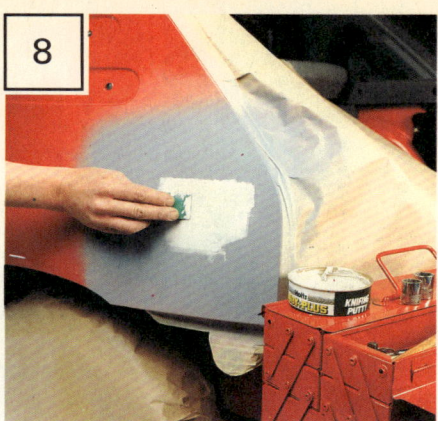

8 Fill any imperfections in the filler surface with a small amount of Holts Body Plus Knifing Putty. Using plenty of clean water, rub down the surface with a fine grade wet-and-dry paper – 400 grade is recommended – until it is really smooth.

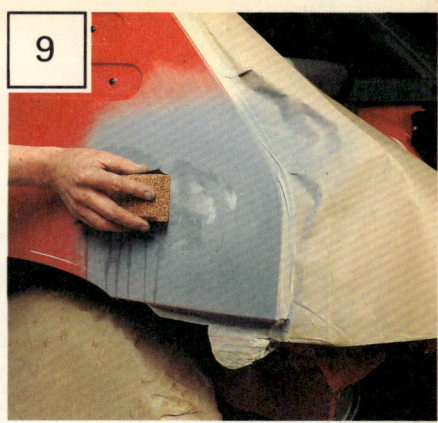

9 Carefully fill any remaining imperfections with knifing putty before applying the last coat of primer. Then rub down the surface with Holts Body Plus Rubbing Compound to ensure a really smooth surface.

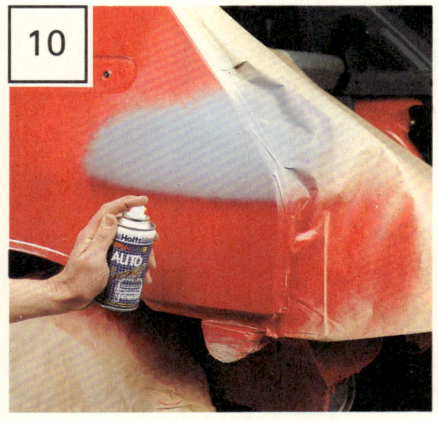

10 Protect surrounding areas from overspray before applying the topcoat in several thin layers. Agitate Holts Dupli-Color aerosol thoroughly. Start at the repair centre, spraying outwards with a side-to-side motion.

10A If the exact colour is not available off the shelf, local Holts Professional Spraymatch Centres will custom fill an aerosol to match perfectly.

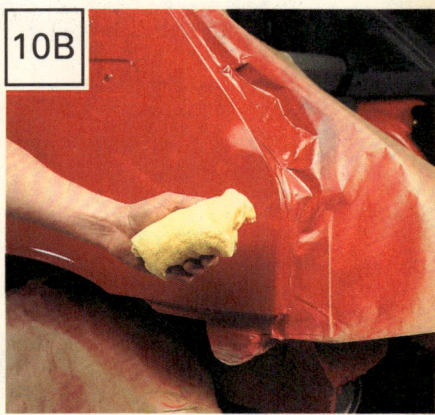

10B To identify whether a lacquer finish is required, rub a painted unrepaired part of the body with wax and a clean cloth.

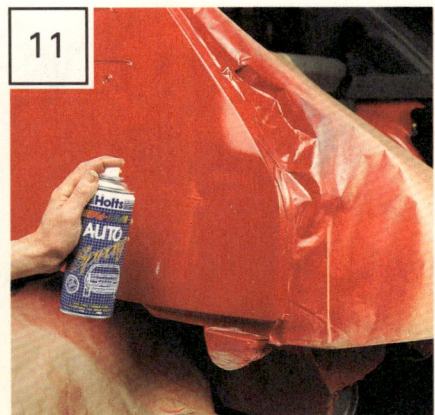

11 If *no* traces of paint appear on the cloth, spray Holts Dupli-Color clear lacquer over the repaired area to achieve the correct gloss level.

12 The paint will take about two weeks to harden fully. After this time it can be 'cut' with a mild cutting compound such as Turtle Wax Minute Cut prior to polishing with a final coating of Turtle Wax Extra.

14 When carrying out bodywork repairs, remember that the quality of the finished job is proportional to the time and effort expended.

HAYNES No1 for DIY

Haynes publish a wide variety of books besides the world famous range of **Haynes Owners Workshop Manuals**. They cover all sorts of DIY jobs. Specialist books such as the **Improve and Modify** series and the **Purchase and DIY Restoration Guides** give you all the information you require to carry out everything from minor modifications to complete restoration on a number of popular cars. In addition there are the publications dealing with specific tasks, such as the **Car Bodywork Repair Manual** and the **In-Car Entertainment Manual**. The **Household DIY** series gives clear step-by-step instructions on how to repair everyday household objects ranging from toasters to washing machines.

Whether it is under the bonnet or around the home there is a Haynes Manual that can help you save money. Available from motor accessory stores and bookshops or direct from the publisher.

39.2B Fitting the baffle pipe retaining bracket

39.3 Lowering the oil sump into position

40.2A Insert the auxiliary drive shaft into the front of the cylinder block

40.2B Auxiliary shaft thrust plate

40.4 Fitting of timing chain and sprockets

41.1 Rear engine plate in position

41.3 Tightening the flywheel retaining bolts

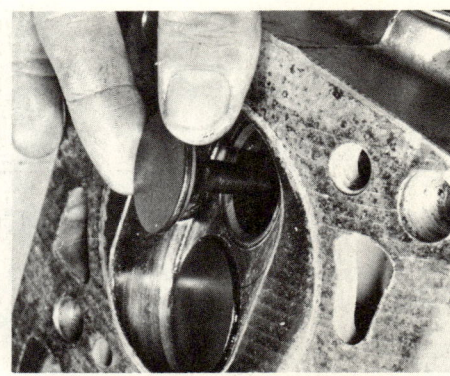
42.3 Fitting a valve into the cylinder head

42.4 Correct location of valve stem oil seal

42.5 Positioning the valve spring lower seat

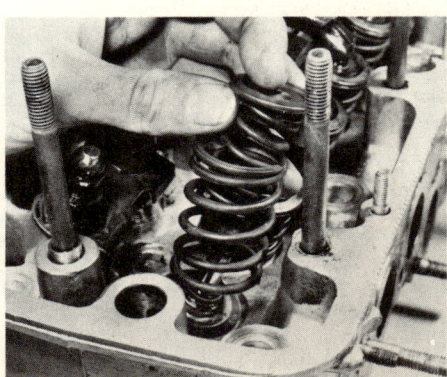
42.6 Fitting the valve springs and cap

42.8 Fitting the retaining collets into the valve spring cap

42.10 Fitting the rockers and retaining clips

43.3 Inserting the camshaft into the housing

44.3 Lowering the cylinder head onto the block

44.5 Tightening the cylinder head bolts

44.6 Fitting the camshaft and housing

44.10A Refitting the timing cover

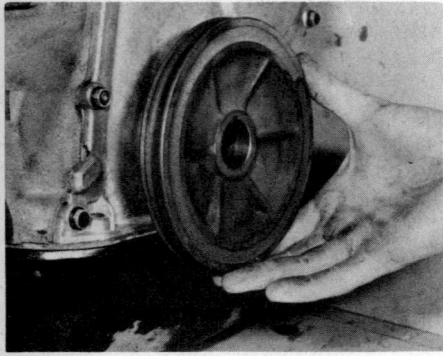
44.10B Fitting the crankshaft pulley

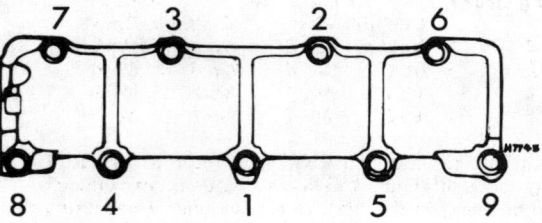
Fig. 1.18 Correct sequence for slackening or tightening camshaft housing nuts (Sec 44)

marks in alignment with their TDC marks. Lower the camshaft and housing onto the cylinder head. Tighten the retaining nuts evenly and progressively (Fig. 1.18).
7   Refit the camshaft sprocket and timing chain as described in Section 12.
8   Fit the chain tensioner and check that the camshaft timing marks are still correctly aligned.
9   Adjust the timing chain tension as described in Section 13, paragraph 9.
10 Finally drift a new seal into the timing cover aperture and fit the timing cover to the cylinder block. Fit the pulley onto the front of the crankshaft and tighten the retaining nut. This nut should be tightened fully after the engine has been refitted in the car (photos).
11 The engine is now ready to have the valve clearances adjusted as described in the following Section.

45 Valve clearances – adjustment

1   Valve clearances should be checked and adjusted if necessary at the intervals specified in Routine Maintenance, or after any operation on the cylinder head which may affect cam to rocker clearance.
2   *If the engine is in the car,* remove the air cleaner as described in Chapter 3. Also remove the rocker cover; if this is done carefully the gasket may not be damaged.
3   The valve clearances are adjusted by slackening the locknut on the rocker pivot stud and screwing the stud in or out until the required clearance is obtained between the top of the rocker and the heel of the cam for both inlet and exhaust valves.
4   The correct clearance for both inlet and exhaust valves is given in the Specifications. Note that the clearance differs between a hot and a

45.4 Check the valve clearances with the special wide feeler gauge

45.6 Tighten the locknut (lower) whilst holding the adjuster nut (upper) still

cold engine. Lada specify the use of a wide feeler gauge (tool no A95111) for checking the gap (photo); if a conventional feeler blade is used, be sure that the rocker pad does not tilt and give a false reading.

5   Turn the engine (with the starting handle if the engine is in the car) until the marks on the camshaft sprocket and camshaft housing are aligned as shown in Fig. 1.3. At this point No 4 piston is at TDC on the firing stroke. Check the clearance of No 4 exhaust valve (cam No 8) and No 3 inlet valve (cam No 6); the rockers for both these valves are on the heels of their respective cams with the engine in this position.

6   The feeler gauge should be a tight sliding fit between the heel of the cam and the rocker arm (gap A in Fig. 1.5). Should adjustment be necessary, slacken the locknut, rotate the adjusting nut until the correct clearance is achieved, then retighten the locknut whilst holding the adjusting nut still (photo). Recheck the clearance.

7   Rotate the crankshaft through 180° in the normal direction of rotation; this will bring No 2 piston to TDC on the firing stroke, and the rockers of exhaust valve No 2 (cam No 4) and inlet valve No 4 (cam No 7) will be on the heels of their cams. Check and adjust those valves as described above.

8   Continue the procedure, referring to the table below for details of which valve to adjust:

Crankshaft position (deg)	Cylinder on firing stroke	Valves to adjust			
		Exhaust		Inlet	
		Cylinder	Cam	Cylinder	Cam
0	No 4	No 4	No 8	No 3	No 6
180	No 2	No 2	No 4	No 4	No 7
360	No 1	No 1	No 1	No 2	No 3
540	No 3	No 3	No 5	No 1	No 2

9   It is very important that the adjusting screw locknuts on No 1 cylinder are fully tightened after adjustment. This is sometimes difficult due to the distributor being in the way. Use a spanner if necessary which has been ground away to tighten these nuts, as if they loosen in service, the valves could impinge on the piston crowns. Alternatively, remove the distributor (see Chapter 4).

10   The importance of correct rocker/cam adjustment cannot be over-stressed. If the gap is too large, the valves will not open fully and the engine 'breathing' will be constricted. If the gap is too small, the valves and pistons may damage each other, and the engine 'breathing' will deteriorate because during the engine cycle both inlet and exhaust valves will open for longer periods than intended. Both effects will reduce engine economy and power.

11   When the valve clearances have been adjusted, the rocker cover may be refitted. If a new gasket is to be used, smear it with a little grease and make sure that all traces of the old gasket are removed from the cover and cylinder head. Position the gasket on the cylinder head, lower the cover into place and tighten the screws.

12   Refit the air cleaner if it was removed.

46 Distributor – refitting

1   Just as it is necessary to synchronise the camshaft with the crankshaft, to ensure that the valves open and close at the correct moments, so it is important to ensure that the distributor is correctly fitted. The spark which ignites the fuel/air charge in the cylinder must arrive at the correct instant to ensure that the maximum amount of energy is extracted from the combustion.

2   The distributor shaft fits in the small splines of the drive gear in the cylinder block and the position of the distributor can be varied quite finely.

3   The distributor must be fitted with the No 1 piston at TDC and the distributor rotor pointing towards the No 1 spark plug HT lead segment in the distributor cap.

4   The correct method of distributor adjustment and ignition timing is fully described in Chapter 4.

5   When the ignition timing is correct the clamp bolt must be tightened.

47.1a Refitting the water pump

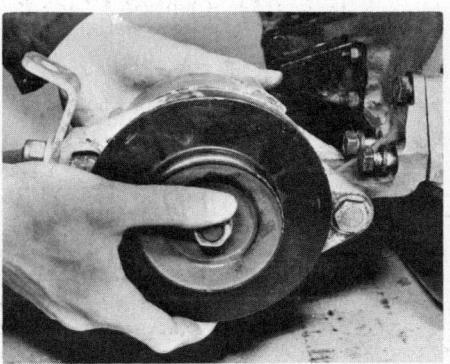

47.1b Fitting the alternator and mounting bracket

47.1c Fitting the fuel pump and spacer

47.1d Fitting the crankcase ventilation cover

47.1e Fitting the exhaust manifold – note the new gasket

47.1f Fitting the inlet manifold with carburettor attached

47.1g Fitting of engine mounting brackets

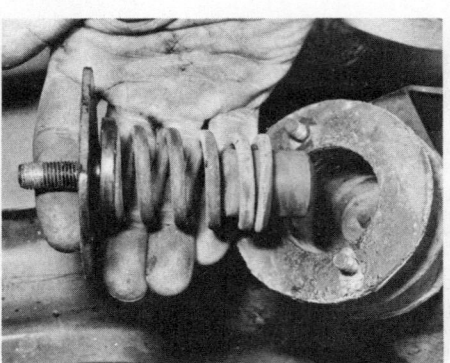
48.2 Correct assembly of left-hand side engine mounting

48.5 Lowering the engine into position

47 Final assembly – general

1 The following components should be assembled onto the engine, before it is refitted into the car. Refer to the relevant Chapters of this manual if you are in any doubt as to points of procedure of reassembly:

   (a) Refit the water pump (photo)
   (b) Refit the alternator and support bracket (photo)
   (c) Refit the spacer and fuel pump (photo)
   (d) Refit the crankcase ventilation cover (photo)
   (e) Fit a new oil filter cartridge
   (f) Refit the exhaust manifold (photo)
   (g) Refit the inlet manifold and carburettor (photo)
   (h) Refit the clutch assembly to the flywheel and centralize it as described in Chapter 5
   (i) Refit the engine support brackets (photo)

48 Engine – refitting

1 If the engine is being refitted to the car with the exhaust and inlet manifolds attached, it is important that the starter motor is in place behind the heat shield prior to lowering the engine into the vehicle.
2 Ensure that the left-hand engine mounting is correctly assembled (photo).
3 Basically the refitting of the engine is the reverse procedure to the removal operation, however, mating the engine to the gearbox can be difficult unless the following method is used.
4 Make sure the clutch is centralised on the flywheel as described in Chapter 5.
5 Carefully lower the engine into the engine compartment using a suitable hoist until the rear of the engine is straight and level with the clutch housing (photo).
6 Push the engine rearwards ensuring the gearbox input shaft enters the clutch assembly in a straight line and not at an angle.
7 If the engine begins to mate up and then stops with a couple of inches still to go, fit a spanner onto the crankshaft starter dog and turn it slowly while pushing the engine rearwards.
8 As soon as the clutch housing mates fully with the rear of the engine, insert one of the gearbox to the engine bolts to hold them together and then lower the engine so that the brackets locate on the engine mountings. Fit the engine mounting nuts and tighten them.
9 Insert the remaining three gearbox retaining bolts and tighten them to the specified torque as described in Chapter 5.
10 Slide the starter motor rearwards into the clutch housing aperture and secure it in position with the three retaining bolts.
11 Finally, lower the hoist completely and remove the engine slings and the jack or blocks from beneath the transmission. Reconnect all the carburettor controls, electrical leads, fuel pipes, exhaust pipe etc, checking each item against the sequence given in Section 5. Also refit the rocker cover if removed for fitting lifting hooks.
12 Do not forget to refill the cooling system, and refill the engine with the recommended grade and quantity of oil.

Chapter 1 Engine

49 Engine – initial start-up after overhaul

1 Make sure that the battery is fully charged and that all lubricants, coolant and fuel are replenished.
2 Prime the carburettor with petrol by operating the lever on the fuel pump.
3 As soon as the engine fires and runs, keep it going at a fast tickover only (no faster), and bring it up to the normal working temperature.
4 As the engine warms up there will be odd smells and some smoke from parts getting hot and burning off oil deposits. The signs to look for are leaks of water or oil which will be obvious if serious. Check also the exhaust pipe and manifold connections, as these do not always 'find' their exact gas tight position until the warmth and vibration have acted on them, and it is almost certain that they will need tightening further. This should be done of course, with the engine stopped.
5 When normal running temperature has been reached adjust the engine idling speed, as described in Chapter 3.
6 Stop the engine and wait a few minutes to see if any lubricant or coolant is dripping out when the engine is stationary.
7 Road test the car to check that the timing is correct and that the engine is giving the necessary smoothness and power. Do not race the engine; if new bearings and/or pistons have been fitted it should be treated as a new engine and run in at a reduced speed for the first 500 miles (800 km).

50 Fault diagnosis – engine

Symptom	Reason/s
Engine fails to turn over when starter control operated	
No current at starter motor	Flat or defective battery
	Loose battery leads
	Defective starter solenoid or switch or broken wiring
	Engine earth strap disconnected
Current at starter motor	Jammed starter motor drive pinion
	Defective starter motor
Engine turns over but will not start	
No spark at spark plug	Ignition damp or wet
	Ignition leads to spark plugs loose
	Shorted or disconnected low tension leads
	Dirty, incorrectly set, or pitted contact breaker points
	Faulty condenser
	Defective ignition switch
	Ignition leads connected wrong way round
	Faulty coil
	Contact breaker point spring earthed or broken
No fuel at carburettor float chamber or at jets	No petrol in petrol tank
	Vapour lock in fuel line (in hot conditions or at high altitude)
	Blocked float chamber needle valve
	Fuel pump filter blocked
	Choked or blocked carburettor jets
	Faulty fuel pump
Engine stalls and will not start	
Excess of petrol in cylinder or carburettor flooding	Too much choke allowing too rich a mixture to wet plugs
	Float damaged or leaking or needle not seating
	Float lever incorrectly adjusted
No spark at spark plug	Ignition failure – sudden
	Ignition failure – misfiring precludes total stoppage
	Ignition failure – in severe rain or after traversing water splash
No fuel at jets	No petrol in petrol tank
	Petrol tank breather choked
	Sudden obstruction in carburettor
	Water in fuel system
Engine misfires or idles unevenly	
Intermittent spark at spark plug	Ignition leads loose
	Battery leads loose on terminals
	Battery earth strap loose on body attachment point
Intermittent sparking at spark plug	Engine earth lead loose
	Low tension leads to (+) and (–) terminals on coil loose
	Low tension lead from (–) terminal side to distributor loose
	Dirty, or incorrectly gapped plugs
	Dirty, incorrectly set, or pitted contact breaker points
	Tracking across inside of distributor cover
	Ignition too retarded
	Faulty coil

Chapter 2 Cooling system

For modifications, and information applicable to later models, see Supplement at end of manual

Contents

Antifreeze and corrosion inhibitors – general ... 11	General description ... 1
Cooling system – draining, flushing and filling ... 2	Radiator – removal, inspection, cleaning and refitting ... 3
Drive belt – adjustment ... 8	Thermostat – removal, testing and refitting ... 4
Drive belt – removal and refitting ... 7	Water pump – dismantling and overhaul ... 6
Electric cooling fan – removal and refitting ... 12	Water pump – removal and refitting ... 5
Expansion tank ... 9	Water temperature gauge – fault finding ... 10
Fault diagnosis – cooling system ... 13	

Specifications

Type
Pressurised; pump assisted circulation with thermostat temperature control

Coolant
Type/specification ... Ethylene glycol based antifreeze, to BS 3151, 3152 or 6580 (Duckhams Universal Antifreeze and Summer Coolant)
Capacity ... 16.8 pints (9.6 litres)

Radiator
Cap opening pressure ... 7.1 lbf/in^2 (0.5 kg f/cm^2)

Thermostat
Type ... Wax filled
Opening temperature ... 80 to 82°C (176 to 179°F)
Fully open ... 90 to 92°C (194 to 197°F)

Water pump
Type ... Centrifugal impeller
Clearance between impeller and body ... 0.035 to 0.051 in (0.9 to 1.3 mm)

Electric cooling fan thermal switch
Location ... Left-hand side of radiator bottom tank
Cut-in temperature ... 198° ± 5°F (92° ± 3°C)
Cut-out temperature ... 189° ± 5°F (87° ± 3°C)

1 General description

The engine cooling water is circulated through the cooling passages in the engine block and radiator by a centrifugal pump driven by the drive belt. A fan forces air through the radiator to assist cooling at low road speeds or high engine revolutions. An electric fan is fitted to the 1500 an 1600 saloon models. The plastic four-blade fan is enclosed in a shroud. The blades being of a twisted aerofoil design. To damp the noise, the blades are unequally spaced around the hub and have rounded rear ends. The fan together with the pulley is held to the hub by three bolts and is V-belt driven from the crankshaft pulley.

The radiator is of the vertical finned-tube type with two rows of tubes and steel tinned fins. The top and bottom tanks and tubes of the radiator are made of brass, the tubes being arranged in a staggered order. The radiator is easily removable, being mounted on two rubber pads and fastened to the front end of the ar body by two bolts passing through steel braces and rubber bushes.

The radiator filler cap contains an outlet valve and vent valve. When the coolant expands due to a temperature rise, it passes into the expansion tank via the vent valve. When the coolant temperature drops, the process is reversed. When the coolant starts to boil or its temperature rises sharply, the capacity of the vent valve proves to be insufficient. The valve closes and isolates the system from the expansion tank. The pressure in the system rises with a resultant rise in the boiling point of the coolant and the radiator head dissipation increases.

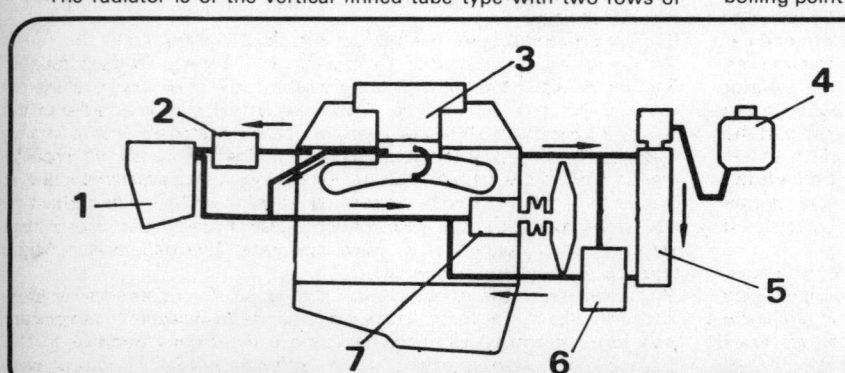

Fig. 2.1 Layout of cooling system (Sec 1)

1 Heater
2 Heater control valve
3 Cylinder head
4 Expansion tank
5 Radiator
6 Thermostat
7 Water pump

Chapter 2 Cooling system

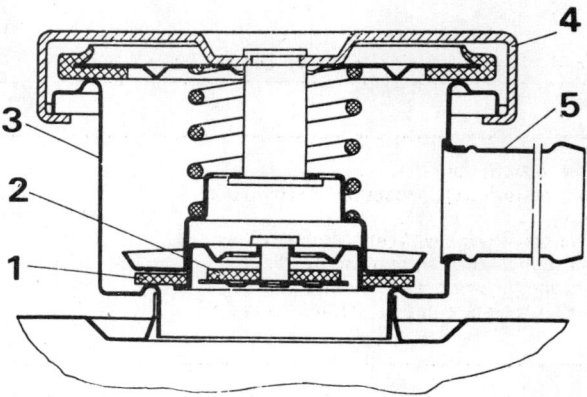

Fig. 2.2 Radiator filler cap (Sec 1)

1 Outlet valve
2 Vent valve
3 Filler pipe
4 Radiator cap
5 Steam outlet (to expansion tank)

When a further rise in temperature brings the pressure in the system to 7.1 lbf/in^2, the outlet valve opens and allows boiling coolant to pass into the expansion tank.

The thermostat maintains the required running temperature of the engine. It has two inlet pipe connections; one being connected to the cylinder head hose and the other being connected to the radiator bottom tank. The one outlet connection is connected to the pump inlet. The temperature-sensitive element of the thermostat consists of a sleeve pressed into the main valve which is held by a spring against its seat. Rolled into the sleeve is a rubber insert which is capable of moving along the piston, the latter being secured in the holder by an adjoining nut. A bypass valve is fitted in the case and supported by a spring bearing on the bottom of the sleeve. The space between the walls of the sleeve and the rubber insert contains solid filler. There are two types of filler, differing in the coefficient of expansion. According to the type of filler, the main valve starts to open at 80°C or 82°C. This temperature is marked on the thermostat bottom.

When the coolant temperature is below 80°C or 82°C, the main valve cuts off the flow of liquid from the radiator, whereas the bypass valve is open thus permitting the liquid to flow from the engine to the pump. When the coolant temperature rises, the filler expands, pushes the piston out of the rubber insert, overrides the spring and lifts the sleeve. The bypass valve, pressed by the bottom of the sleeve through the spring against the end of the inlet pipe connection (cylinder head hose), cuts off direct circulation of coolant from the engine to the pump. As the coolant temperature rises above 90°C to 92°C (depending on the type of thermostat) the main valve opens widely thus permitting the coolant to circulate through the radiator. At intermediate temperatures the coolant circulates both through main and bypass valves so that cold and hot liquids are gradually mixed thus providing optimum temperature conditions of the engine.

The heater which is standard on all models, is fed by a hot water pipe from the cylinder head water jacket. The cooling system circulation is as follows: From the cylinder head, coolant passes through the top hose to the radiator header tank, it then passes down the radiator core and is drawn into the pump via the thermostat. The pump forces coolant into the cylinder block where it passes up to the head and the cycle repeats itself. When the engine is first started from cold the thermostat is closed, preventing the coolant from circulating through the radiator and thus ensuring rapid warm-up. As soon as the engine reaches normal operating temperature the thermostat opens and enables the coolant to circulate through the radiator.

The thermostat is housed in the three-way junction and when cold blocks the bottom radiator hose. When hot it blocks the by-pass pipe from the head. When the water heats up and the thermostat valve opens the water circulates through the top radiator hose, down through the radiator and up to the other side of the thermostat and to the pump.

2 Cooling system – draining, flushing and filling

1 With the car on level ground, drain the system as follows:
2 If the engine is cold, remove the filler cap from the radiator by turning the cap anti-clockwise. If the engine is hot, having just been run, turn the filler cap very slightly until the pressure in the system has had time to disperse. Use a rag over the cap to protect your hand from escaping steam. If, with the engine very hot, the cap is released suddenly, the drop in pressure can result in the water boiling. With the pressure released, the cap can be removed.
3 If antifreeze is in the radiator drain it into a clean bucket or bowl for re-use.
4 Open the drain plug at the bottom of the radiator and the drain plug on the left-hand side of the cylinder block.
5 When the water has finished running, probe the drain plug orifices with a short piece of wire to dislodge any particles of rust or sediment which may be blocking the taps and preventing all the water draining out. **Note**: *Opening only the radiator tap will not drain the cylinder block.*
6 If neglected the cooling system will gradually lose its efficiency as the radiator becomes choked with rust scales, deposits from water and other sediment. To clean the system out, remove the radiator cap and drain tap and leave a hose running in the radiator cap orifice for ten to fifteen minutes.
7 Where the system is suspected of being choked with scale of one sort or another, it is best to give it a treatment with one of the many proprietary chemical descaling compounds which are on the market. Check that whichever one you choose is suitable for use with aluminium cylinder heads.
8 When filling the system use clean water (mixed with antifreeze as appropriate – see Section 11). Some may take the trouble to use only distilled water or rain water. This will greatly reduce the rate of deposit formation. It is recommended that the following procedure is followed when filling the system:

(a) *Remove the hose from the heater outlet pipe and open the heater valve*
(b) *Fill the radiator with coolant until it starts flowing out of the heater outlet pipe. Fit the hose on the pipe and secure it with the clamp without stopping the filling operation, so as to minimise an air lock*
(c) *Fill the radiator until it overflows and replace the radiator cap. Add coolant to the expansion tank until the level reaches 3 cm above the minimum mark*

3 Radiator – removal, inspection, cleaning and refitting

1 Drain the cooling system as described in Section 2.
2 Undo the clip which holds the top water hose to the header tank and pull it off. Unclip and pull off the bottom hose.
3 Slacken the expansion tank pipe clamp and pull the pipe off the radiator filler neck connection.
4 Remove the securing bolts and detach the fan shroud from the radiator (if fitted with the two-piece shroud, first disconnect the halves). On 1500 and 1600 models, disconnect and unbolt the electric fan assembly as described in Section 12.
5 Remove the two bolts, one on each side at the top of the radiator, that secure it to the body (photo). Take care of the washers, spacers and buffer pads.
6 Lift the radiator out (the bottom is located in a channel).
7 Clean out the inside of the radiator by flushing, as described in Section 2. When the radiator is out of the car, it is well worthwhile to invert it for reverse flushing. Clean the exterior of the radiator by hosing down the matrix (honeycomb cooling material) with a strong water jet to clear away embedded dirt and insects which will impede the air flow. If it is thought that the radiator may be partially blocked, it is best to first flush it with a chemical solution while still in the car. If this does not have any effect, flush water through and check that water flows through it at a reasonable rate. Five gallons should go through in about half a minute.
8 Leaks which are on the exterior can usually be repaired by soldering but leaks in the vertical tubes are not so accessible. A temporary leak stopper can be obtained by using a proprietary additive to the cooling liquid or by blocking up the offending area of radiator with mastic or filler paste. Beyond this through there is little alternative than

Chapter 2 Cooling system

3.5 One of the radiator retaining bolts

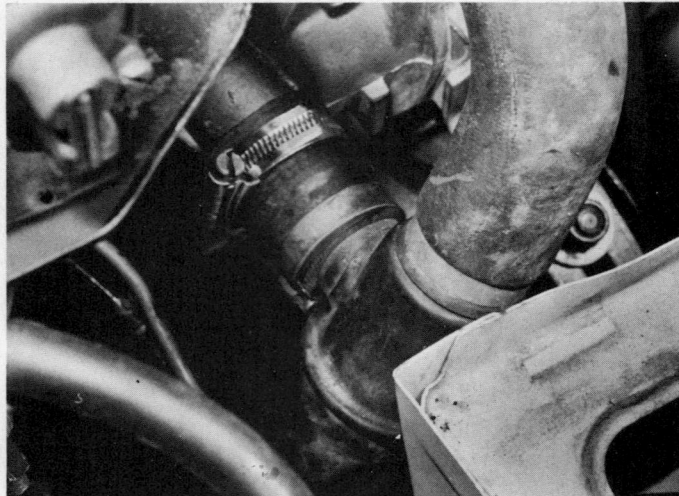

4.1 Hose connections to thermostat housing

a new radiator or a professional repair.
9 Replacement of the radiator is a reversal of the removal procedure. Check that the hoses are not cracked or brittle with age. Fit new ones if they are. Some owners prefer to fit screw type worm drive hose clips in place of the type fitted. Whichever is used do not overtighten them.

4 Thermostat – removal, testing and refitting

1 To remove the thermostat, partially drain the radiator, slacken the clips and remove the three hoses from the thermostat housing (photo).
2 Remove the thermostat and suspend it by a piece of string in a saucepan of cold water together with a thermometer. Neither the thermostat nor the thermometer should touch the bottom of the saucepan, to ensure a false reading is not given.
3 Heat the water, stirring it gently with the thermometer to ensure temperature uniformity and note when the thermostat begins to open. The temperature at which this should happen is given in the Specifications.
4 Discard the thermostat if it opens too early. Continue heating the

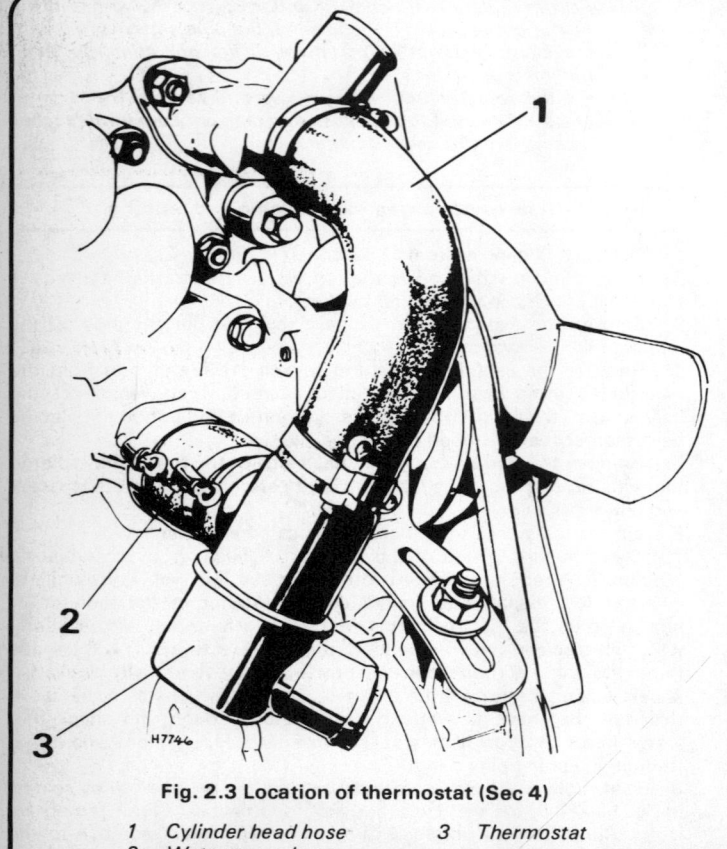

Fig. 2.3 Location of thermostat (Sec 4)

1 Cylinder head hose
2 Water pump hose
3 Thermostat

Fig. 2.4 Sectional view through thermostat housing (Sec 4)

1 Inlet pipe connection (from engine)
2 Bypass valve
3 Bypass valve spring
4 Sleeve
5 Rubber insert
6 Outlet pipe connection
7 Main valve spring
8 Main valve seat
9 Main valve
10 Holder
11 Adjusting nut
12 Piston
13 Inlet pipe connection (from radiator)
14 Filler
15 Case
16 Coolant from engine
17 Coolant from radiator
18 Coolant to pump
19 Travel of bypass valve

water until the thermostat is fully open then let is cool down naturally. If the thermostat will not open fully in boiling water, or does not close down as the water cools, then it must be renewed.
5 Replace the thermostat assembly using the reverse procedure to removal.

5 Water pump – removal and refitting

1 Refer to Section 3 and remove the radiator.
2 Slacken the top fixing bolt on the alternator so that it can be moved towards the engine and the drive belt loosened. The drivebelt and the alternator adjusting strap may now be removed.
3 Detach the hoses from the pump body and undo also the bolts holding the fan and pulley hub to the shaft (photo). On 1500 and 1600 models with electric fan fitted, no fan blades are attached to the water pump bolts.
4 Undo the two nuts that secure the heater return pipe flange (photo) and take out the three bolts securing the body to the cylinder block.
5 Draw the pump off the cylinder block.
6 Refitting is the reverse sequence to removal, but the following additional points should be noted:
7 Clean the mating faces of the pump and cylinder block free of all fragments of old gasket to ensure a good water tight joint.
8 Always use a new gasket.
9 Adjust fan belt tension as described in Section 8.

6 Water pump – dismantling and overhaul

1 If the water pump starts to leak, shows excessive movement of the spindle, or is noisy during operation, the pump may be dismantled and overhauled. Make sure before you begin that the suspected faulty parts are readily available as spares. It is general policy nowadays not to hold pump parts but to supply the whole unit ready to fit in the engine.
2 Assuming that you have obtained the spare parts necessary to renovate the pump, it may be dismantled as follows:
3 After having removed the four nuts that hold the water pump body

5.3 Removing the cooling fan (not 1500 and 1600 models)

5.4 Removing the heater pipe from the rear of the water pump

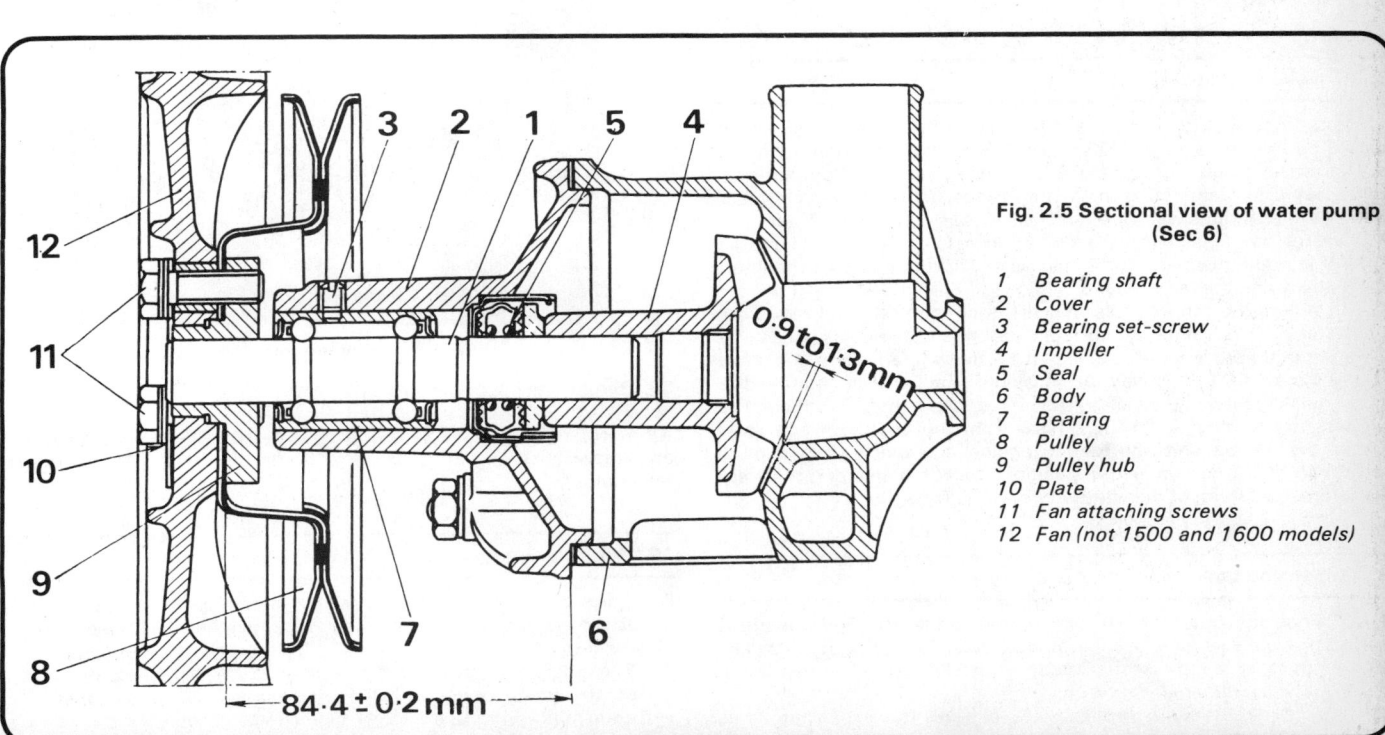

Fig. 2.5 Sectional view of water pump (Sec 6)

1 Bearing shaft
2 Cover
3 Bearing set-screw
4 Impeller
5 Seal
6 Body
7 Bearing
8 Pulley
9 Pulley hub
10 Plate
11 Fan attaching screws
12 Fan (not 1500 and 1600 models)

in its housing and removed the pump from the car, hold the pump body lightly in a vice and with either a conventional puller or the special Lada tool A 40026 withdraw the pump impeller from the shaft.

4 Next the set-screw which locates the bearing in the pump body is removed, to allow the shaft together with the pulley hub and bearings to be gently driven out.

5 The water seal may be tapped out of the pump body with an ordinary drift.

6 Again using a hub puller, the pulley hub may be removed from the shaft if necessary.

7 Clean all the components and inspect them thoroughly for signs of wear or damage. Replace parts as necessary.

8 To assemble the pump, first carefully insert the water seal into the pump body. Do not use excessive force, and be careful not to damage the carbon seal block and face. The seal block is fragile and if scratched or damaged will allow water to escape past the mating face of the pump impeller into the bearing assembly in the pump.

9 When offering up the shaft assembly to the body, ensure that the hole in the bearing outer race is aligned to accept the set-screw which locates the shaft assembly in position in the pump. Carefully drive the shaft assembly into the pump and screw in the set-screw. Press the fan pulley hub onto the shaft, ensuring the dimensions 84.4 ± 0.2 mm (Fig. 2.5).

10 Finally press the impeller onto the shaft until the free running clearance between the impeller vanes and the pump body is 0.035 to 0.051 in (0.9 to 1.3 mm).

11 The pump is now ready to be fitted to the engine, see Section 5 of this Chapter.

7 Drivebelt – removal and refitting

If the drivebelt is worn or has over stretched it should be renewed. The most usual reason for replacement is that the belt has broken. It is therefore recommended that a spare belt is always carried in the car. Replacement is a reversal of the removal procedure, but if a replacement is due to breakage:

1 Loosen the alternator pivot and slotted link bolts and move the alternator towards the engine.

2 Carefully fit the belt over the crankshaft, water pump and alternator pulleys.

3 Adjust the belt as described in Section 8 and tighten the alternator mounting bolts. **Note**: *After fitting a new belt it will require adjustment 250 miles (400 km) later.*

8 Drivebelt – adjustment

1 It is important to keep the drive belt correctly adjusted and it should be checked every 6000 miles (9600 km) or 6 months. If the belt is loose it will slip, wear rapidly and cause the alternator and water pump to malfunction. If the belt is too tight the alternator and water pump bearings will wear rapidly and cause premature failure.

2 The fan belt tension is correct when there is 0.4 to 0.7 in (10 to 15 mm) of lateral movement at the mid point position between the alternator pulley and the fan pulley.

3 To adjust the fan belt, slacken the securing bolts and move the alternator in or out until the correct tension is obtained. It is easier if the alternator bolts are only slackened a little so it requires some effort to move the unit. In this way the tension of the belt can be arrived at more quickly than by making frequent adjustments. If difficulty is experienced in moving the unit away from the engine, a tyre lever placed behind the unit and resting against the block gives a good control so that it can be held in position whilst the securing bolts are tightened. Be careful of the alternator cover, it is fragile.

9 Expansion tank

1 The coolant expansion tank is mounted on the left-hand side of the radiator in the engine compartment, and does not require any maintenance. It is important that the expansion tank filler cap is not removed whilst the engine is hot.

2 Should it be found necessary to remove the expansion tank, disconnect the radiator to expansion tank hose from the radiator union.

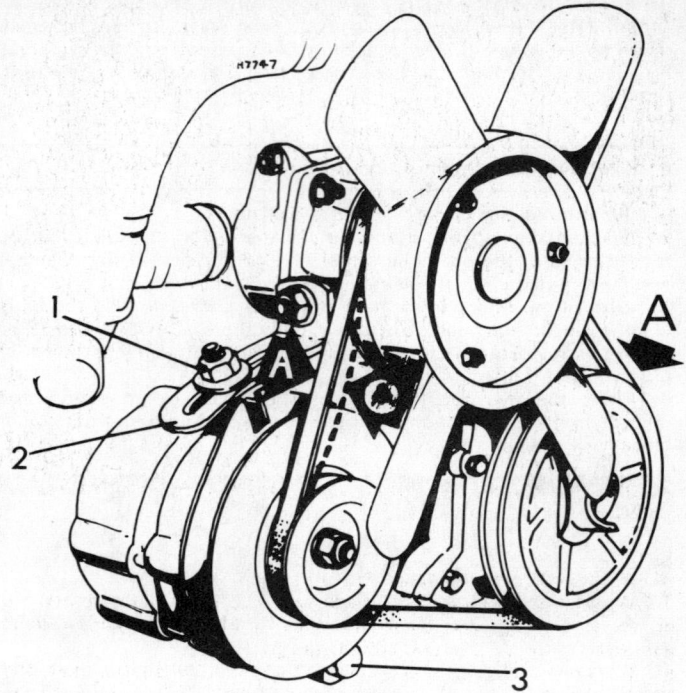

Fig. 2.6 Adjusting the drive belt (Sec 8)

1 Adjusting nut
2 Sliding bracket
3 Securing bolt

A Belt tension point 0.5 in (12.7 mm)

Fig. 2.7 Expansion tank (Sec 9)

3 Remove the bracket nuts and carefully lift away the tank and its hose.

4 Refitting is the reverse sequence to removal. Add either water or anti-freeze solution until it is up to the level mark of 30 mm above minimum.

10 Water temperature gauge – fault finding

1 Correct operation of the water temperature gauge is very important as the engine can otherwise overheat without it being observed.

2 The gauge is an electrically operated instrument comprising a transmitter unit screwed into the left-hand side of the cylinder head and transmitting through a Lucar type connector and cable to the dial mounted on the fascia instrument panel. The instrument only operates

Chapter 2 Cooling system

10.2 Temperature gauge transmitter unit

when the ignition is switched on (photo).
3 Where the water temperature gauge reads high/low intermittently, or not at all, then first check the security of the connecting cable between the transmitter unit and the gauge.
4 Disconnect the Lucar connector from the transmitter unit and switch on the ignition. The gauge should read COLD. Now earth the cable to the engine block; the gauge needle should indicate HOT. This test proves the gauge to be functional and the fault must therefore lie in the cable or transmitter unit. Renew as appropriate.
5 If the fuel gauge shows signs of malfunction at the same time as the water temperature gauge then a blown fuse (9) may be the cause.

11 Antifreeze and corrosion inhibitors – general

1 In circumstances where it is likely that the temperature will drop below freezing, it is essential that some of the water is drained and an adequate amount of ethylene glycol antifreeze is added to the cooling system. If antifreeze is not used, it is essential to use a corrosion inhibitor in the cooling system in the proportion recommended by the inhibitor manufacturer.
2 Any antifreeze which has an ethylene glycol base can be used. Never use an antifreeze with an alcohol base as evaporation is too high.
3 Most products with an anti-corrosion additive can be left in the cooling system for up to two years, but after six months it is advisable to have the specific gravity of the coolant checked at your local garage, and thereafter once every three months.
4 The table below gives the proportion of antifreeze and degree of protection:

Antifreeze	Commences to freeze		Frozen	
%	°C	°F	°C	°F
25	-13	9	-26	-15
33	-19	-2	-36	-33
50	-36	-33	-48	-53

Note: *Never use antifreeze in the windscreen washer reservoir as it will cause damage to the paintwork.*
Warning: *Ethylene glycol is poisonous, keep out of the reach of children.*

12 Electric cooling fan – removal and refitting

1 The electric cooling fan is attached to the radiator cowl. To remove the fan, detach the cowl from the radiator, unplug the fan and remove fan and cowl together. The fan can then be detached from the cowl.
2 The fan cannot be repaired, and if defective must be renewed. Check first that the fault is not in the thermal switch or the wiring, by applying 12 volts directly to the fan motor terminals.
3 Refitting is the reverse of the removal procedure.

13 Fault diagnosis – cooling system

Symptom	Reason/s
Loss of coolant	Leak in system
Loss of coolant from radiator into expansion bottle	Defective radiator pressure cap Overheating Blown cylinder head gasket, causing excess pressure in cooling system forcing coolant past radiator cap into expansion bottle Cracked block or head due to freezing
Overheating	Insufficient coolant in system Water pump not turning properly due to slack fan belt Kinked or collapsed water hoses Electric cooling fan defective (when fitted) Faulty thermostat (not opening properly) Engine out of tune Blocked radiator either internally or externally Cylinder head gasket blown forcing coolant out of system New engine not run-in
Engine running too cool	Faulty thermostat
Coolant leaking into engine oil	Cylinder head gasket leaking Cylinder head plugs leaking Cylinder head cracked Cylinder block cracked or porous

Chapter 3 Fuel and exhaust systems

For modifications, and information applicable to later models, see Supplement at end of manual

Contents

Air cleaner and element – removal and refitting	2
Carburettor – adjustment	10
Carburettor – dismantling, inspection and reassembly	8
Carburettor – general description	6
Carburettor – removal and refitting	7
Choke – adjustment	9
Closed circuit crankcase ventilation system	16
Fault diagnosis – fuel and exhaust systems	17
Fuel gauge sender unit - fault finding	14
Fuel pipes and lines – general inspection	13
Fuel pump – dismantling, inspection and reassembly	5
Fuel pump – removal and refitting	3
Fuel pump – testing	4
Fuel tank – cleaning and repair	12
Fuel tank – removal and refitting	11
General description	1
Manifolds and exhaust system – general	15

Specifications

Fuel pump
Type .... Mechanically operated by a cam on the distributor/oil pump drive shaft

Filters
Air .... Champion W106
Fuel .... Champion L101

Carburettor
Type .... Twin choke downdraught
Parameters **1200 and 1300 models (Lada 2101, 2102, 21011)**

	2101–1107010		2101–1107010–01		2101-1107010-02	
No of barrel/choke	I	II	I	II	I	II
Barrel diameter mm	32	32	32	32	32	32
Larger venturi diameter mm	23	23	23	23	23	23
Smaller venturi diameter mm	10.5	10.5	10.5	10.5	8	10.5
Atomiser calibration	4.5	4.5	4	4	4	4.5
Diameter of main fuel jet mm	1.35	1.25	1.3	1.3	1.3	1.25
Diameter of main air jet mm	1.7	1.9	1.5	2.0	1.5	1.9
Emulsion tube calibration	15	15	15	15	15	15
Diameter of idling fuel jet mm	0.45	0.60	0.45	0.60	0.50	0.45
Diameter of idling air jet mm	1.80	0.70	1.70	0.70	1.70	1.70
Diameter of acceleration pump atomiser orifice mm	0.40		0.40		0.40	
Diameter of acceleration pump bypass jet mm	0.40		0.40		0.40	
Capacity of acceleration pump (10 full strokes) cc	7 ± 25%		7 ± 25%		7 ± 25%	
Diameter of enrichment device fuel jet mm	1.50		1.50		1.50	
Diameter of enrichment device air jet mm	0.90		1.20		0.90	
Diameter of enrichment device emulsion jet mm	1.70		1.50		1.70	
Diameter of choke mechanism air jet mm	0.70		0.70		0.70	
Distance from float to carburettor cover with gasket mm	7.5 ± 0.25		6.5 ± 0.25		7.50	
Diameter of hole in needle valve seat mm	1.75		1.75		1.75	

1500 and 1600 models (Lada 2103 and 2106)

	2103–1107010		2103–1107010–01	
No of barrel/choke	I	II	I	II
Barrel diameter mm	32	32	32	32
Larger venturi diameter mm	23	24	23	24
Smaller venturi diameter mm	10.5	10.5	8	8
Atomiser calibration	4.5	4.5	4	4
Diameter of main fuel jet mm	1.35	1.4	1.3	1.4
Diameter of main air jet mm	1.7	1.9	1.5	1.5
Emulsion tube calibration	15	15	15	15
Diameter of idling fuel jet mm	0.50	0.80	0.45	0.60
Diameter of idling air jet mm	1.70	0.70	1.70	0.70
Diameter of acceleration pump atomiser orifice mm	0.50		0.40	
Diameter of acceleration pump bypass jet mm	0.40		0.40	
Capacity of acceleration pump (10 full strokes) cc	7 ± 25%		7 ± 25%	
Diameter of enrichment device fuel jet mm	1.80			
Diameter of enrichment device air jet mm	1.20			

Chapter 3 Fuel and exhaust systems

Diameter of enrichment device emulsion jet mm	1.60	
Diameter of choke mechanism air jet mm	0.70	0.70
Distance from float to carburettor cover with gasket mm	7.5 ± 0.25	6.5 ± 0.25
Diameter of hole in needle valve seat mm	1.75	1.75

Adjustment data
Idling speed .... 750 to 800 rpm
CO content of exhaust gas at idle .... 1.5 to 2.5%

Fuel tank capacity
Saloon .... 8.5 gallons (38.6 litres)
Estate .... 10.0 gallons (45.5 litres)

Torque wrench setting
	lbf ft	kgf m
Inlet and exhaust manifold stud nut	15 to 22	2.1 to 2.6

1 General description

The basic layout of the fuel system on all models comprises a fuel tank mounted in the boot, a fuel pump mechanically operated by a cam on the distributor driveshaft and a carburettor and inlet manifold. The interconnecting fuel lines are small bore metal pipes with flexible hose connections where necessary.

The carburettor is a twin-choke down-draught type, fitted with a water heated jacket to ensure more efficient vaporisation of the petrol. The carburettor air inlet is protected by an air cleaner containing a replaceable filter element.

2 Air cleaner and element – removal and refitting

1   To gain access to the filter element, remove the three nuts and lift off the top cover plate from the air cleaner assembly (photo).
2   Remove the filter element and wipe out the interior of the air cleaner housing. Fit a new filter element, replace the fabric disc and top cover and tighten the nuts.
3   To remove the complete air cleaner assembly, remove the filter element as described, undo the four retaining nuts and remove the stiffener plate if fitted (photo).
4   Detach the hot air inlet pipe and crankcase breather pipe and lift off the air cleaner assembly.
5   Refit the air cleaner using the reverse procedure.
6   The air cleaner can be adjusted to draw in either hot air from around the exhaust manifold or cool air from the engine compartment.
7   In the summer, remove the three nuts and rotate the top cover so that the blue spot is aligned with the arrow on the air inlet spout. In winter turn the cover to align the red spot with the arrow (Fig. 3.1).
8   When the cover is in the correct position refit and tighten the nuts.

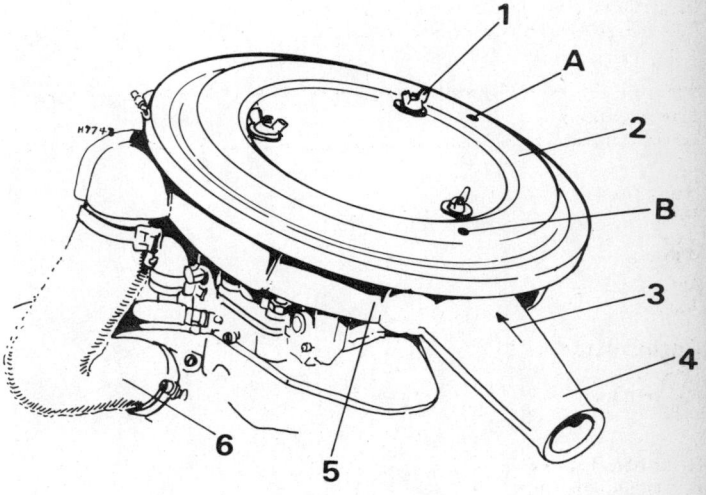

Fig. 3.1 Air cleaner assembly (Sec 2)

1 Wing nut
2 Cover
3 Indicating arrow
4 Air intake (summer operation)
5 Housing
6 Air inlet hose (heated for winter operation)
A Blue mark
B Red mark

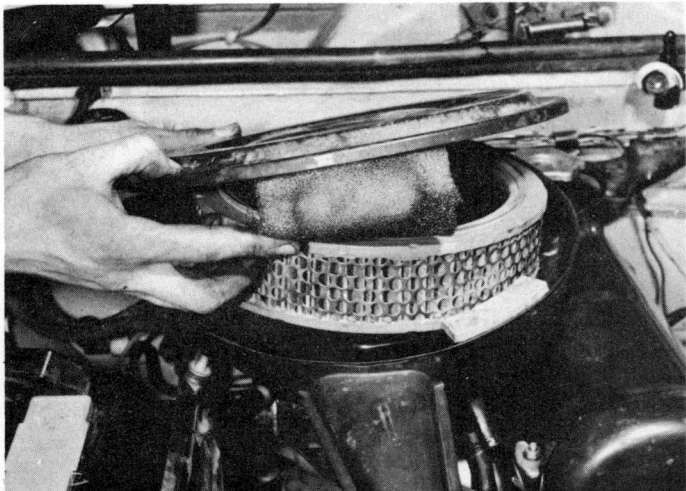

2.1 Removing the air cleaner cover plate

2.3 Interior view of air cleaner assembly

3 Fuel pump – removal and refitting

1 The mechanical fuel pump is mounted on the left-hand side of the cylinder block. It is operated by a pushrod which is actuated by a cam on the distributor/oil pump drive shaft (photo).
2 To remove the pump, disconnect the two fuel pipes and if necessary, plug the end of the pipe from the petrol tank to prevent loss of fuel.
3 Remove the two retaining nuts and remove the pump, insulating spacer and gaskets from the cylinder block. Discard the two gaskets.
4 Check that the pushrod is still in place in the fuel pump and has not been withdrawn when the pump was removed.
5 To refit the pump, first rotate the engine until the pushrod is in the fully retracted position (on the heel of the cam).
6 Now fit the insulating spacer and two new gaskets and check that the pushrod protrudes beyond the outer gasket by 0.031 to 0.051 in (0.8 to 1.3 mm), see Fig. 3.2 (photo).
7 To obtain this dimension your Lada dealer can supply fuel pump gaskets in thicknesses of 0.011 in (0.28 mm), 0.029 in (0.74 mm) and 0.040 in (1.2 mm).
8 When the correct pushrod dimension is obtained, refit the pump to the cylinder block and tighten the retaining nuts.

4 Fuel pump – testing

The fuel pump can be easily tested by disconnecting the fuel inlet pipe from the carburettor and operating the lever on the side of the pump. With each stroke of the pump lever a strong jet of petrol should spurt out of the fuel pipe. Note that it may be necessary to turn the engine over on the starting handle just slightly before the pump lever can be operated.

5 Fuel pump – dismantling, inspection and reassembly

1 Before deciding to overhaul a fuel pump, it will be wise to consider

3.1 Location of fuel pump

3.6 Fuel pump spacer and gaskets in position

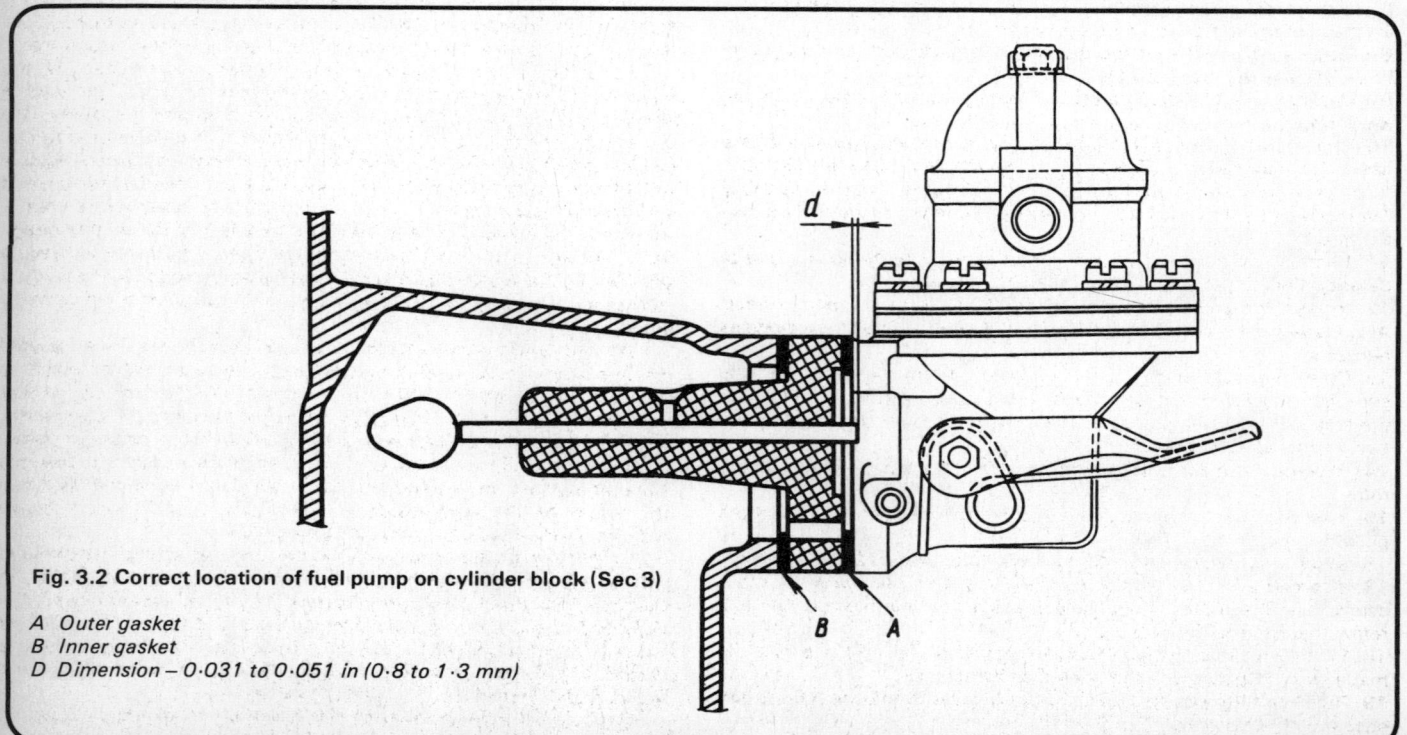

Fig. 3.2 Correct location of fuel pump on cylinder block (Sec 3)

A *Outer gasket*
B *Inner gasket*
D *Dimension – 0·031 to 0·051 in (0·8 to 1·3 mm)*

Chapter 3 Fuel and exhaust systems

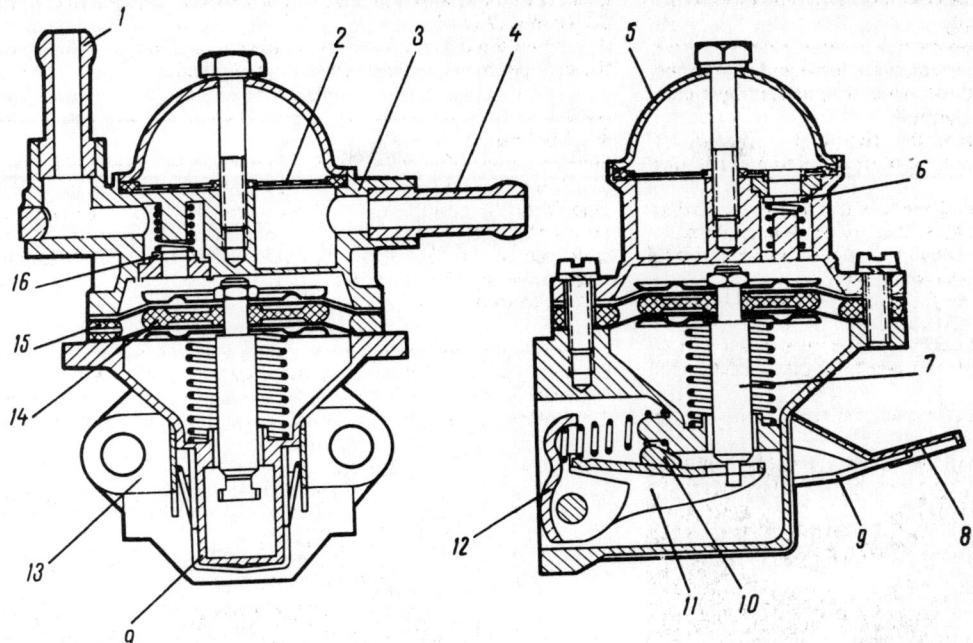

Fig. 3.3 Sectional views of fuel pump (Sec 5)

1 Delivery union
2 Strainer
3 Body
4 Suction union
5 Cover
6 Suction valve
7 Pullrod
8 Priming lever
9 Spring
10 Eccentric
11 Rocker
12 Rocker arm
13 Bottom cover
14 Spacer (inner)
15 Spacer (outer)
16 Delivery valve

a factory exchange unit, particularly where the unit has been in service for a considerable mileage. If you decide to overhaul the pump, obtain a repair kit which contains all necessary renewable items.

2 The pump cover (which can be removed with the pump in situ if required) is secured with a hexagon headed screw. Remove the screw and top cover and lift out the filter screen, noting its relation to the pump body. Should further investigation be required, remove the pump from the cylinder block.

3 Make a mark across the upper and lower bodies so that they may be refitted in their original positions.

4 Undo and remove the screws and spring washers that secure the two body halves together.

5 Carefully lift off the upper body. It is possible for the diaphragm to stick to the mating flanges. If this is the case free it with a sharp knife.

6 Using a parallel pin punch carefully tap out the rocker arm pin.

7 Lift out the rocker arm and recover the spring.

8 Depress the centre of the diaphragm and detach the rocker arm link. Lift away the rocker arm link.

9 Carefully lift the diaphragm and spring up and away from the lower body.

10 The valves should not normally require renewal but in extreme cases this may be done. On some models the valves are retained with two screws. These should be removed. On other models carefully remove the peening and lift out the valves. Note which way round they are fitted.

11 Carefully examine the diaphragm for signs of splitting, cracking or deterioration. Obtain a new one if suspect.

12 Inspect the pump bodies for signs of cracks or stripped threads. Also inspect the rocker arm, link and pin for wear. Obtain new parts as necessary.

13 Clean the recesses for the two valves and refit the valves, making sure they are the correct way round. Use a sharp centre punch to peen the edges of the casting or retain them with the H-shaped retainer and two screws as applicable.

14 Replace the diaphragm spring and insert the diaphragm and pull-rod.

15 Insert and hook the end of the rocker arm link onto the end of the pullrod.

16 Replace the rocker arm into the lower body and locate the spring.

17 Insert the rocker arm pin, lining up the hole with a small screwdriver. Peen over the pin hole edges with a sharp centre punch to retain the pin in position.

18 Fit the upper body to the lower body and line up the previously made marks. Secure with the screws and washers.

19 Refit the filter screen, gasket and top cover to the upper body and secure with the screw.

20 The fuel pump is now ready for refitting to the engine.

6 Carburettor – general description

The carburettor fitted to all Lada models is of the dual barrel fixed choke type. The term 'fixed choke' relates to the throat (venturi) into which the main petrol jet sprays the fuel, and because on these carburettors this venturi is a fixed size, several other petrol jets are incorporated in the carburettor to enrich the fuel/air mixture passing into the engine as and when necessary. All fuel jets take the form of inserts screwed into the carburettor body.

The carburettor functions as follows: Petrol is pumped into the float chamber and is regulated by the needle valve actuated by the float. As air is sucked into the engine a slight vacuum (proportional to air flow) is created in the venturi of the carburettor; this vacuum draws a corresponding amount of fuel from the float chamber through the emulsion tube where it mixes with the small amount of air coming from the fuel correction jet. The fuel air emulsion passes on through the main jet and into the main air stream in the venturi of the carburettor.

The flow of air through the carburettor is controlled by the butterfly valve operated by the accelerator linkage. The engine develops power in proportion to the amount of air and fuel drawn into the engine and the economy is dependant on the relative proportions of fuel and air taken into the engine. On these fixed choke carburettors the relative proportions of the fuel and air are controlled by the sizes of the main jet, carburettor throat (venturi) and the other minor enrichment jets. All these sizes are fixed and decided by the engine designers. The only adjustment or trim to the fuel/air mixture available is provided by the accelerator's butterfly valve stop and the idling mixture control screw. Section 10 sets down the carburettor adjustment procedure.

As mentioned earlier, the fixed jet carburettors require additional fuel jets to enrich the fuel/air mixture when necessary. When power is required for acceleration the engine needs an enriched mixture of fuel and air. An acceleration jet is provided in the throat of the carburettor and it is supplied with fuel under pressure from the accelerator pump. The pump is activated by levers connected to the accelerator linkage. There is also a compressing jet which introduces additional fuel into the throat of the carburettor to create the slightly rich mixture necessary to sustain the engine at high speeds.

In addition to the accelerator jet and compensating jets, there is the idling jet. The flow of fuel through this jet is adjustable. When the engine is idling and the accelerator butterfly valve almost closed, the slight vacuum in the inlet manifold sucks air and fuel via the pilot jet through the idling jet orifice. The three minor jets therefore provide the mixture enrichment necessary when accelerating, maintaining high speed, and engine idling.

Arrangements for cold starting consists of a choke flap above the main jet and venturi. To start the engine when cold, appreciably more fuel is required to overcome the losses due to condensation of the fuel vapour in the inlet manifold ducts. By closing the choke valve an increased vacuum is created in the carburettor barrel and this sucks a greater amount of fuel out of the main jet system. The choke flap is mounted eccentrically on the support spindle, so that when the engine demands more air the flap will partially open itself against a control spring to admit the air into the carburettor and engine.

A vacuum-operated diaphragm valve is connected to the choke linkage and controls the position of the choke flap during starting and idling to prevent excessive enrichment or weakening of the mixture.

A crankcase ventilation valve is connected to the throttle spindle on some carburettors and controls the amount of crankcase vapour to be drawn back into the carburettor throughout the throttle opening range.

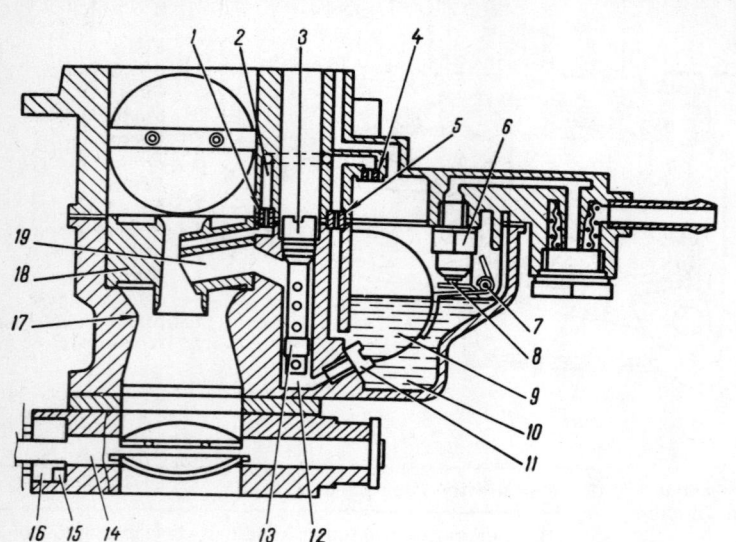

Fig. 3.4 Sectional view of carburettor showing the fuel metering system (Sec 6)

1 Econostat emulsion jet
2 Econostat emulsion passage
3 Air jet (main metering system)
4 Econostat air jet
5 Econostat fuel jet
6 Needle valve
7 Float pin
8 Needle ball
9 Float
10 Float chamber
11 Main jet
12 Emulsion well
13 Emulsion tube
14 Primary throttle shaft
15 Crankcase vapour valve groove
16 Rotary crankcase vapour valve
17 Venturi (large)
18 Venturi (small)
19 Nozzle

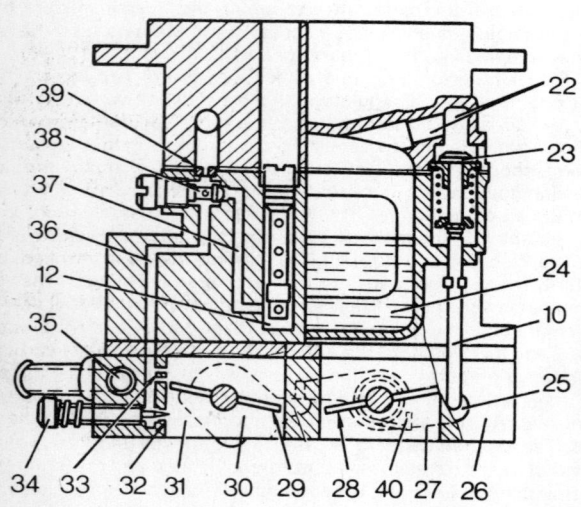

Fig. 3.5 Sectional view of carburettor showing idling system (Sec 6)

10 Float chamber
12 Emulsion well
22 Float chamber vent
23 Vent shut-off valve
24 Rod
25 Idler lever
26 Body
27 Secondary barrel
28 Secondary throttle
29 Primary throttle
30 Accelerator pump lever (with abutment)
31 Primary barrel
32 Adjustable discharge hole
33 Progression holes
34 Adjusting screw
35 Throttle body jacket passage
36 Idling emulsion passage
37 Idling fuel passage
38 Idling emulsion jet
39 Idling air jet
40 Vent shut-off valve idler lever spring

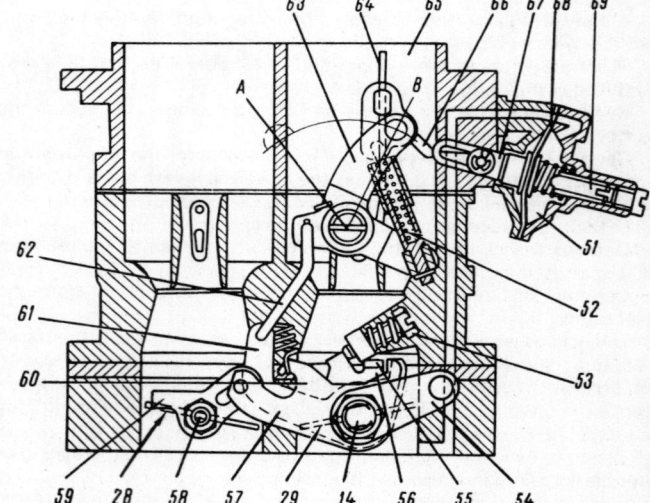

Fig. 3.6 Carburettor throttle and choke linkage (Sec 6)

14 Primary throttle shaft
28 Secondary throttle
29 Primary throttle
51 Depression chamber
52 Telescopic link
53 Adjusting screw (primary throttle)
54 Throttle operating lever
55 Projection
56 Sector
57 Lever (primary throttle shaft)
58 Secondary throttle shaft
59 Lever
60 Lever projection
61 Lever
62 Link
63 Choke operating lever
64 Choke/lever assembly
65 Primary barrel air horn
66 Link
67 Rod
68 Diaphragm
69 Adjusting screw (choke control unit)
A Position of lever 63 at starting
B Position of lever 63 during engine running

Chapter 3 Fuel and exhaust systems

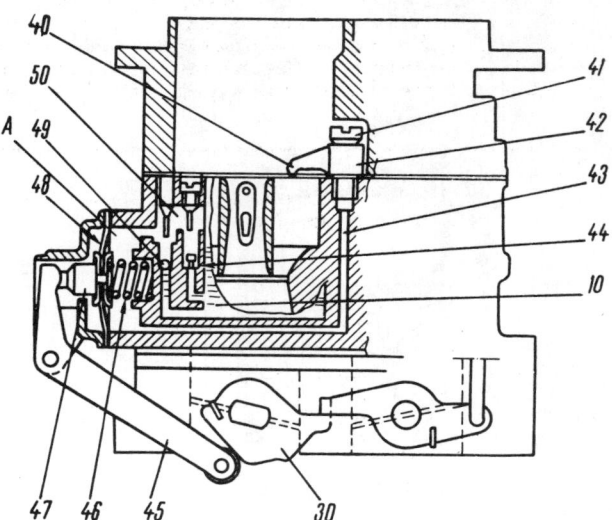

Fig. 3.7 Sectional view of carburettor accelerator pump (Sec 6)

- 10 Float chamber
- 30 Accelerator pump lever
- 40 Nozzle
- 41 Ball valve (outlet)
- 42 Accelerator pump nozzle
- 43 Fuel passage
- 44 Bypass jet
- 45 Accelerator pump rocker arm
- 46 Pump return spring
- 47 Diaphragm seat
- 48 Pump diaphragm
- 49 Ball valve (inlet)
- 50 Vapour chamber
- A Pump working chamber

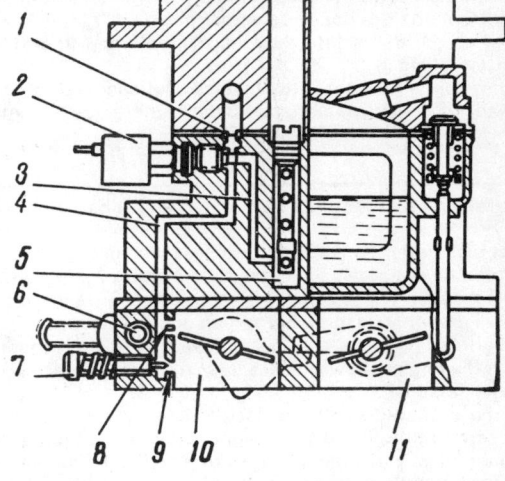

Fig. 3.8 Idle system (1500, 1600 carburettors) (Sec 6)

- 1 Idle air jet
- 2 Fuel shut-off valve
- 3 Idle fuel passage
- 4 Idle emulsion passage
- 5 Emulsion well
- 6 Water heated jacket
- 7 Mixture adjusting screw
- 8 Progression holes
- 9 Bleed hole
- 10 Primary barrel
- 11 Secondary barrel

The carburettor used on 1500 and 1600 models incorporates a solenoid type fuel shut-off valve. This electrically operated device prevents the engine running on after the ignition has been switched off. This valve does not require adjusting but in the event of a fault developing renew the valve complete.

7 Carburettor – removal and refitting

1 First disconnect the crankcase ventilation hoses and remove the aircleaner as described in Section 2.
2 Disconnect the fuel pipe from the carburettor union (photo). Tape over the end to prevent dirt ingress.
3 Disconnect the choke cable and throttle linkage from the side of the carburettor (photos).
4 Disconnect the vacuum lines from the carburettor. On 1500 and 1600 models, disconnect the electrical lead from the fuel shut off valve.
5 Undo and remove the four nuts and spring washers that secure the carburettor to the inlet manifold. Lift away the carburettor. The water heated insulator will be removed with the carburettor after the cooling

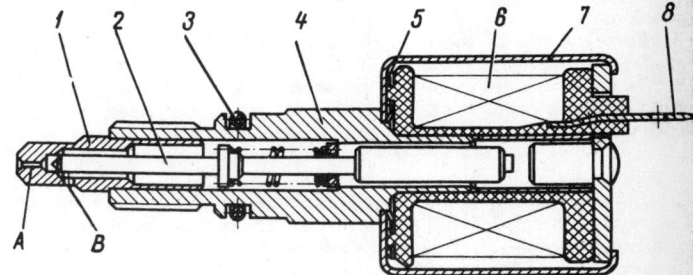

Fig. 3.9 Fuel shut off valve (1500, 1600 carburettors) (Sec 6)

- 1 Idle jet
- 2 Needle
- 3 Seal
- 4 Valve body
- 5 Plate
- 6 Coil (solenoid)
- 7 Casing
- 8 Lead
- A Calibrated jet
- B Fuel orifice

7.2 Disconnecting the fuel pipe from the carburettor

7.3a Choke cable connection to carburettor

7.3b Throttle linkage connection to carburettor

system has been drained and the insulator hoses disconnected. The insulator is connected to the carburettor by two screws from under the base of the insulator into the carburettor body.

6   Place a wedge of rag into the inlet manifold aperture to stop any foreign matter falling in.

7   Refitting the carburettor is the reverse sequence to removal. It will be necessary to check the various adjustments as described later in this Chapter. Always use a new gasket.

8 Carburettor – dismantling, inspection and reassembly

1   Do not dismantle the carburettor unless it is absolutely necessary, when systematic diagnosis indicates that there is a fault with it. The internal mechanism is delicate and finely balanced and unnecessary tinkering will probably do more harm than good.

2   Although certain parts may be removed with the carburettor still attached to the engine, nevertheless it is considered safer to remove it and work over a bench.

3   With the carburettor off the car disassembly is as follows (this covers all carburettors fitted with slight variations, which, because of their simple and obvious nature, are not specifically mentioned each time, but it does not cover the automatic choke working of these carburettors so fitted): Remove the top of the carburettor by undoing its fixing screws with the correct size screwdriver. These screws are made of comparatively soft metal and will damage very easily. With the top will come the choke mechanism, fuel inlet pipe and filter, needle valve and float.

4   Remove the gasket and the float retaining spindle and then the float itself.

5   Remove any jets and their washers which have screwdriver cuts in their heads and are removable from the top and inside of the carburettor.

6   Unscrew all the external adjusting screws and the float chamber drain plugs. Retain all washers and springs and code for their relevant positions.

7   Turn the top of the carburettor upside down and remove the needle valve and its washer.

8   There is no point under any circumstances in removing any more parts from the carburettor. If any of these parts are in need of attention, then a complete new carburettor is needed. It is safer and more efficient if you have reached this stage of need of repair to replace the complete unit.

9   Carefully wash all the parts with clean petrol and blow out the fuel passages with compressed air if available. Clean the carburettor body with an old toothbrush dipped in petrol and dry with a fluff-free cloth.

10   Inspect the carburettor body for cracks or damage and the throttle and choke flap valves and shafts for excessive wear. If worn, it is advisable to obtain a new or reconditioned carburettor.

11   Check the jets and ball valves for dirt and ensure that the accelerator pump and diaphragm type power piston plunger operate smoothly.

12   Reassemble the gasket first, the needle valve and float and with the top section of the carburettor held vertically, check the distance between the top of the float and the carburettor body is correct (see Fig. 3.12). If not, carefully bend the float hinge to rectify.

13   Check the needle valve for correct seating by holding the carburettor top section upside-down and trying to suck air through the petrol inlet union. If the seating is good no air will pass through.

14   Check all mating surfaces with a straight-edge and all gaskets for damage; renew if necessary.

15   Reassemble the carburettor using the reverse sequence to dismantling. Providing that the location of each part was noted and laid out in the right order, no major problems should be encountered.

16   Refit the carburettor, as described in Section 7.

9 Choke – adjustment

1   Ensure that the choke cable moves freely; if necessary, lightly oil the inner cable.

2   Check the operation of the choke flap on the carburettor, move the lever to the open position and ensure that the flap snaps to the fully closed position when the operating lever is released.

3   With the choke knob in the fully closed position fit the cable onto the choke control lever. Pull the choke knob right out and check that

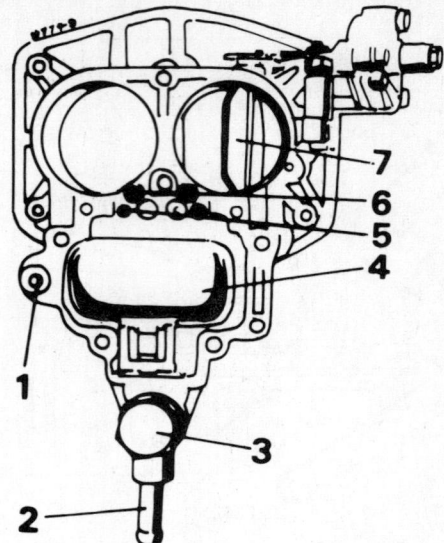

Fig. 3.10 Carburettor top cover and associated components (Sec 8)

1   Float chamber vent shut-off valve
2   Fuel inlet connection
3   Fuel filter cover
4   Float
5   Econostat fuel jet
6   Econostat emulsion jet
7   Choke

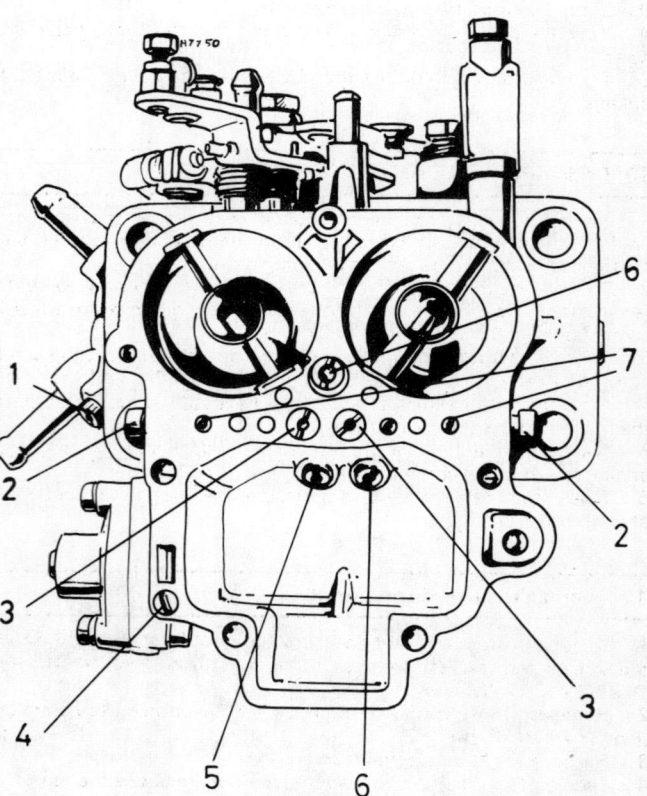

Fig. 3.11 Carburettor main body and jets (Sec 8)

1   Idle mixture adjusting screw
2   Idling jet body
3   Main air jets
4   Accelerator pump valve plug
5   Main jets
6   Accelerator pump nozzle valve
7   Idling air jets

Chapter 3 Fuel and exhaust systems

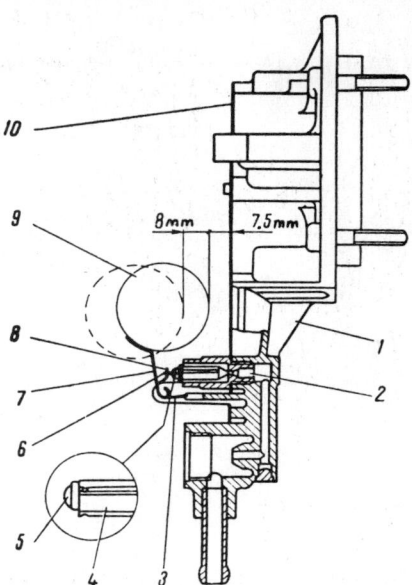

Fig. 3.12 Carburettor needle valve and float mechanism (Sec 8)

1 Carburettor cover
2 Float needle seat
3 Slip
4 Needle
5 Needle ball
6 Needle yoke
7 Float arm
8 Tab
9 Float
10 Gasket

the choke flap is in the fully closed position.
4 Push the choke knob right in and check that the choke flap is wide open.
5 Do not attempt to adjust the choke vacuum valve or associated linkage.

10 Carburettor – adjustment

1 Before adjusting the carburettor it is essential that the other engine adjustments are correct; this includes ignition timing, valve clearances, contact breaker gap and plug gaps.
2 First start the engine and bring it up to normal operating temperature. Now locate the throttle stop screw and mixture adjusting screw (see Fig. 3.13).
3 Turn the throttle stop screw in the appropriate direction until a slow but steady idling is achieved (approximately 600 rpm).
4 Now turn the mixture screw in the necessary direction to obtain the maximum possible engine speed.
5 Next turn the throttle stop screw anti-clockwise to reduce the engine speed to the correct idling rpm (see Specifications).
6 Finally road test the car and check the operation of the carburettor at all throttle openings.

11 Fuel tank – removal and refitting

1 Remove the tool bag and peel back the boot floor covering. Extract the screws and take off the plastic trim cover from over the fuel tank. Disconnect the battery.
2 Remove the tank cap and drain the fuel by undoing the drain plug (Figs. 3.14 and 3.15).
3 Remove the strap clamping screw.
4 Take off the strap and remove the guard plate above the tank.
5 Disconnect the fuel pipe and breather pipe.
6 Disconnect the earth wire and fuel gauge wire and lift the tank out, working the filler neck out of the rubber sealing grommet.
7 If the tank is suspected of leaking do not be tempted to solder it. Have it repaired professionally or buy a new tank. If the tank is persistently contaminating the fuel because of deterioration of the interior then the tank should be replaced.

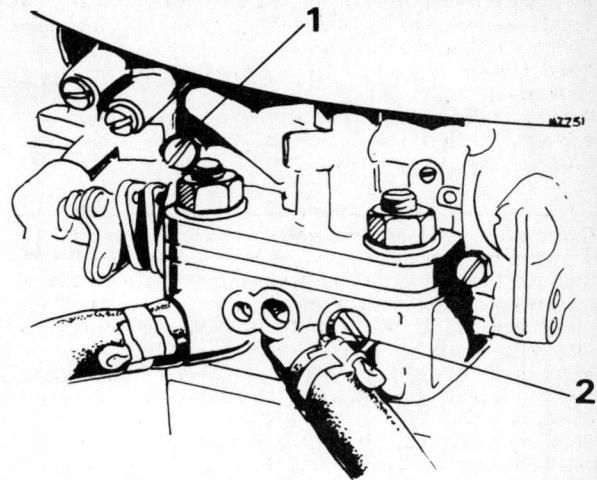

Fig. 3.13 Carburettor adjustment screws (Sec 10)

1 Throttle stop screw
2 Mixture adjustment screw

8 Estate car tanks are fitted flat at the rear floor and are secured by screws to the floor panel.
9 Replacement is a reversal of the removal procedure.

12 Fuel tank – cleaning and repair

1 With time it is likely that sediment will collect in the bottom of the fuel tank. Condensation, resulting in rust and other impurities is sometimes found in the fuel tank.
2 With the tank removed it should be vigorously flushed out and turned upside down, and if facilities are available, steam cleaned.
3 Temporary repairs to the fuel tank to stop leaks are best carried out using resin adhesive and hardeners as supplied by most accessory shops. In cases of repairs made to large areas, glass fibre mats or perforated zinc sheet may be required to give the area support. If any soldering, welding or brazing is contemplated, the tank must be steamed out to remove any traces of petroleum vapour. It is extremely dangerous to use naked flames on a fuel tank without this, even though it may have been lying empty for a considerable time.

13 Fuel pipes and lines – general inspection

1 Check all flexible hoses for signs of perishing, cracking or damage and replace if necessary.
2 Carefully inspect all metal fuel pipes for signs of corrosion, cracking, kinking or distortion and replace any pipe that is suspect. These pipes are clipped to the underbody.

14 Fuel gauge sender unit – fault finding

1 The sender unit is mounted on the fuel tank and access is straightforward.
2 If the fuel gauge does not work correctly then the fault is either in the sender unit, the gauge in the instrument cluster or the wiring.
3 To check the sender unit disconnect the wire from the unit at the connector. Switch on the ignition and the gauge should read 'Empty'. Now connect the lead to earth and the gauge should read 'Full'. Allow 30 seconds for each reading.
4 If both the situations are correct then the fault lies in the sender unit.
5 If the gauge does not read 'Empty' with the wire disconnected from the sender unit, the wire should then also be disconnected from the gauge to the sender unit.
6 If not, the gauge is faulty and should be replaced (see Chapter 10).
7 With the wire disconnected from the sender unit and earthed, if

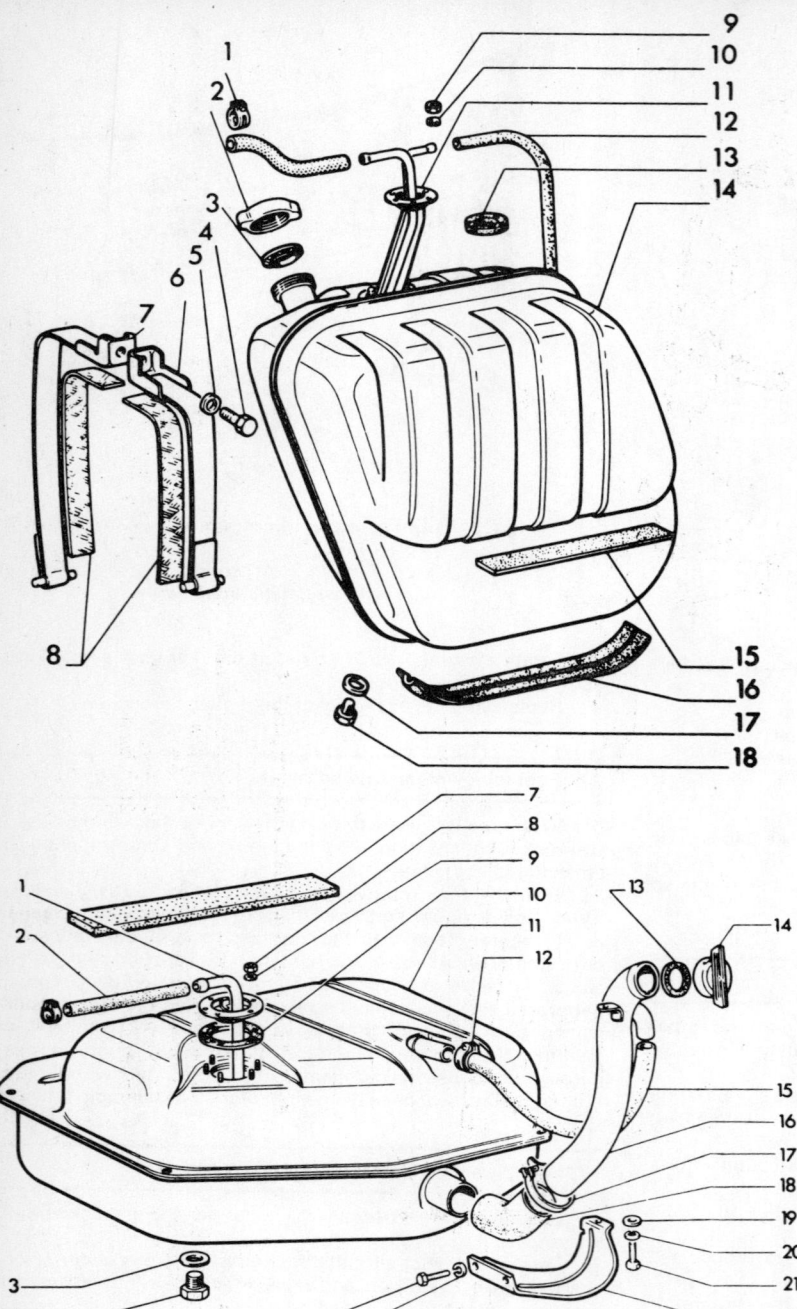

Fig. 3.14 Fuel tank and fittings (saloon models) (Sec 11)

1 Outlet pipe clip
2 Filler cap
3 Washer
4 Strap bolt
5 Washer
6 Short strap
7 Long strap
8 Insulation
9 Nut
10 Spring washer
11 Outlet and vent pipe assembly
12 Vent pipe extension
13 Gasket
14 Tank
15 Padding strip
16 Insulation strip
17 Washer
18 Drain plug

Fig. 3.15 Fuel tank and fittings (estate models) (Sec 11)

1 Outlet pipe
2 Connecting hose
3 Washer
4 Drain plug
5 Screw
6 Spring washer
7 Padding strip
8 Nut
9 Washer
10 Gasket
11 Tank
12 Hose clip
13 Washer
14 Filler cap
15 Vent pipe
16 Filler pipe
17 Hose clip
18 Filler elbow
19 Washer
20 Spring washer
21 Screw
22 Support bracket

the gauge reads anything other than 'Full' check the rest of the circuit (see Chapter 10 for the wiring diagram).
8 To remove the unit, first remove the tank as described in Section 11.
9 Undo and remove the screws and spring washers that secure the unit to the tank. Lift away the unit taking care not to bend the wire arm. Recover the gasket.
10 Refitting the sender unit is the reverse sequence to removal. Always use a new gasket.

15 Manifolds and exhaust system – general

1 The inlet manifold (aluminium) and the exhaust manifold (cast iron) are two separate castings bolted to the right-hand side of the cylinder head (photo).
2 The exhaust system is in three sections. The front section consists of a twin outlet pipe from the exhaust manifold which converges into one. It is supported on a bracket at the rear end just before it joins the middle section. The middle section includes an expansion box at the rear end near where the pipe joins the tail section. This joint is just in front of the rear axle where the pipe rises to go over it. The middle section has no support points along its length. The rear section consists of the silencer and tail pipe. This is supported by a rubber sling at the front end of the silencer and a flexible mounting on the tail pipe.
3 It is important for any exhaust system to see that the support hangers are in good condition. If any one is broken it should be renewed immediately, otherwise the unsupported weight of the system will strain and fracture the pipe and other hangers.
4 If a new section is to be fitted the whole system should be

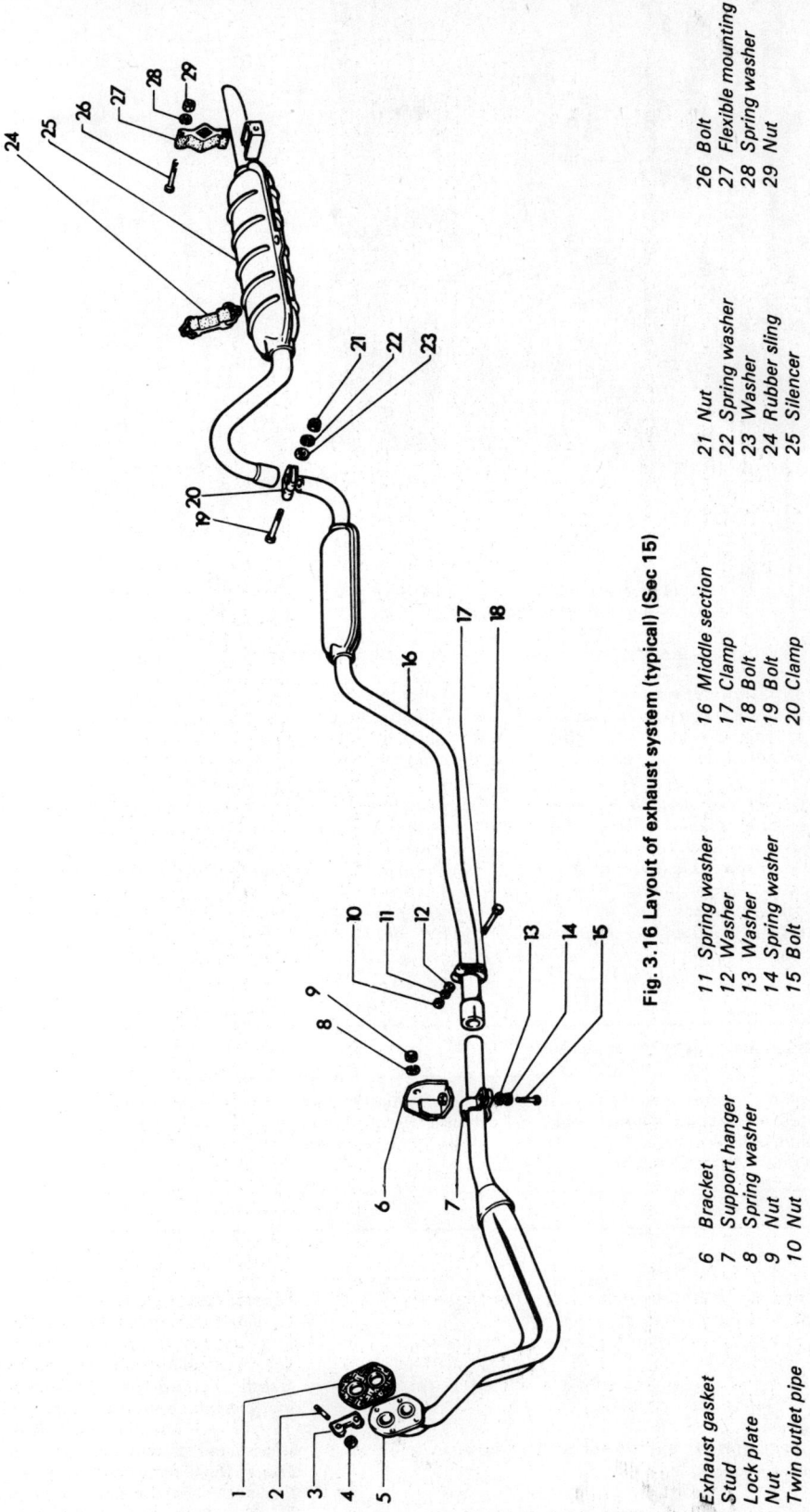

Fig. 3.16 Layout of exhaust system (typical) (Sec 15)

1 Exhaust gasket
2 Stud
3 Lock plate
4 Nut
5 Twin outlet pipe
6 Bracket
7 Support hanger
8 Spring washer
9 Nut
10 Nut
11 Spring washer
12 Washer
13 Washer
14 Spring washer
15 Bolt
16 Middle section
17 Clamp
18 Bolt
19 Bolt
20 Clamp
21 Nut
22 Spring washer
23 Washer
24 Rubber sling
25 Silencer
26 Bolt
27 Flexible mounting
28 Spring washer
29 Nut

Chapter 3 Fuel and exhaust systems

15.1 Removing the exhaust manifold

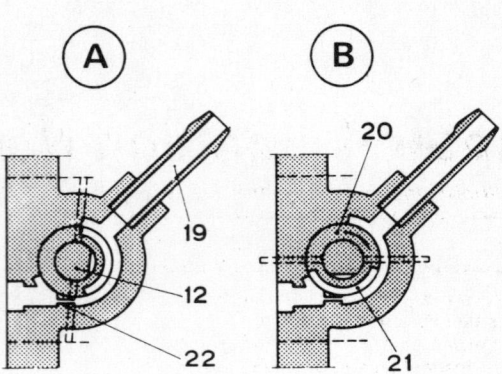

Fig. 3.17 Operation of crankcase ventilation valve (Sec 16)

A Starting position
B Running position
12 Throttle spindle
19 Gas inlet
20 Rotary valve
21 Valve slot
22 Metering orifice

removed unless it is easy to separate the parts in-situ. Otherwise it is easy to damage or strain another section whilst struggling to get one off.
5 Assemble the system together loosely before replacing it on the car. Always fit new clamps.
6 Arrange the pipes to follow the lines of the body and tighten the manifold flange, with a new gasket, before anything else. Then fit and tighten the three hangers and finally tighten the two pipe joining clamps. Do not overtighten the pipe clamps or the tube will distort and weaken.
Note: *Difficulty will be experienced in removing the front twin pipe section from the exhaust manifold flange due to the restricted space between the engine and body member. To increase the clearance, remove the heat shield from the starter motor and then disconnect the engine right-hand mounting by unscrewing the upper and lower nuts. On no account unscrew the two small nuts which secure the plate to the mounting as this retains a heavy coil spring which could fly out and cause injury. Place a jack under the sump pan with a block of wood as* an insulator, and raise the engine until the exhaust downpipe can be removed.

16 Closed circuit crankcase ventilation system

1 All models are fitted with a positive crankcase ventilation/circulation system which feeds unburnt fuel and oil vapours back through the combustion chambers, thereby reducing the amount of pollution released into the atmosphere.
2 The hoses connecting the crankcase and timing chain cover to the air cleaner should be removed periodically and cleaned with petrol.
3 On some models a valve is fitted to the throttle spindle on the carburettor and controls the amount of crankcase vapour entering the inlet manifold in relation to throttle opening.
4 The valve disc is particularly susceptible to gummy deposits which could affect slow running. It is easily taken out for cleaning after removing the circlip and washers from the end of the spindle. Make sure it is put back the right way round (see Fig. 3.17).

17 Fault diagnosis – fuel and exhaust systems

Unsatisfactory engine performance and excessive fuel consumption are not necessarily the fault of the fuel system or carburettor. In fact they more commonly occur as a result of ignition and timing faults. Before acting on the following it is necessary to check the ignition system first. Even though a fault may lie in the fuel system it will be difficult to trace unless the ignition is correct. The faults below, therefore, assume that this has been attended to first (where appropriate).

Symptom	Reason/s
Smell of petrol when engine is stopped	Leaking fuel lines or unions Leaking fuel tank
Smell of petrol when engine is idling	Leaking fuel line unions between pump and carburettor Overflow of fuel from float chamber due to wrong level setting, ineffective needle valve or punctured float
Excessive fuel consumption for reasons not covered by leaks or float chamber faults	Worn jets Over-rich setting Sticking mechanism
Difficult starting, uneven running, lack of power, cutting out	One or more jets blocked or restricted Float chamber fuel level too low or needle valve sticking Fuel pump not delivering sufficient fuel Faulty solenoid fuel shut-off valve

Chapter 4 Ignition system

For modifications, and information applicable to later models, see Supplement at end of manual

Contents

Condenser – removal, testing and refitting 4	Fault diagnosis – ignition system 9
Contact breaker points – adjustment 2	General description 1
Contact breaker points – removal and refitting 3	Ignition timing 7
Distributor – inspection and overhaul 6	Spark plugs and high tension leads 8
Distributor – removal and refitting 5	

Specifications

Spark plugs
Make and type .... Champion RN9YC or RN9YCC
Electrode gap:
    RN9YC .... 0.024 in (0.6 mm)
    RN9YCC .... 0.032 in (0.8 mm)

Coil
Rated primary circuit voltage .... 12 volts
Primary winding resistance at 20°C .... 3.2 ohms ± 4%
Secondary winding resistance at 20°C .... 6000 ohms ± 10%
Resistance of insulation to earth at 500 volts dc .... 50 M ohms

Distributor
Rotation .... Clockwise
Initial advance setting:
    1200 and 1300 cc models .... 5 to 7° BTDC
    1500 and 1600 cc models .... 3 to 5° BTDC
Vernier adjuster range – Octane selector (manual adjustment) .. ± 2.5° retard or advance
Automatic advance .... 30°
Breaker point gap .... 0.014 to 0.017 in (0.36 to 0.43 mm)
Dwell angle .... 55°
Insulating resistance between terminals and earth at
500 volts dc .... 10 M ohms
Condenser capacity at 50 to 1000 Hz .... 0.20 to 0.25 microfarads
Condenser insulation resistance at 100 ± 2°C, 100 volts dc ... 1 M ohm/microfarad
Oil type/specification .... Multigrade engine oil, viscosity range SAE 10W/30 to 20W/50, to API SE or SF (Duckhams QXR, Hypergrade, or 10W/40 Motor Oil)

Torque wrench setting

	lbf ft	kgf m
Spark plugs:		
Nominal	27.5	3.8
Minimum	23.2	3.2
Maximum	29.0	4.0

1 General description

In order that the engine can run correctly it is necessary for an electrical spark to ignite the fuel/air mixture in the combustion chamber at exactly the right moment in relation to engine speed and load. The ignition system is based on feeding low tension voltage from the battery to the coil where it is converted to high tension voltage. The high tension voltage is powerful enough to jump the spark plug gap in the cylinders many times a second under high compression pressures, providing that the system is in good condition and that all adjustments are correct.

The ignition system is divided into two circuits, the low tension circuit and the high tension circuit.

The low tension (sometimes known as the primary) circuit consists of the battery, lead to the control box, lead to the ignition switch, lead from the ignition switch to the low tension or primary coil windings, and the lead from the low tension coil windings to the contact breaker points and condenser in the distributor.

The high tension circuit consists of the high tension or secondary coil windings, the heavy ignition lead from the centre of the coil to the centre of the distributor cap, the rotor arm, and the spark plug leads and spark plugs.

The system functions in the following manner: High tension voltage is generated in the coil by the interruption of the low tension circuit. The interruption is effected by the opening of the contact breaker points in this low tension circuit. High tension voltage is fed via the carbon brush in the centre of the distributor cap to the rotor arm of the distributor.

The rotor arm revolves clockwise at half engine speed inside the distributor cap. Each time it comes in line with one of the four metal segments in the cap, which are connected to the spark plug

Chapter 4 Ignition system

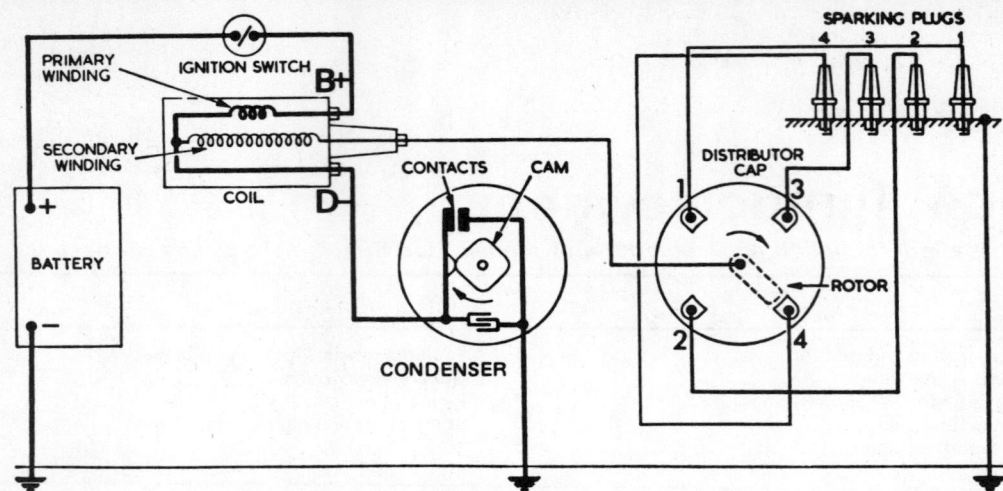

Fig. 4.1 Schematic diagram of ignition circuit (Sec 1)

leads, the opening and closing of the contact breaker points causes the high tension voltage to build up, jump the gap from the rotor arm to the appropriate metal segment, and so via the spark plug lead to the spark plug, where it finally jumps the spark plug gap before going to earth.

The ignition is advanced and retarded automatically, to ensure the spark occurs at just the right instant for the particular load at the prevailing engine speed.

The ignition advance is controlled mechanically. The mechanical governor mechanism comprises two weights, which move out from the distributor shaft as the engine speed rises, due to centrifugal force. As they move outwards they rotate the cam relative to the distributor shaft, and so advance the spark. The weights are held in position by two springs and it is the tension of the springs which is largely responsible for correct spark advancement.

An octane selector on the side of the distributor enables the timing to be advanced or retarded to a maximum of $\pm 2.5°$ either way for fine timing adjustment.

2 Contact breaker points – adjustment

1 To adjust the contact breaker points to the correct gap, first pull off the two clips that secure the distributor cap to the distributor body, and lift away the cap. Clean the cap inside and out with a dry cloth. It is unlikely that the four segments will be badly burned or scored, but if they are, the cap will have to be renewed.
2 Check the carbon brush located in the top of the cap to make sure that it is not broken or missing.
3 Gently prise the contact breaker points open to examine the condition of their faces. If they are rough, pitted or dirty, it wil be necessary to remove them for resurfacing, or for replacement points to be fitted.
4 Presuming the points are satisfactory, or that they have been cleaned and replaced, measure the gap between the points by turning the engine over until the contact breaker arm is on the peak of one of the four cam lobes.
5 A feeler gauge of thickness 0·014 to 0·017 in (0.36 to 0.43 mm) should fit between the points whilst lightly touching each face (photo).
6 If the gap varies from this amount, slacken the contact securing screw.
7 Adjust the contact gap by inserting a screwdriver in the nick in the side of the fixed plate and lever it in the required direction.
8 Refit the distributor cap and clip it into place.
Note: *The points gap should be checked by a dwell meter when possible.*
9 Adjustment of the contact breaker points gap by feeler blades must be regarded as an initial setting in order to get the engine running. A final and more precise adjustment should now be made by checking the dwell angle on a dwell meter.
10 The dwell angle is the number of degrees through which the distributor cam turns between the instant of closure and opening of

2.5 Setting the contact breaker gap

the contact breaker points. This can only be checked on a dwell meter connected in accordance with the maker's instructions.
11 The correct dwell angle is as shown in the Specifications. If the angle is too large, increase the points gap; if too small, reduce the points gap.

3 Contact breaker points – removal and refitting

1 Remove the distributor cap.
2 Remove the two screws that secure the contact points assembly to the distributor and slacken the insulated terminal in the side of the body sufficiently to release the low tension lead.
3 Lift the complete contact set assembly off the pivot pin of the mounting plate.
4 If the condition of the points is not too bad, they can be reconditioned by rubbing the contacts clean with fine emery cloth or a fine carborundum stone. It is important that the faces are rubbed flat and parallel to each other so that there will be complete face to face contact when the points are closed. One of the points will be pitted and the other will have deposits on it.
5 It is necessary to completely remove the built up deposits, but not necessary to rub the pitted point right down to the stage where all the pitting has disappeared, though obviously if this is done it will prolong the time before the operation of refacing the points has to be repeated.

Chapter 4 Ignition system

6   Wipe the points clean and dry and refit the assembly with the two screws. Reconnect the low tension lead to the terminal.
7   Adjust the points gap as described in the previous Section.

4 Condenser – removal, testing and refitting

1   The purpose of the condenser (sometimes known as capacitor) is to ensure that when the contact breaker points open there is no sparking across them which would waste voltage and cause rapid deterioration of the points.
2   The condenser is fitted in parallel with the contact breaker points. If it develops a short circuit, it will cause ignition failure as the points will be prevented from interrupting the low tension circuit.
3   If the engine becomes very difficult to start or begins to misfire whilst running and the breaker points show signs of excessive burning, then the condition of the condenser must be suspect. A further test can be made by separating the points by hand with the ignition switched on. If this is accompanied by a bright spark at the contact points it is indicative that the condenser has failed.
4   Without special test equipment the only sure way to diagnose condenser trouble is to replace a suspected unit with a new one and note if there is any improvement.
5   To remove the condenser from the distributor, take out the screw which secures it to the underside of the distributor body and slacken the insulated terminal nut enough to remove the wire connection tag.
6   When fitting the condenser it is vital to ensure that the fixing screw is secure and the condenser tightly held. The lead must be secure on the terminal with no chance of short circuiting.

5 Distributor – removal and refitting

1   Unless particular care is taken the ignition timing will be disturbed when removing the distributor. If this is to be prevented first remove the cap and turn the engine so that the rotor points towards an accurate reference point. Then mark the bottom flange of the distributor body and the adjacent block with punch marks so they may be realigned.
2   Undo the clamp securing nut and take away the clamp.
3   Lift the distributor straight out (photo).
4   Provided the engine is not turned the distributor may be replaced with the rotor and marks aligned and no further adjustment is needed. Otherwise the ignition timing must be reset as described in Section 7. When refitting the distributor do not forget the oil sealing ring if fitted.

6 Distributor – inspection and overhaul

1   Apart from the contact points, the other parts of a distributor which deteriorate with age and use are the cap, the rotor, the shaft bushes and the bob weight springs.
2   The cap must have no flaws or cracks and the 4 high tension terminal contacts should not be severely corroded. The centre spring loaded carbon contact is replaceable. If in any doubt about the cap buy a new one.
3   The rotor deteriorates minimally but with age the metal conductor tip may corrode. It should not be cracked or chipped and the metal conductor must not be loose. If in doubt renew it. Always fit a new rotor if fitting a new cap.
4   The rotor is held in position by two screws and when removed, the centrifugal advance mechanism is exposed. There is no way to test the bob weight springs other than by checking the performance of the distributor on special test equipment, so if in doubt fit new springs anyway. If the springs are loose where they loop over the posts it is more than possible that the post grooves are worn in which case the various parts which include the shaft will need renewal. Wear to this extent would mean that a new distributor is probably the best solution in the long run. Be sure to make an exact note of both the engine number and any serial number on the distributor when ordering. When refitting the rotor, note that there are two positioning lugs and recesses, one round, one square, to make sure it is replaced the proper way round.
5   If the main shaft is slack in its bushes allowing even the slightest sideways play, it means that the contact points gap setting can only be a compromise because the cam position relative to the cam follower

5.3 Lifting out the distributor

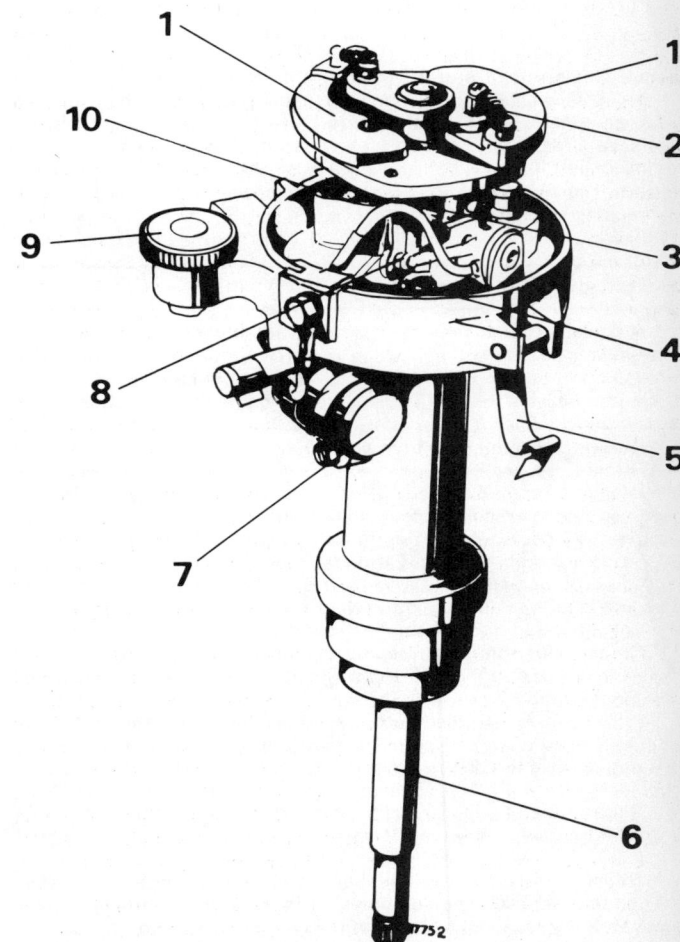

Fig. 4.2 Main components of distributor (Sec 5)

1   Governor weights
2   Weight springs
3   Weight shaft with cams
4   Breaker point gap adjusting screw
5   Cap clip
6   Drive shaft
7   Condenser and bracket
8   Breaker points
9   Vernier adjuster nut
10 Oil wick

Chapter 4 Ignition system

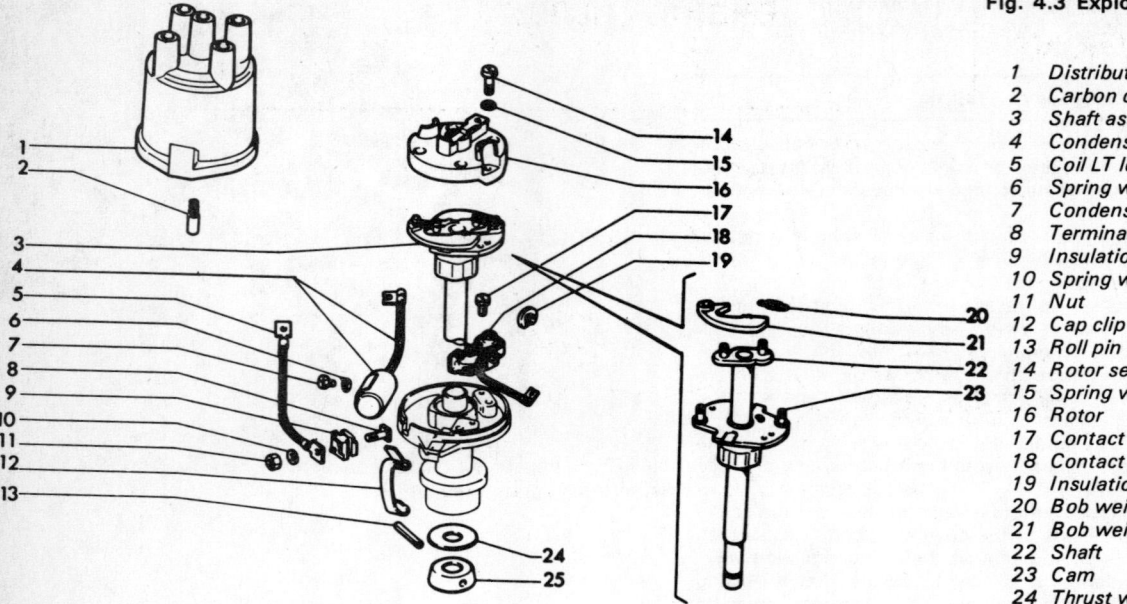

Fig. 4.3 Exploded view of distributor (Sec 6)

1 Distributor cap
2 Carbon contact
3 Shaft assembly
4 Condenser
5 Coil LT lead
6 Spring washer
7 Condenser securing screw
8 Terminal screw
9 Insulation block
10 Spring washer
11 Nut
12 Cap clip
13 Roll pin
14 Rotor securing screw
15 Spring washer
16 Rotor
17 Contact points securing screw
18 Contact points assembly
19 Insulation
20 Bob weight spring
21 Bob weight
22 Shaft
23 Cam
24 Thrust washer
25 Retainer (oil slinger)

on the moving point arm is not constant. If the top of the shaft can move 0·003 in sideways it means the points gap can vary by about the same amount. It is not practical to re-bush the distributor body unless you have a friend who can bore and bush it for you. The shaft can be removed by driving out the roll pin from the retaining collar at the bottom. The collar also acts as an oil slinger to prevent excess engine oil creeping up the shaft.

7 Ignition timing

1 It is necessary to time the ignition when it has been upset due to overhauling or dismantling, which may have altered the relationship between the position of the pistons and the moment at which the distributor delivers the spark. Also, if maladjustments have affected the engine performance it is very desirable, although not always essential, to reset the timing starting from scratch. In the following procedures it is assumed that the intention is to obtain standard performance from the standard engine which is in reasonable condition. It is also assumed that the recommended fuel octane rating is used.
2 Set the transmission to neutral and remove all four spark plugs.
3 Place a thumb over No 1 cylinder spark plug hole (front cylinder) and rotate the engine clockwise by means of the starting handle until pressure is felt building up in the No 1 cylinder. This indicates that the No 1 piston is approaching top dead centre (TDC) on the firing stroke.
4 Continue to rotate the engine until the notch in the crankshaft pulley is directly opposite the appropriate mark on the timing cover (see Specifications).
5 Now remove the distributor cap and check that the distributor rotor is aligned with the No 1 HT lead segment in the distributor cap. If it is not, remove the distributor and rotate the rotor to the correct position. Hold the rotor in this position and refit the distributor to the block.
6 Slacken the distributor clamp bolt and rotate the distributor body until the contact breaker points are just opening. Tighten the clamp bolt.
7 Difficulty is sometimes experienced in determining exactly when the contact breaker points open. This can be ascertained most accurately by connecting a 12v bulb in parallel with the contact breaker points (one lead to earth and the other from the distributor low tension terminal). Switch on the ignition and turn the distributor body until the bulb lights up, indicating the points have just opened.
8 As a final check on the ignition timing the best method is to use a strobe lamp.
9 Put a spot of white paint on the notch in the crankshaft pulley and the centre pointer on the timing cover and connect the strobe light into the No 1 cylinder HT circuit.
10 Run the engine at idling speed and point the strobe lamp at the timing marks. At idling speed the white paint marks should be immediately opposite each other; open the throttle slightly and check that as the engine revolutions rise the spot on the crankshaft pulley moves anti-clockwise. This indicates the advance mechanism is operating correctly.
11 Fine adjustment of the ignition timing can be made by rotating the vernier adjuster on the side of the distributor in the appropriate direction (+ for advance, – for retard).

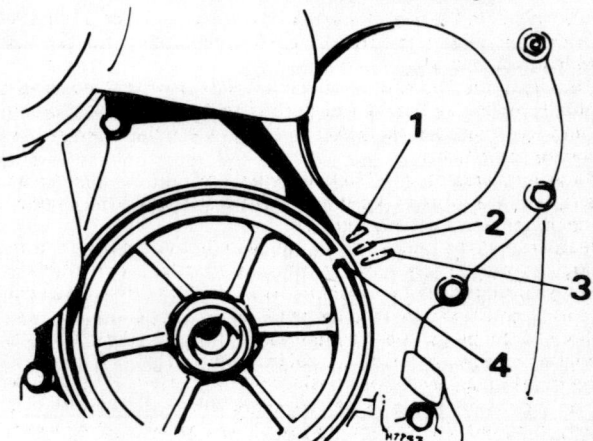

Fig. 4.4 Ignition timing marks (Sec 7)

1 10° BTDC mark
2 5° BTDC mark
3 TDC mark
4 Notch in crankshaft pulley

8 Spark plugs and high tension leads

1 The correct functioning of the spark plugs is vital for the correct running and efficiency of the engine. It is essential that the plugs fitted are appropriate for the engine, and the suitable type is specified at the beginning of this chapter. If this type is used and the engine is in good condition, the spark plugs should not need attention between scheduled replacement intervals. Spark plug cleaning is rarely necessary and should not be attempted unless specialised equipment is available as damage can easily be caused to the firing ends.
2 The condition of the spark plug will also tell much about the overall condition of the engine.
3 If the insulator nose of the spark plug is clean and white, with no

Chapter 4 Ignition system

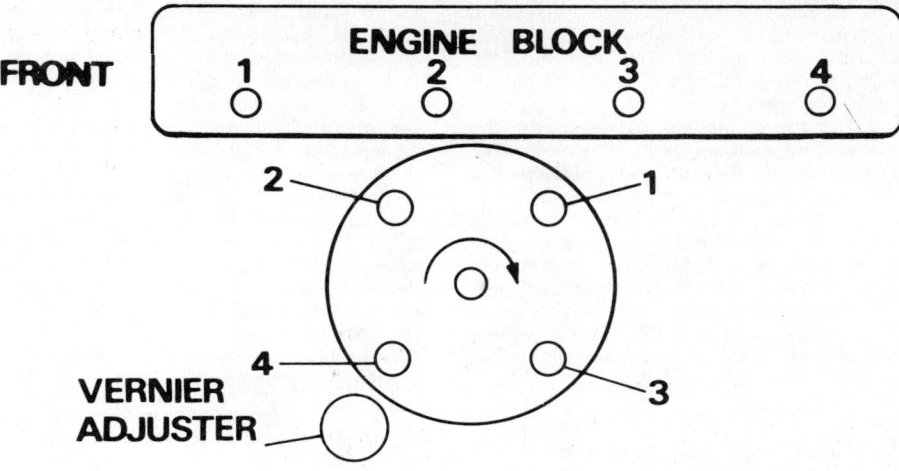

Fig. 4.5 Spark plug lead connection diagram (Sec 8)

deposits, this is indicative of a weak mixture, or too hot a plug. A hot plug transfers heat away from the electrode slowly, a cold plugs transfers it away quickly.

4 If the tip of the insulator nose is covered with sooty black deposits, then this is indicative that the mixture is too rich. Should the plug be black and oily, then it is likely that the engine is fairly worn, as well as the mixture being rich.

5 If the insulator nose is covered with light tan to greyish brown deposits, then the mixture is correct and it is likely that the engine is in good condition and correctly tuned.

6 Clean around the plug seats in the cylinder head before removing the plugs to prevent dirt getting into the cylinders. Plugs should be cleaned by a sand blasting machine, which will free them from carbon more thoroughly than cleaning by hand. The machine will also test the behaviour of the plugs under compression. Any plug that fails to spark at the recommended pressure should be renewed.

7 The spark plug gap is of considerable importance, as if it is too large or too small the size of the spark and its efficiency will be seriously impaired. The spark plug gap should be set to the gap shown in the Specifications for the best results.

8 To set it, measure the gap with a feeler gauge, and then bend open, or close, the outer plug electrode until the correct gap is achieved. The centre electrode should never be bent as this may crack the insulation and cause plug failure, if nothing worse.

9 When replacing the plug see that the washer is intact and carbon free, also the shoulder of the plug under the washer. Make sure also that the plug seat in the cylinder head is quite clean.

10 Replace the leads from the distributor in the correct firing order, which is 1–3–4–2; No 1 cylinder being the one nearest the radiator.

11 The plug leads require no routine attention other than being kept clean and wiped over regularly. At intervals, say twice yearly, pull each lead off the plug in turn and also from the distributor cap. Water can seep down into these joints giving rise to a white corrosive deposit which must be carefully removed from the brass connectors at the end of each cable.

12 If an ohmmeter is available, test each HT lead for condition. If the leads are of carbon core type the indicated resistance should not exceed 25 000 ohms, otherwise renew the leads.

9 Fault diagnosis – ignition system

By far the majority of breakdown and running troubles are caused by faults in the ignition system either in the low tension or high tension circuits.

There are two main symptoms indicating faults. Either the engine will not start or fire, or the engine is difficult to start and misfires. If it is a regular misfire (ie the engine is running on only two or three cylinders), the fault is almost sure to be in the secondary or high tension circuit. If the misfiring is intermittent the fault could be in either the high or low tension circuits. If the car stops suddenly, or will not start at all, it is likely that the fault is in the low tension circuit. Loss of power and overheating, apart from faulty carburation settings, are normally due to faults in the distributor or to incorrect ignition timing.

Engine fails to start

1 If the engine fails to start and the car was running normally when it was last used, first check there is fuel in the petrol tank. If the engine turns over normally on the starter motor and the battery is evidently well charged, then the fault may be in either the high or low tension circuits. First check the HT circuit. **Note**: *If the battery is known to be fully charged, the ignition light comes on, and the starter motor fails to turn the engine check the tightness of the leads on the battery terminals and also the secureness of the earth lead to its connection to the body.* It is quite common for the leads to have worked loose, even if they look and feel secure. If one of the battery terminal posts gets very hot when trying to work the starter motor this is a sure indication of a faulty connection to that terminal.

2 One of the commonest reasons for bad starting is wet or damp spark plug leads and distributor. Remove the distributor cap. If condensation is visible internally dry the cap with a rag and also wipe over the leads. Refit the cap.

3 If the engine still fails to start, check that voltage is reaching the plugs by disconnecting each plug lead in turn at the spark plug end, and holding the end of the cable about $\frac{1}{4}$ in (6 mm) away from the cylinder block. Spin the engine on the starter motor.

4 Sparking between the end of the cable and the block should be fairly strong with a strong regular blue spark. Hold the lead with rubber to avoid electric shocks. If voltage is reaching the plugs, then remove them and clean and regap them. The engine should now start.

5 If there is no spark at the plug leads, take off the HT lead from the centre of the distributor cap and hold it to the block as before. Spin the engine on the starter once more. A rapid succession of blue sparks between the end of the lead and the block indicate that the coil is in order and that the distributor cap is cracked, the rotor arm is faulty, or the carbon brush into the top of the distributor cap is not making good contact with the spring on the rotor arm. Possibly, the points are in bad condition. Clean and reset them as described in this Chapter, Section 2 or 3.

6 If there are no sparks from the end of the lead from the coil, check the connections at the coil end of the lead. If it is in order start checking the low tension circuit.

7 Use a 12v voltmeter or a 12v bulb and two lengths of wire. With the ignition switch on and the points open, test between the low

tension wire to the coil (it is marked 15 or +) and earth. No reading indicates a break in the supply from the ignition switch. Check the connections at the switch to see if any are loose. Refit them and the engine should run. A reading shows a faulty coil or condenser, or broken lead between the coil and the distributor.

8 Take the condenser wire off the points assembly and with the points open test between the moving point and earth. If there now is a reading then the fault is in the condenser. Fit a new one and the fault is cleared.

9 With no reading from the moving point to earth, take a reading between earth and the − or 1 terminal of the coil. A reading here shows a broken wire which will need to be replaced between the coil and distributor. No reading confirms that the coil has failed and must be replaced, after which the engine will run once more. Remember to refit the condenser wire to the points assembly. For these tests it is sufficient to separate the points with a piece of dry paper while testing with the points open.

Engine misfires

10 If the engine misfires regularly, run it at a fast idling speed, pull off each of the plug caps in turn and listen to the note of the engine. Hold the plug cap in a dry cloth or with a rubber glove as additional protection against a shock from HT supply.

11 No difference in engine running will be noticed when the lead from the defective circuit is removed. Removing the lead from one of the good cylinders will accentuate the misfire.

12 Remove the lead from the defective circuit and hold it about $\frac{3}{16}$ in (5 mm) away from the block. Re-start the engine. If the sparking is fairly strong and regular, the fault must lie in the spark plug.

13 The plug may be loose, the insulation may be cracked, or the points may have burnt away giving a wide gap for the spark to jump. Worse still, one of the points may have broken off. Either renew the plug, or clean it, reset the gap, and then test it.

14 If there is no spark at the end of the plug lead, or if it is weak and intermittent, check the ignition lead from the distributor to the plug. If the insulation is cracked or perished, renew the lead. Check the connections at the distributor cap.

15 If there is still no spark, examine the distributor cap carefully for tracking. This can be recognised by a very thin black line running between two or more electrodes, or between an electrode and some other part of the distributor. These lines are paths which now conduct electricity across the cap thus letting it run to earth. The only answer is a new distributor cap.

16 Apart from the ignition timing being incorrect, other causes of misfiring have already been dealt with under the Section dealing with the failure of the engine to start. To recap, these are that:

- (a) The coil may be faulty giving an intermittent misfire
- (b) There may be a damaged wire or loose connection in the low tension circuit
- (c) The condenser may be faulty
- (d) There may be a mechanical fault in the distributor (broken driving spindle or contact breaker spring)

17 If the ignition timing is too far retarded, it should be noted that the engine will tend to overheat and there will be a quite noticeable drop in power. If the engine is overheating, the power is down and the ignition timing is correct, then the carburettor should be checked, as it is likely that this is where the fault lies.

Chapter 5 Clutch

For modifications, and information applicable to later models, see Supplement at end of manual

Contents

Clutch – adjustment 2	General description 1
Clutch assembly – removal 8	Hydraulic system – bleeding 5
Clutch – inspection and renovation 9	Master cylinder – removal, dismantling and reassembly 6
Clutch pedal – adjustment 3	Release bearing – removal and refitting 10
Clutch pedal – removal and refitting 4	Slave (operating) cylinder – removal, dismantling and reassembly 7
Clutch – refitting 11	
Fault diagnosis – clutch 12	

Specifications

Type . . . . Single dry plate, diaphragm spring

Driven plate
Diameter:
   1200 and 1300 models . . . . 7·87 in (200 mm)
   1500 and 1600 models . . . . 8·1 in (206 mm)
Thickness . . . . 0·130 in (3·3 mm)

Release mechanism . . . . Hydraulic, master and slave cylinder
Master cylinder diameter . . . . 0·75 in (19·05 $^{+\,0·025}_{-\,0·015}$ mm)

Slave cylinder diameter . . . . 0·75 in (19·05 $^{+\,0·025}_{-\,0·015}$ mm)

Hydraulic fluid
Type/specification . . . . Hydraulic fluid to SAE J1703 or FMVSS DOT 3 or DOT 4 (Duckhams Universal Brake and Clutch Fluid)

Clutch adjustment
Pedal free travel . . . . 0·78 to 1·18 in (20 to 30 mm)
Release lever free travel . . . . 0·157 to 0·236 in (4 to 6 mm)

Torque wrench settings	**lbf ft**	**kgf m**
Cover to flywheel bolts	21·7	3·0
Slave cylinder bolts	10·9	1·5
Flexible hose connections	15	2·1
Clutch bellhousing to engine bolts	61·6	8·5

1 General description

Note: The diameter of the clutch plate will depend on the model of car (see Specifications). The clutch is of the single dry plate, diaphragm spring type. The unit comprises a pressed steel cover which is dowelled to the rear face of the flywheel and bolted to it and contains the pressure plate, pressure plate diaphragm spring and the fulcrum rings.

The clutch disc is free to slide along the splined first motion shaft and is held in position between the flywheel and the pressure plate by the pressure of the pressure plate spring. Friction lining material is riveted to the clutch disc and it has a spring cushioned hub to absorb transmission shocks and to help ensure a smooth take-off.

The circular diaphragm spring is mounted on shouldered pins held in place in the cover by two fulcrum rings. The spring is also held to the pressure plate by the three spring steel clips which are riveted in position.

The clutch is actuated hydraulically by a master and slave cylinder. The pendant type clutch pedal is connected to the clutch master cylinder and hydraulic fluid reservoir by a short pushrod. The master cylinder and hydraulic reservoir are mounted on the engine side of the bulkhead in front of the driver. Depressing the clutch pedal moves the piston in the master cylinder forwards, so forcing hydraulic fluid through the clutch hydraulic pipe to the slave cylinder. The piston in the slave cylinder moves forward on the entry of the fluid and actuates the clutch release arm by means of a short pushrod.

The release arm pushes the release bearing forwards to bear against the release plate, so moving the centre of the diaphragm spring inwards. The spring is sandwiched between two annular rings which act as fulcrum points. As the centre of the spring is pushed in, the outside of the spring is pushed out, so moving the pressure plate backwards and disengaging the pressure plate from the clutch disc.

When the clutch pedal is released, the diaphragm spring forces the pressure plate into contact with the high friction linings on the clutch disc and at the same time pushes the clutch disc a fraction of an inch forwards on its splines so engaging the clutch disc with the flywheel. The clutch disc is now firmly sandwiched between the pressure plate and the flywheel so the drive is taken up. As the friction linings on the clutch disc wear the pressure plate automatically moves closer to the disc to compensate. There is therefore no need to periodically adjust the clutch.

Chapter 5 Clutch

2 Clutch – adjustment

1 A clearance of 0·157 to 0·236 in (4 to 6 mm) must be maintained between the slave cylinder pushrod nut and the withdrawal lever.

2 Carry out any required adjustment by releasing the locknut on the pushrod and turning the adjuster nut (see Fig. 5.2).

3 When the clearance at the withdrawal lever is correct, there should be a free movement at the centre of the pedal pad of approximately 1·180 in (30 mm).

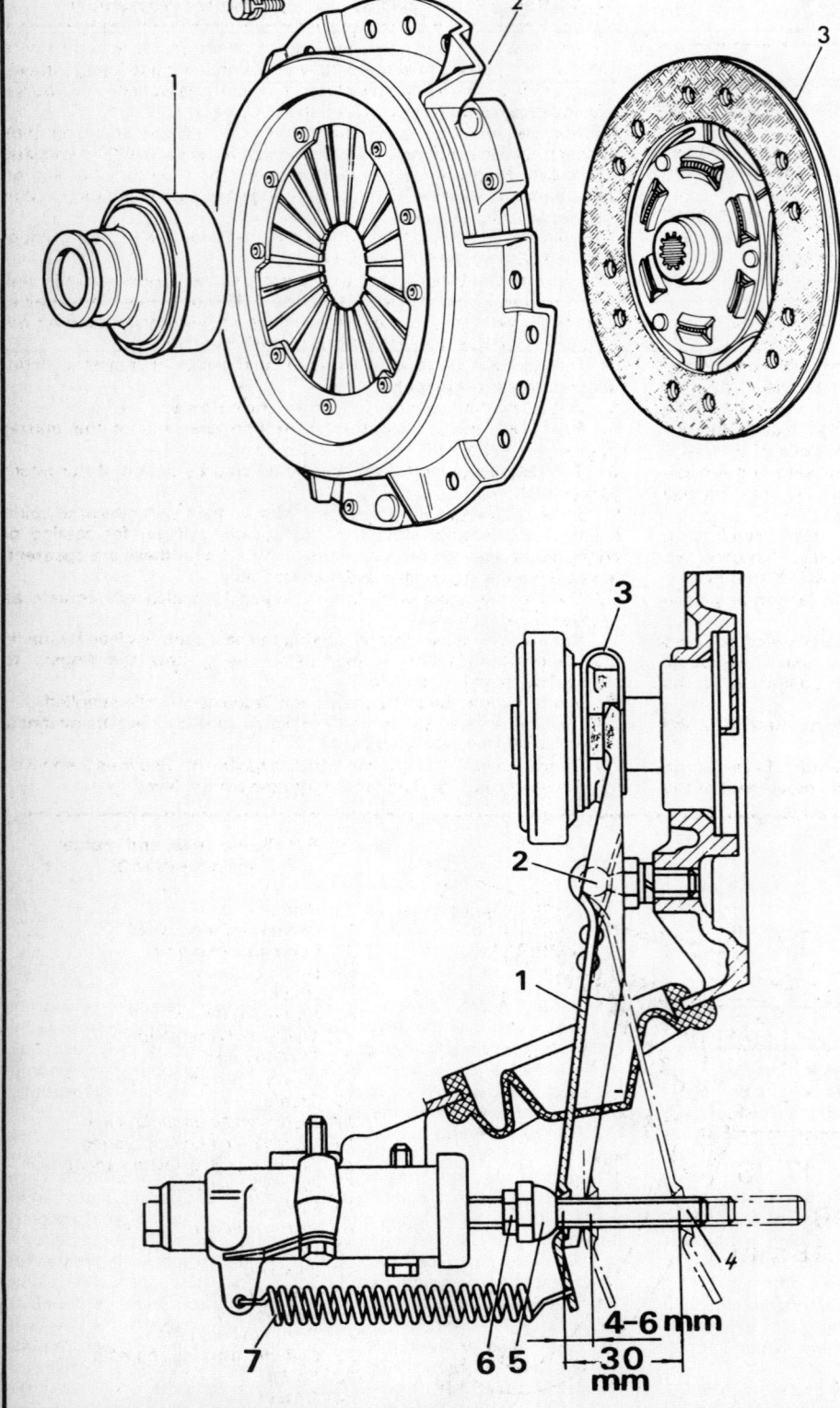

Fig. 5.1 Main clutch components (Sec 1)

1 Release bearing
2 Cover plate/diaphragm spring assembly
3 Driven plate

Fig. 5.2 Clutch release arm and slave cylinder (Sec 2)

1 Clutch release fork
2 Ball fulcrum
3 Spring
4 Pushrod
5 Adjusting nut
6 Locknut
7 Return spring

Chapter 5 Clutch

3 Clutch pedal – adjustment

The clutch pedal height is correctly set at the factory and will not normally require adjustment until the car has covered a considerable mileage. The pedal stop bolt, (Fig. 5.3) should be adjusted so that there is 0·003 to 0·019 in (0·07 to 0·48 mm) of free play between the end of the clutch pedal pushrod and the master cylinder piston. This gives a clutch pedal height of approximately 5·5 in (140 mm).

4 Clutch pedal – removal and refitting

1 To remove the clutch pedal it is required that the complete clutch and brake pedal assembly be removed.
2 Remove both master cylinders and steering column. Remove pedal springs and bracket retaining bolts. Remove pedal bracket with pedals complete.
3 To remove the pedals from the bracket, remove the bolt running through the pedal pivots.
4 Refitting is a reversal of removal. Grease the mating surfaces.
5 Check the adjustments as described in Sections 2 and 3.

5 Hydraulic system – bleeding

1 The need for bleeding the cylinders and fluid line arises when air gets into it. Air gets in whenever a joint or seal leaks or part has to be dismantled. Bleeding is simply the process of venting the air out again.
2 Make sure the reservoir is filled and obtain a piece of $\frac{3}{16}$ in (4·8 mm) bore diameter rubber tube about 2ft (60 cm) long and a clean glass jar. A small quantity of fresh, clean hydraulic fluid is also necessary.
3 Detach the cap (if fitted) on the bleed nipple at the clutch slave cylinder and clean up the nipple and surrounding area. Unscrew the nipple $\frac{3}{4}$ turn and fit the tube over it. Put about 1 in (25·4 mm) of fluid in the jar and put the other end of the pipe in it. The jar can be placed on the ground under the car.
4 The clutch pedal should then be depressed quickly and released slowly until no more air bubbles come from the pipe. Quick pedal action carries the air along rather than leaves it behind. Keep the reservoir topped up.
5 When the air bubbles stop, tighten the nipple at the end of the down stroke.
6 Check that the operation of the clutch is satisfactory. Even though there may be no exterior leaks, it is possible that the movement of the pushrod from the clutch cylinder is inadequate because fluid is leaking internally past the seals in the master cylinder. If this is the case, it is best to replace all seals in both cylinders.
7 Always use clean hydraulic fluid which has been stored in an airtight container and has remained unshaken for the preceding 24 hours.

6 Master cylinder – removal, dismantling and reassembly

1 The master cylinder and fluid reservoir are joined together by a hose and indications of something wrong with it are if the pedal travels down without operating the clutch efficiently (assuming, of course, that the system has been bled and there are no leaks).
2 Remove the unit from the car, clamp off the adjoining pipe between the reservoir and the master cylinder to reduce fluid wastage whilst dismantling the pipes. Alternatively, the fluid may be bled off from the clutch reservoir into a jar through the pipe when removed at the master cylinder end.
3 Disconnect the fluid line which runs between the master cylinder and the slave (operating) cylinder.
4 If the master cylinder has a separate reservoir, disconnect the pipe from the master cylinder and drain the fluid into a suitable container (photo). *Note: Care must be taken not to spill the brake fluid on the bodywork as it has the same effect as paint remover.*
5 Unscrew and remove the two bolts that secure the master cylinder to the engine rear bulkhead.
6 Withdraw the master cylinder from the bulkhead.
7 Peel back the rubber dust cover from the end of the master cylinder and extract the circlip and stop ring.
8 The piston assembly will now be ejected by action of the return spring.
9 Clean all components in clean hydraulic fluid or methylated spirit. Examine the internal surface of the master cylinder for scoring or bright areas; also the surface of the piston. Where these are apparent, renew the complete master cylinder assembly.
10 Discard the seals and obtain a repair kit which will contain all necessary components.
11 Commence reassembly by dipping the new seals in clean hydraulic fluid and fitting them to the piston, using only the fingers to manipulate them into position.
12 Insert the internal components and fit the new circlip supplied.
13 Bolt the master cylinder to the bulkhead and connect the hydraulic pipe. Bleed the system (Section 5).
14 Check the pedal height and clutch adjustment (Sections 2 and 3).
15 Finally top-up the fluid reservoir to the correct level.

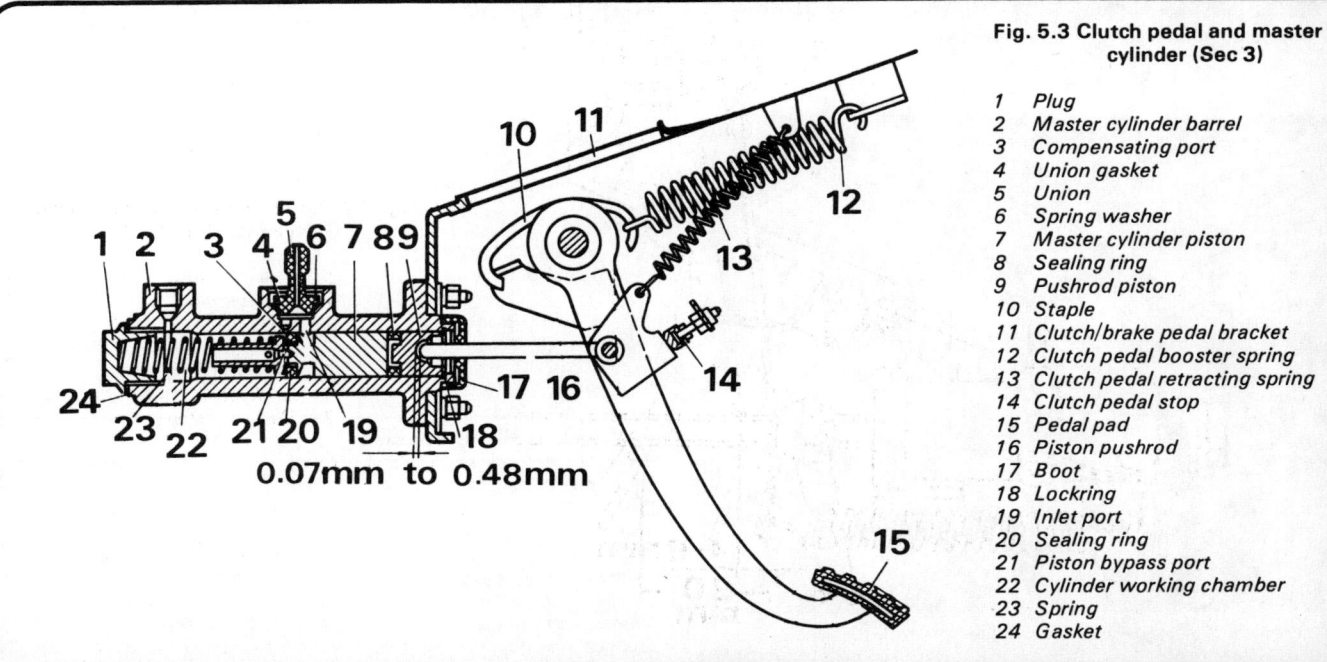

Fig. 5.3 Clutch pedal and master cylinder (Sec 3)

1 Plug
2 Master cylinder barrel
3 Compensating port
4 Union gasket
5 Union
6 Spring washer
7 Master cylinder piston
8 Sealing ring
9 Pushrod piston
10 Staple
11 Clutch/brake pedal bracket
12 Clutch pedal booster spring
13 Clutch pedal retracting spring
14 Clutch pedal stop
15 Pedal pad
16 Piston pushrod
17 Boot
18 Lockring
19 Inlet port
20 Sealing ring
21 Piston bypass port
22 Cylinder working chamber
23 Spring
24 Gasket

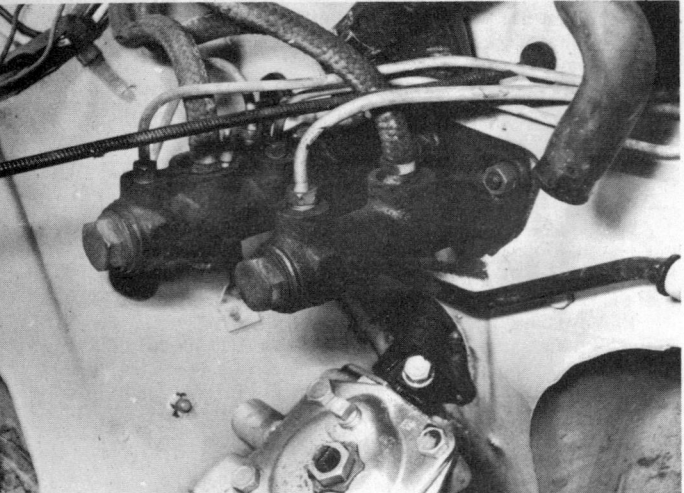

6.4 Clutch master cylinder with separate reservoir

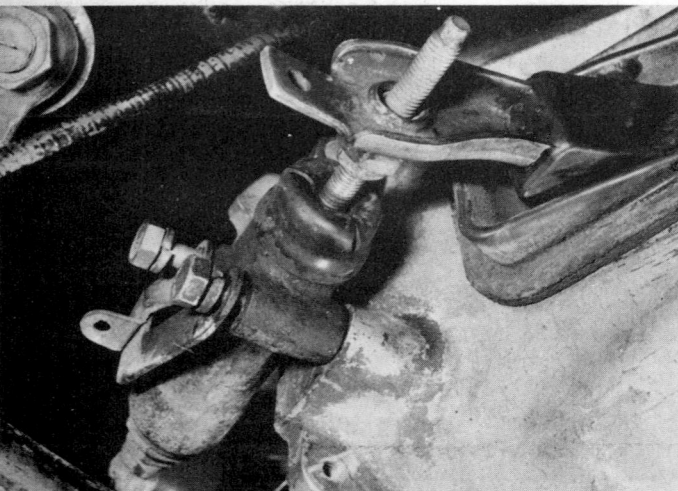

7.4 Removing the clutch slave cylinder retaining bolts

7.11 Correct refitting of clutch slave cylinder and return spring

7 Slave (operating) cylinder – removal, dismantling and re-assembly

1 Disconnect the rigid hydraulic pipe from the flexible hose at the support bracket on the body side frame.
2 Remove the locking clip from the bracket and release the flexible hose. Now unscrew the flexible hose from the slave cylinder and plug the end to prevent loss of fluid.
3 Detach the return spring from the slave cylinder. Remove the split-pin from the end of the pushrod.
4 Unscrew and remove the two bolts that secure the cylinder to the clutch bellhousing and remove it (photo).
5 Peel off the rubber dust excluder and pull out the pushrod.
6 Extract the piston assembly either by tapping the cylinder on a piece of wood or applying air pressure to the fluid inlet hole.
7 Wash all components in clean hydraulic fluid or methylated spirit. Discard the seals and examine the piston and cylinder bore surfaces for scoring or bright areas. Where these are evident, renew the complete assembly.
8 Obtain a repair kit which will contain all necessary components.
9 Commence reassembly by dipping the new seals in clean hydraulic fluid and fitting them to the piston, using only the fingers to manipulate them into position.
10 Dip the piston in clean fluid and insert it into the cylinder bore, fit the new dust cover supplied and refit the pushrod.
11 Bolt the cylinder to the clutch bellhousing (photo).
12 Reconnect the flexible hose taking care not to twist it or to route it near the exhaust pipe.
13 Bleed the system as described in Section 5.
14 Check the clutch adjustment (Sec 2).
15 Finally top-up the fluid reservoir to the correct level.

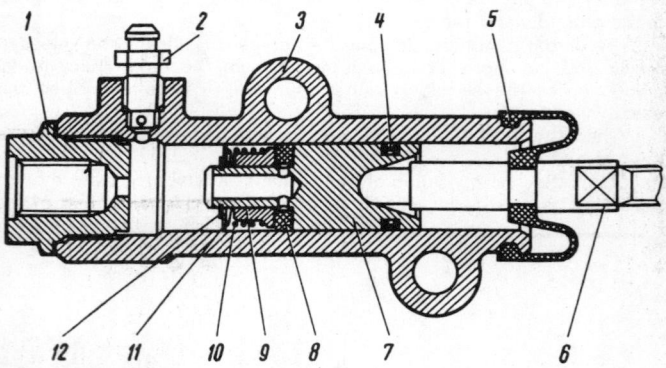

Fig. 5.4 Sectional view of slave cylinder (Sec 7)

1 Plug
2 Bleed nipple
3 Barrel
4 Sealing ring
5 Boot
6 Release fork
7 Piston
8 Seal
9 Sleeve
10 Piston spring
11 Spring disc
12 Lockring

8 Clutch assembly – removal

1 Remove the gearbox (see Chapter 6).
2 Mark the position of the clutch cover relative to the flywheel (photo).
3 Slacken off the bolts holding the cover to the flywheel in a diagonal sequence, undoing each bolt a little at a time. This keeps the pressure even all round the diaphragm spring and prevents distortion. When all the pressure is released on the bolts, remove them, lift the cover off the dowel pegs and take it off together with the friction disc which is between it and the flywheel.

Chapter 5 Clutch

8.2 Clutch assembly prior to removal

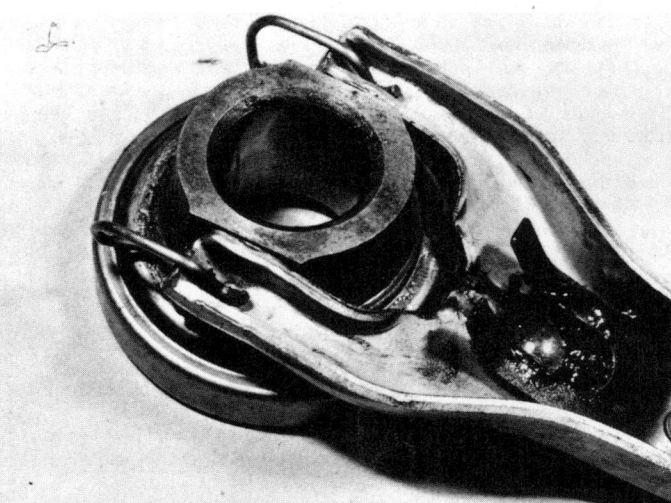
10.4 Clutch release bearing clips

10.9 Correct assembly of clutch release lever and bearing

9 Clutch – inspection and renovation

1 Examine the clutch disc friction linings for wear and loose rivets and the disc for rim distortion, cracks, broken hub springs and worn splines. The surface of the friction linings may be highly glazed, but as long as the clutch material can be clearly seen this is satisfactory. Compare the amount of lining wear with a new clutch disc at the stores in your local garage and if the linings are more than three quarters worn replace the disc.
2 It is always best to renew the clutch driven plate as an assembly to preclude further trouble. The manufacturers do not advise that only the linings are renewed. Personal experience dictates that it is far more satisfactory to renew the driven plate complete than to try and economize by only fitting new friction linings.
3 Check the machined faces of the flywheel and the pressure plate. If either is grooved it should be machined until smooth or renewed.
4 If the pressure plate is cracked or split, or if the pressure of the diaphragm spring is suspect, it is essential that an exchange unit is fitted.
5 Check the release bearing for smoothness of operation. There should be no harshness and no slackness in it. It should spin reasonably freely, bearing in mind it has been prepacked with grease.

10 Release bearing – removal and refitting

1 The release bearing is of the ball bearing, grease sealed type and although designed for long life it is worth renewing at the same time as the other clutch components are being renewed or serviced.
2 Deterioration of the bearing should be suspected when there are signs of grease leakage or the unit is noisy when spun with the fingers.
3 Remove the rubber dust excluder which surrounds the release lever at the bellhousing aperture.
4 Remove the spring clips that secure the release bearing to the lever (photo).
5 Withdraw the release bearing/hub assembly from the input shaft.
6 If necessary, remove the release lever from its ball pivot.
7 Apply a dab of wheel bearing grease to the tip of the release lever pivot ball and pack the internal recess of the bearing retainer hub with the same type of lubricant.
8 Fit the bearing assembly to the release lever.
9 Refit the release lever and bearing assembly, ensuring the lever retaining clip is correctly located on the pivot stud (photo).
10 Check that the release bearing turns freely.

11 Clutch – refitting

1 To refit the clutch, first place the driven plate (friction disc) against the flywheel with the shorter side of the centre hub facing towards the flywheel. Some clutch discs are marked 'flywheel side' to ensure correct refitting (photo). On no account should the clutch disc be replaced the wrong way round as it will be found quite impossible to operate the clutch with the friction disc incorrectly fitted.
2 Fit the clutch cover assembly on the flywheel. Refit the bolts and spring washers and tighten them finger tight so that the clutch disc is gripped but can still be moved.
3 The clutch disc must now be centralised so that when the engine and gearbox are mated, the gearbox input shaft splines will pass through the splines in the centre of the hub.
4 Centralisation can be carried out by using a clutch aligning tool, whether it be a universal type or the manufacturers type.
5 Tighten the clutch bolts firmly in a diagonal sequence to ensure that the cover plate is pulled down evenly without distortion of the flange. Tighten the bolts to the torque wrench setting given in the Specifications.
6 Reconnect the gearbox to the engine. Do this by supporting the gearbox and engaging the input shaft with the driven plate hub splines and the flywheel spigot bush. Keep the input shaft and gearbox perfectly square during the refitting operation and do not allow the weight of the gearbox to hang, even momentarily, upon the input shaft while it is only partially engaged with the driven plate, otherwise damage to the clutch components may result.
7 Insert the clutch bellhousing to engine crankcase securing bolts

11.1 Refitting the clutch assembly – engine out of car

and tighten them to the specified torque wrench setting.
8 Bleed the hydraulic system and adjust the clutch as described earlier in this Chapter.

12 Fault diagnosis – clutch

There are four main faults to which the clutch and release mechanism are prone. They may occur by themselves or in conjunction with any of the other faults. They are clutch squeal, slip, spin and judder.

Clutch squeal – diagnosis and cure

1 If on taking up the drive or when changing gear, the clutch squeals, this is a sure indication of a badly worn clutch release bearing.
2 As well as regular wear due to normal use, wear of the clutch release bearing is much accentuated if the clutch is ridden, or held down for long periods in gear, with the engine running. To minimise wear of this component the car should always be taken out of gear at traffic lights and for similar holdups.

Clutch slip – diagnosis and cure

1 Clutch slip is a self evident condition which occurs when the clutch friction plate is badly worn, when oil or grease have got onto the flywheel or pressure plate faces, or when the pressure plate itself is faulty.
2 The reason for clutch slip is that, due to one of the faults listed above, there is either insufficient pressure from the pressure plate, or insufficient friction from the friction plate to ensure solid drive.
3 If small amounts of oil get onto the clutch, they will be burnt off under the heat of clutch engagement, and in the process, gradually darken the linings. Excessive oil on the clutch will burn off leaving a carbon deposit which can cause quite bad slip, or fierceness, spin and judder.
4 If clutch slip is suspected, and confirmation of this condition is required, there are several tests which can be made.
5 With the engine in top gear and pulling lightly up a moderate incline sudden depression of the accelerator pedal may cause the engine to increase its speed without any increase in road speed.
6 In extreme cases of clutch slip the engine will race under normal acceleration conditions.
7 If slip is due to oil or grease on the linings a temporary cure can sometimes be effected by squirting carbon tetrachloride into the clutch. The permanent cure is, of course, to renew the clutch driven plate and trace and rectify the oil leak.

Clutch spin – diagnosis and cure

1 Clutch spin is a condition which occurs when the release arm travel is excessive, there is an obstruction in the clutch either on the primary gear splines or in the operating lever itself, or the oil may have partially burnt off the clutch linings and have left a resinous deposit which is causing the clutch to stick to the pressure plate or flywheel. Air in the hydraulic circuit may also cause this fault.
2 The reason for clutch spin is that due to any, or a combination of the faults just listed, the clutch pressure plate is not completely freeing from the centre plate even with the clutch pedal fully depressed.
3 If clutch spin is suspected, the condition can be confirmed by extreme difficulty in engaging first gear from rest, difficulty in changing gear, and very sudden take up of the clutch drive at the fully depressed end of the clutch pedal travel as the clutch is released.
4 Check that the clutch free travel is correctly adjusted and, if in order, the fault lies internally in the clutch. It will then be necessary to remove the clutch for examination and to check the gearbox input shaft.

Clutch judder – diagnosis

1 Clutch judder is a self evident condition which occurs when the gearbox or engine mountings are loose or too flexible, when there is oil on the faces of the clutch friction plate, or when the clutch pressure plate has been incorrectly adjusted during assembly.
2 The reason for clutch judder is that due to one of the faults just listed, the clutch pressure plate is not freeing smoothly from the friction disc and is snatching.
3 Clutch judder normally occurs when the clutch pedal is released in first or reverse gears, and the whole car shudders as it moves backwards or forwards.

Chapter 6 Gearbox

For modifications, and information applicable to later models, see Supplement at end of manual

Contents

Fault diagnosis – gearbox ... 9	General description ... 1
Gearbox – dismantling ... 3	Mainshaft – dismantling ... 4
Gearbox – inspection and renewal of components ... 5	Mainshaft – reassembly ... 6
Gearbox – reassembly ... 7	Speedometer drive gear ... 8
Gearbox – removal and refitting ... 2	

Specifications

Number of gears .... Four forward, one reverse

Type .... Constant mesh, helical spur gears with sliding baulk ring synchromesh on all forward speeds

Lubrication
Oil type/specification .... Hypoid gear oil, viscosity SAE 90 or 75W/90, to API GL5 (Duckhams Hypoid 75W/90S)
Capacity .... 2.36 pints (1.35 litres)

Ratios

	1200, 1300, 1500 models	1600 model
First	3·75 : 1	3·242 : 1
Second	2·30 : 1	1·989 : 1
Third	1·49 : 1	1·289 : 1
Fourth	1 : 1	1 : 1
Reverse	3·87 : 1	3·340 : 1

Tolerances

	in	mm
Gear backlash	0·004	0·1
Main gears to shafts	0·002 to 0·004	0·05 to 0·10
Limit	0·006	0·15
Reverse gear to shaft	0·002 to 0·004	0·05 to 0·10
Limit	0·006	0·15
Sleeve to hub spline flanks	0·003 to 0·006	0·07 to 0·16
Limit	0·010	0·25
Radial clearance on bearings	0·002 max	0·05 max
Axial clearance on bearings	0·020	0·50
Maximum permissible shaft run-out	0·001	0·025

Torque wrench settings

	lbf ft	kgf m
Gearbox to engine mounting screws	62	8·5
Casing stud nuts M10	36	5·0
Casing stud nuts M8	18	2·5
Layshaft front bearing bolt	69	9·5
Mainshaft flexible coupling nut	58	8·0
Detent spring cover retaining nuts	18	2·5
Gear lever support stud nuts	7	1·0

1 General description

All models are fitted with a four-speed manual gearbox equipped with synchromesh on all gears except reverse. The synchromesh mechanism comprises (basically) spring-loaded cone type clutches contained in two splined hubs. When a gear is selected the cones have a braking effect which synchronises the speed of the gears prior to engagement and ensures that the gears mesh smoothly. The synchromesh components will last for a considerable time but as they begin to wear gearchanging will become difficult and noisy. However before assuming the worst, check the clutch adjustment as described in Chapter 5.

The input shaft, mainshaft and laygear are more or less conventional. The gearbox cover is at the bottom of the box and the selector mechanism consists of forks held to rails by lock bolts. The ends of the rails protrude from the rear of the main casing and the gear lever, which is mounted in the rear extension housing, engages directly with the rails. There is therefore, a very positive gearchange action. The selector rails are controlled and located by conventional detent balls and springs with transverse interlock plungers.

Reversing lamps are controlled by a switch which is mounted in the rear extension housing. **Note:** *2101 models up to approximately April 1978 have no provision for this reversing lamp switch.*

Chapter 6 Gearbox

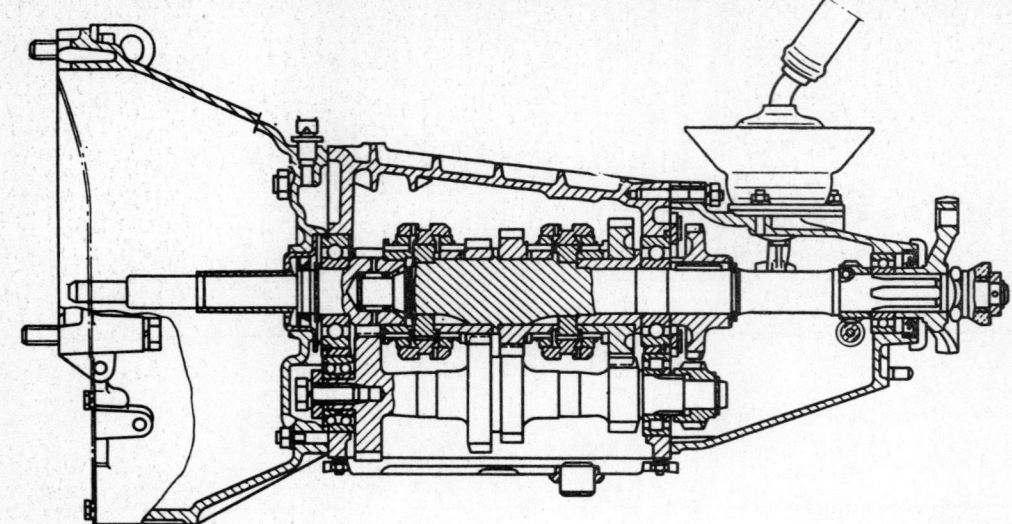

2 Gearbox – removal and refitting

The gearbox may be removed without disturbing the engine significantly. A special tool to undo the top bolts that secure the bellhousing to the engine is available, but the bolts can be undone with a long reach socket spanner quite satisfactorily. The procedure for gearbox removal is as follows.

Note: *If a centre console is fitted, it must be removed in order to improve access when taking out the slotted nylon ring (4 in Fig. 6.3) to release the gear lever.*

1 Raise the car so that there is approximately a 2 ft high space underneath the engine and gearbox region. Put chassis stands underneath the front of the car and chock the back wheels. It is essential to ensure that the car is safe in the raised position because when you will be working underneath it, the efforts made to loosen bolts and the like could easily topple an inadequately supported vehicle.
2 From inside the car, set the gearbox lever in neutral and remove the retaining plate and boot from around the base of the gear lever (photo).
3 The top section of the lever is held to the lower section by a nylon

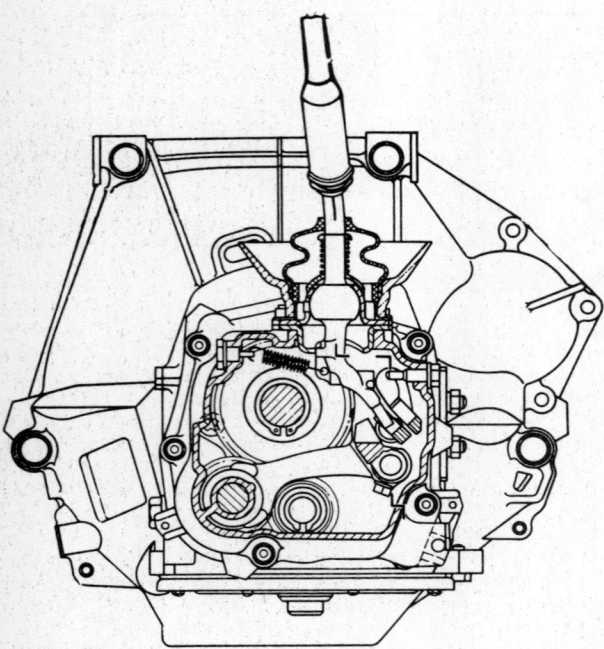

Fig. 6.1 Sectional views of gearbox (Sec 1)

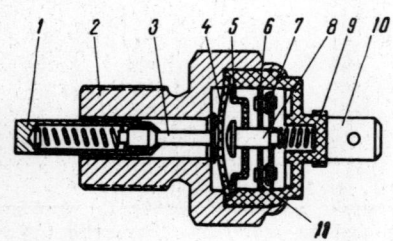

Fig. 6.2 Gearbox reverse lamp switch (Sec 1)

1 Plunger	7 Movable contact
2 Switch body	8 Tappet
3 Piston	9 Cover
4 Diaphragm	10 Spade terminal
5 Disc	11 Contact
6 Contact	

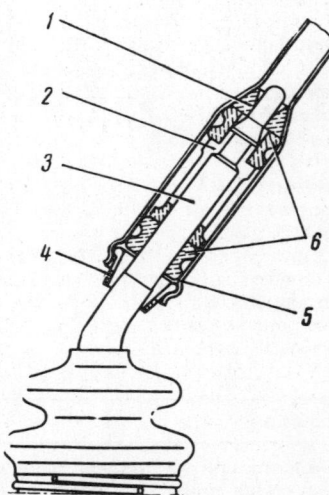

Fig. 6.3 Gearbox extension joint (Sec 2)

1 Cone	4 Slotted ring
2 Spacer sleeve	5 Gear lever extension
3 Gear lever	6 Flexible bush

Chapter 6 Gearbox

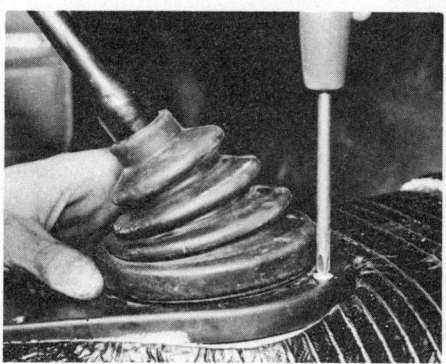
2.2 Removing the gearchange lever floor plate

2.3 Removing the gearchange lever retaining ring

2.7 Front exhaust pipe clamp

2.9 Speedometer cable retaining nut

2.21 Gearbox rear support member

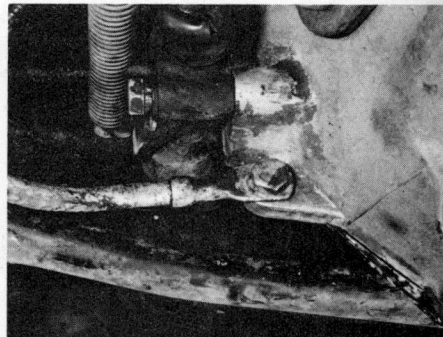
2.22 Location of earthing wire

slotted ring which can be prised out of the lower end of the tube with a screwdriver (photo).
4 The lever top section may then be lifted off.
5 Drain the gearbox oil into a suitable container.
6 Unhook the clutch lever return spring, remove the slave cylinder retaining bolts and tie the cylinder out of the way. Note that it is not necessary to disconnect the flexible hose.
7 Slacken the clamp that secures the front exhaust pipe to the rear pipe and silencer (photo). Disconnect the rear silencer bracket and remove the rear pipe and silencer assembly.
8 Remove the clamp that secures the front pipe to the gearbox and remove the nuts that secure the front pipe to the exhaust manifold to enable the pipe to be positioned out of the way.
9 Unscrew the knurled nut holding the speedometer cable to the gearbox (photo). Disconnect the reverse lamp switch leads.
10 Remove the screws that secure the cover plate to the front of the clutch housing and detach the plate.
11 Remove the heat shield from above the starter motor. This is held by two nuts to the exhaust manifold. Undo and remove the three starter motor securing bolts. The starter should remain in position.
12 Disconnect the propeller shaft. This means undoing the bolts holding the propeller shaft spider to the flexible front coupling (see Chapter 7). Remove the crosspiece under the front section of shaft, followed by the centre bearing mounting. The shaft may then be moved to one side well out of the way.
13 Some support for the engine should be devised now. A chassis stand placed underneath the sump and firmly wedged into position should serve adequately. Alternatively if a winch is available, a sling may be passed around the engine and attached to the winch overhead. The engine weight may then be taken by the winch.
14 A trolley jack should now be placed under the gearbox and raised to accept the weight of the box. Once the jack is in position undo the two nuts that secure the rear support crossmember of the bodyshell.
15 The four main bolts that join the clutch bellhousing and engine block can now be undone with the Lada tool No A55035 or a sufficiently long reach socket spanner.
16 The gearbox assembly is free to be moved away from the engine. Make sure that during its first movement rearwards, when the gearbox drive input shaft is still engaged with the clutch mechanism and flywheel, the gearbox is kept aligned to the engine. The gearbox input shaft and the clutch friction plate can easily be seriously damaged if the two are allowed to become misaligned.
17 Having cleared the gearbox assembly from the engine, it may now be lowered from the car and wheeled away to a work bench ready for dismantling.
18 Having removed the gearbox it is worthwhile examining the amount of wear left in the clutch friction disc whilst the opportunity arises (see Chapter 5).
19 When refitting the gearbox on to the engine the clutch friction plate must be centred if it has been disturbed. When juggling the gearbox into position do not let any strain be imposed on the input shaft.
20 If the support crossmember has been detached from the gearbox refit it before getting the gearbox up into position on the engine.
21 Bolt the bellhousing to the engine before refitting the support crossmember back up to the bodywork (photo).
2 When refitting the flywheel lower cover plate do not forget the earth strap bolted on the left side (photo).
23 When refitting the gear lever upper section, make sure that the nylon snap ring is pushed securely home inside the housing.
24 Do not forget to adjust the clutch pedal travel.
25 Fill the gearbox with oil.

3 Gearbox – dismantling

1 Before taking the gearbox apart remember that individual parts are expensive and may take time to acquire. If the gearbox is in very bad condition the economies of fitting a new gearbox must be considered. If the problem is bearing failure early action in replacing them will be worthwhile. If, however, the gearbox has been used for sometime with failed bearings the wear patterns of the gears will have altered (and increased). The fitting of new bearings will not then necessarily cure all the noise that may have developed. If the problem is one of jumping out of gear or very sloppy feel to the gearchange it is possible that the detent springs may be the cause. These can be renewed without

Chapter 6 Gearbox

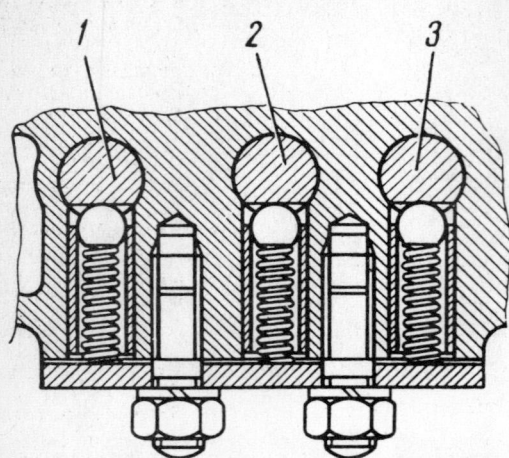

Fig. 6.4 Gearchange selector rod detent balls and springs (Sec 3)

1. 1st and 2nd selector rod
2. 3rd and 4th selector rod
3. Reverse shift fork

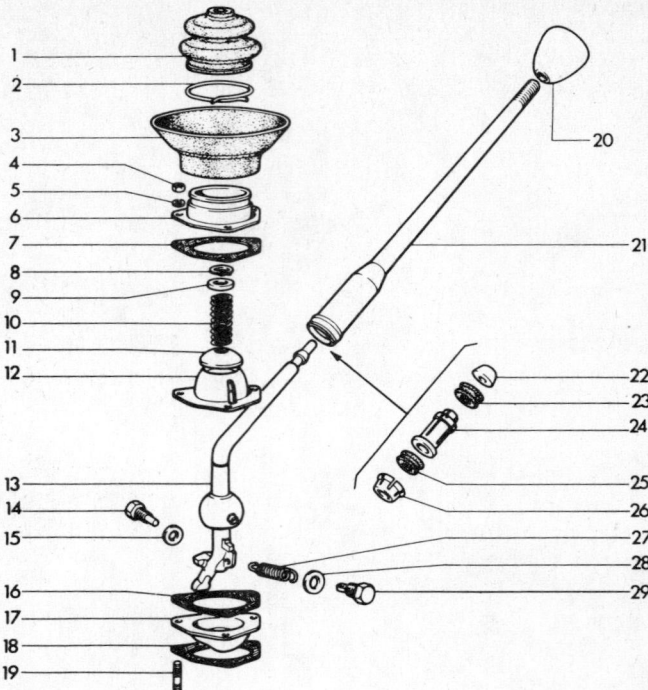

Fig. 6.5 Gearchange lever and components (Sec 3)

1. Rubber boot
2. Clip
3. Draught excluder
4. Nut
5. Spring washer
6. Mounting flange
7. Gasket
8. Circlip
9. Washer
10. Spring
11. Cap
12. Ball socket
13. Change lever
14. Stop screw
15. Copper washer
16. Gasket
17. Plate
18. Gasket
19. Stud
20. Grip
21. Change lever extension
22. Pad
23. Rubber bush
24. Spacer
25. Rubber bush
26. Nylon securing collar
27. Spring
28. Washer
29. Spring retainer screw

taking the gearbox from the car. Simply remove the detent balls and springs cover from the front right-hand edge of the main casing when access to the springs (which can break) and balls (which can wear) is possible (see Fig. 6.4).

2 Preparation for dismantling should ensure that a good firm flat clean working space is available. Unless there are adequate facilities for laying parts out when removed the job will be many times more difficult.

3 Thoroughly clean off the exterior with paraffin and hose it down with water.

4 Remove the seven nuts that secure the bellhousing to the front of the main casing and take it off complete with clutch operating arm and thrust bearing. Retrieve the convex washer from between the housing and input shaft bearing.

5 Stand the gearbox upside-down and remove the bottom cover plate screws and gasket. From now on all dismantling and assembly work is carried out with the gearbox inverted so when references are made to the top or bottom of the box it means when the box is in position in the car. So remember, the cover plate is on the bottom of the box!

6 Remove the circlip from the rear end of the mainshaft and then take off the sleeve (a small puller may be required) and rubber ring. Undo the lockwasher tab and take off the nut followed by the universal joint spider. In order to lock the mainshaft when doing this it will be necessary to lock the spider with a rod through one of the bolt holes.

7 Remove the speedometer cable drive unit after removing the securing nut.

8 Undo the two nuts that secure the detent ball and spring retaining plate on the right rear of the main casing. Take care not to lose the three springs and balls which will fall out from the holes. Note that the one for the reverse rail is different from the others (the one nearest the bottom).

9 To take off the rear extension cover and gear lever assembly first remove the stop screw on the right-hand side which restricts the sideways movement of the lever. Then unscrew the six nuts that secure the extension to the main casing.

10 Move the gear lever over to the left and the cover can be drawn off. Remove the reversing lamp switch to avoid damaging it.

11 From the rear end of the mainshaft draw off and collect the locating ball and speedometer drive gear. Do not lose the locking ball in the shaft.

12 Reverse gear selector fork and rail (the bottom one) can now be drawn out of the casing, at the same time drawing the gear off the spindle with it.

13 With expanding circlip pliers remove the circlip holding reverse driving gear on the end of the layshaft. Remove the dished washer and gear and the washer behind the gear.

14 Undo the circlip that secures the reverse gear on to the main shaft and remove the dished washer and gear.

15 Take the large Woodruff key out of its groove in the shaft.

16 Lock the shafts by engaging two gears simultaneously (move the shift forks to do this) and from the front of the layshaft remove the front bearing retaining bolt and flat washer.

17 The front and rear layshaft bearings can be tapped out of the casing from the inside. Tilt the layshaft to lift it up and out.

18 Draw the centre selector rail (3rd/4th gears) out of the casing, releasing the fork while doing so. Do not lose the small interlock plunger from the rail itself or the larger one which will be seen in the bore in the corner of the casing inside (Fig. 6.6).

19 Remove the 1st/2nd gear selector fork locking bolt and draw out the rail in the same way as before. Lift out the two forks.

20 The third interlock plunger can be retrieved now from the bottom of the bore in the casing.

21 The input shaft which has been left in position till now, can be taken out complete with bearing from the front of the casing. The needle rollers which carry the mainshaft in the counterbore of the shaft may or may not be in a cage. If they are loose and fall out recover them without delay. There should be 23.

22 Undo the three countersunk cross-head screws that secure the mainshaft bearing retaining plate and the rear of the casing. The bearing can then be taken out. The bearing retainer plate also locks into a slot in the reverse idler gear shaft which can now also be withdrawn.

23 Tilt the mainshaft inside the casing and carefully withdraw it complete with gears and hubs.

Chapter 6 Gearbox

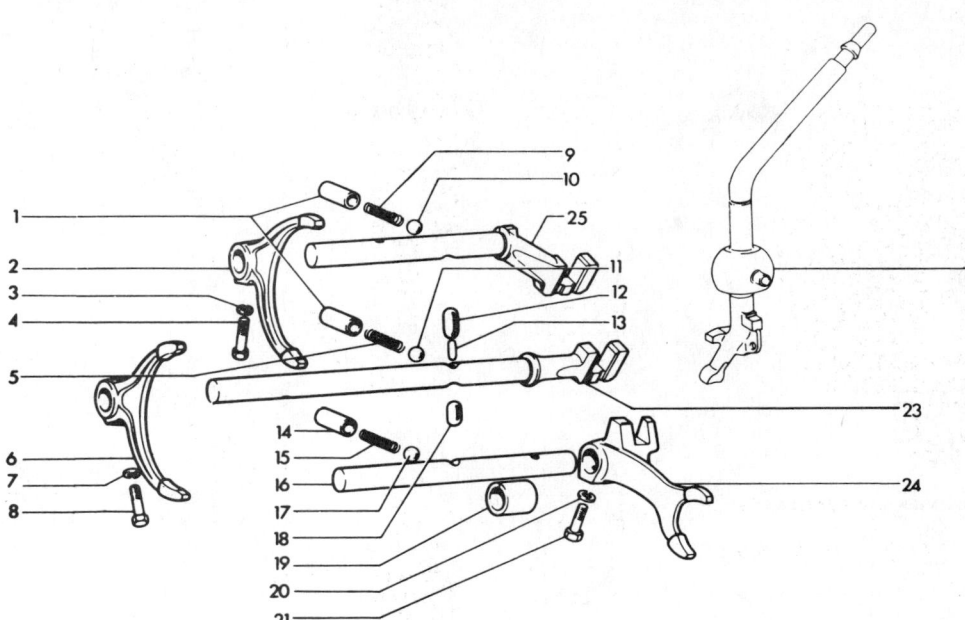

Fig. 6.6 Gearchange selector rods and forks (Sec 3)

1. Bushes (in casing)
2. 1st/2nd selector fork
3. Spring washer
4. Fork locking bolt
5. Detent spring
6. 3rd/4th selector fork
7. Spring washer
8. Fork locking bolt
9. Detent spring
10. Detent ball
11. Detent ball
12. Interlock plunger
13. Interlock plunger
14. Bush
15. Detent spring
16. Reverse selector rod
17. Detent ball
18. Interlock plunger
19. Sleeve
20. Spring washer
21. Fork locking bolt
22. Gear lever assembly
23. 3rd/4th selector rod
24. Reverse shift fork
25. 1st/2nd selector rod

4 Mainshaft – dismantling

1 The mainshaft will have to be dismantled in order to renew any of the gears or synchromesh components fitted to it. Before starting work make careful reference to Fig. 6.9 and make sure that the various components are kept in the correct order of removal.

2 Remove the circlip from the forward end of the mainshaft, so that the spring spacer, synchronizing sleeve hub, 3rd gear wheel and synchromesh assembly can be slid off the mainshaft.

3 The 1st gear wheel and its synchromesh baulk ring assembly may be slid off the rear end of the mainshaft next, followed by the 1st and 2nd gear synchronizing sleeve and hub. Finally the 2nd gear wheel and its synchromesh baulk ring assembly may be slid off the shaft.

4 It will be seen that each synchronizing baulk ring assembly is located on a seating on the gear wheel itself. The assembly is retained by a wide circlip. Once this circlip is removed the synchronizing baulk ring, thrust spring and spring seat ring can be removed from the gear hub.

5 Having separated all the parts of the mainshaft assembly they can be closely inspected for wear and tear.

6 The radial and endplay on the ball races used in the gearbox can be checked with a dial gauge, but even if partially worn, it is advisable to renew the bearings if new ones are readily available. See the Specification at the beginning of this Chapter. The radial play of the 1st, 2nd, 3rd and reverse gears on their respective bushing and the seating on the mainshaft can be similarly checked against clearances given in the Specification.

7 If necessary, the input shaft bearing can be driven off after first removing the circlip and convex washer.

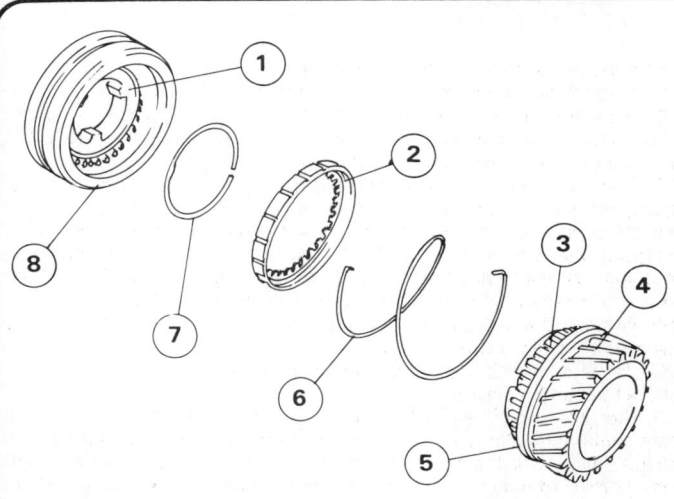

Fig. 6.7 Exploded view of a synchromesh assembly (Sec 4)

1. Hub
2. Baulk ring
3. Dog teeth
4. 3rd speed gear
5. Spring seat
6. Spring
7. Lockring
8. Sleeve

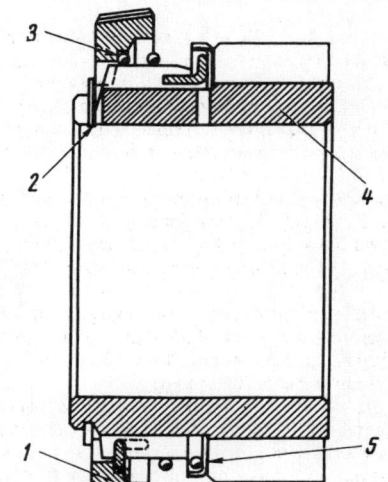

Fig. 6.8 Sectional view of a synchromesh assembly (Sec 4)

1. Baulk ring
2. Lockring
3. Spring
4. 3rd speed gear
5. Spring seat

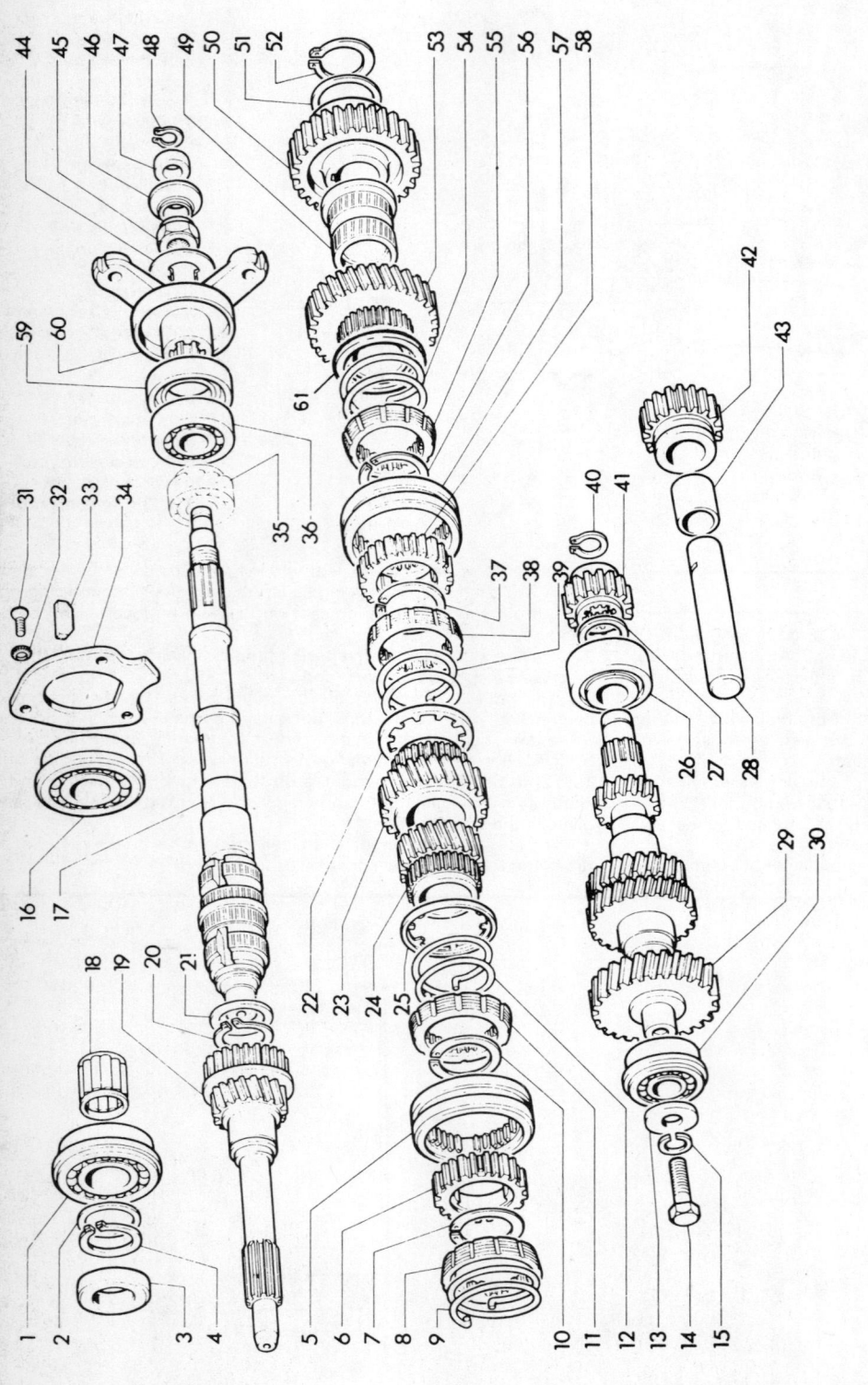

Fig. 6.9 Exploded view of gearbox (Sec 4)

1 Input shaft bearing
2 Convex washer
3 Oil seal
4 Circlip
5 3rd/4th synchro unit outer sleeve
6 3rd/4th synchro unit hub
7 Circlip
8 Baulk ring (4th gear)
9 Spring
10 Circlip
11 Baulk ring (3rd gear)
12 Spring
13 Washer
14 Bolt
15 Spring washer
16 Mainshaft bearing
17 Mainshaft
18 Needle roller bearing
19 Input shaft with 4th gear
20 Circlip
21 Convex washer
22 Toothed washer
23 2nd gear
24 3rd gear
25 Toothed washer
26 Laygear needle roller bearing
27 Convex washer
28 Reverse idler gear shaft
29 Laygear
30 Laygear double ball bearing
31 Countersunk set screw
32 Lockwasher
33 Woodruff key
34 Bearing retainer plate
35 Speedo drive gear
36 Mainshaft rear bearing
37 Circlip
38 Baulk ring (2nd gear)
39 Spring
40 Circlip
41 Reverse idler gear
42 Sleeve
43 Tab washer
44 Nut
45 Dust cover
46 Centring ring
47 Circlip
48 1st gear bush
49 Mainshaft rear bearing
50 Reverse mainshaft gear
51 Convex washer
52 Circlip
53 1st gear
54 Spring
55 Baulk ring (1st gear)
56 Circlip
57 1st/2nd synchro unit outer sleeve
58 1st/2nd synchro unit hub
59 Oil seal
60 Coupling spider
61 Washer

Chapter 6 Gearbox

5 Gearbox – inspection and renewal of components

1 Thoroughly wash all parts in paraffin to remove oil residue and lay them out to drain and dry on sheets of clean paper.
2 Examine the casing and covers for any signs of cracks and see that the mating faces are free from scratches and burrs. The lip type oil seals in the bellhousing and rear covers should be renewed as a matter of routine.
3 The seal can be levered out with a screwdriver and a new one tapped in square. The condition of the shafts, gears and synchro hubs is a question of degree. The mating of the baulk rings in the synchromesh should be such that no rock is present. Renewal of the baulk rings should be made if there is any significant backlash in the splines or wear on the cone face. Serious backlash (more than 0.010 in (0.25 mm)) in the hubs between the inner and outer sleeves means that the whole unit should be renewed and this can be expensive.
4 The gears themselves should not have chipped or excessively worn teeth. Backlash between gear teeth should not exceed 0.008 in (0.20 mm)
5 The clearance between the gears and mainshaft or bush should not exceed 0.006 in (0.15 mm).
6 The selector rails should slide easily in the transmission casing but without play which might cause jamming tendencies. The detent springs should not be weak and the balls not worn or pitted. The selector forks must be at right angles to the rails when fitted and the tongues which engage in the synchro sleeve grooves must not be worn on either face which would give rise to excessive clearance and consequent lost motion in gear engagement. This fault can cause gears to jump out of engagement.
7 Finally check the needle roller bearing that supports the rear end of the input shaft. If signs of wear are evident it must be renewed.

6 Mainshaft – reassembly

1 If the synchro baulk rings have been removed from the gears, refit these first and ensure that the circlip is fully engaged in the groove (photo).
2 Starting at the front of the shaft, fit 3rd gear (the smallest of the three loose gears) over the shaft, gear teeth first (photo).
3 Fit the hub of 3rd/4th synchro unit, engaging the dogs into the shaft cutouts. It may go on either way round (photo).
4 Put the convex washer over the end of the shaft, convex side towards the shaft end (photo).
5 Put the circlip into position as near as possible. To get it into the groove a piece of pipe of suitable diameter ($1\frac{1}{8}$ in) will be needed to drive the circlip down so it will engage in the groove (photo).
6 Turn to the other end of the shaft and place 2nd gear (the middle size one) over the shaft, gear teeth first (photo).
7 Refit the hub section of the 1st/2nd gear synchro unit, it can go on either way round (photo).
8 Fit the outer sleeves or both synchro hub units noting that they have special master locating splines with a flat section which must engage together (photo).
9 Put the bush into 1st gear from the gear teeth side (photo).
10 Fit 1st gear together with its bush on the shaft with the synchro baulk ring leading (photo).
11 The mainshaft is now ready for reassembly into the gearbox.

7 Gearbox – reassembly

1 Refit the mainshaft carefully into the casing making sure you get it the proper way round (photo).
2 Fit the rear bearing over the tail end of the shaft and into its recess in the casing. If a new bearing is fitted remember to refit the circlip in the outer annular groove. The circlip is towards the rear (photo).
3 Put the reverse idler gear shaft spindle into its hole in the casing and engage the bearing retaining plate in the slot, and push it fully home (photo).
4 Refit the retaining plate screws and use an impact screwdriver to tighten them. Alternatively use a pair of self-grips on a screwdriver to obtain extra leverage (photo). It is a good idea to stake them when tight.
5 Take up the input shaft and place the needle rollers in the counter-bore (photo). If the rollers are not caged note that there is a spacer ring in front of and behind them. Hold them in position with grease and make sure that all 23 are fitted.
6 Fit the input shaft into the casing, engaging the needle bearings over the front of the mainshaft (photo).
7 With the input shaft and bearing fully home the two shafts should rotate independently and together smoothly (photo).
8 Take up the shorter of the two selector rails (which have integral lugs at one end) and put it into the top hole of the three at the rear of the casing. The single transverse cutout in the rail faces the bottom of the box and the other three cutouts face the side (photo).
9 When the selector rail is partially inserted fit 1st/2nd gear selector fork into the grooves of the synchro hub sleeve and pass the rail through the fork boss (photo).
10 Refit and tighten the fork securing bolt.
11 Fit one of the two large interlock plungers into the drilling in the casing (photo). If the selector is at the neutral position the end of the plunger will fit the transverse recess in the rail.
12 Fit the second selector rail into the centre hole and fit the 3rd/4th selector fork in the same manner as for the other one. Before pushing the rail right in fit the small interlock plunger through the hole in the shaft and then trap it by pushing the rail into the casing (photo). Make sure the rear end of the rail lines up alongside the other with the gear lever cutouts together.
13 Tighten the selector fork locking bolt (photo).
14 Fit the laygear into the box with the larger gear towards the front (photo).
15 Fit the double ball bearing into the front of the casing over the nose of the laygear (photo) and tap it home.
16 Refit the securing bolt into the laygear with the spacer and lockwasher (photo).
17 Refit the needle roller bearing into the casing at the rear, over the laygear (photo).
18 After tapping the laygear home fit the spacer ring (photo).
19 Refit the convex washer, convex side out (photo).
20 Refit the reverse gear onto the splines of the layshaft (photo) and then fit the circlip as near as possible to the groove in which it goes (photo).

6.1 Fitting a synchro baulk ring retaining circlip

6.2 Placing 3rd gear on the front of the mainshaft

6.3 Sliding 3rd/4th synchro hub into position

6.4 Fitting the convex washer to the mainshaft

6.5 Driving the circlip into its groove

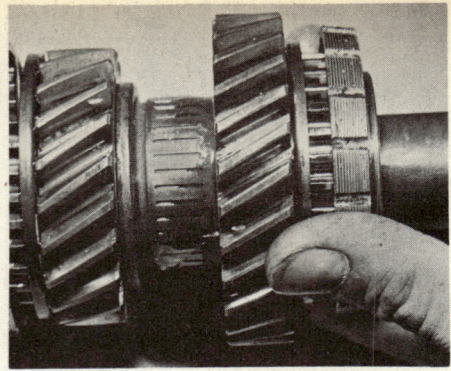
6.6 Sliding 2nd gear into position from the rear of the mainshaft

6.7 Placing 1st/2nd synchro hub into position on the mainshaft

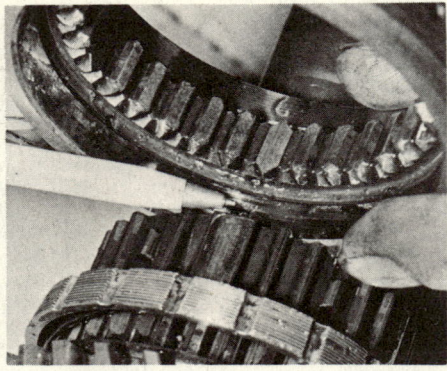
6.8 Lining up the master splines on the synchro hub and outer sleeve

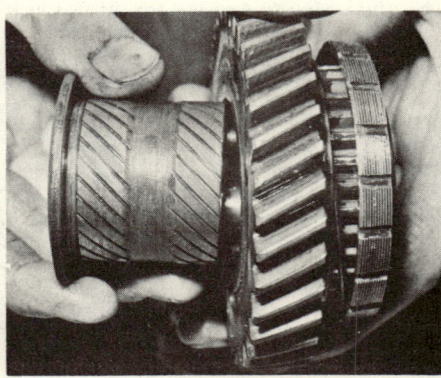
6.9 Fitting the bush into the 1st gear

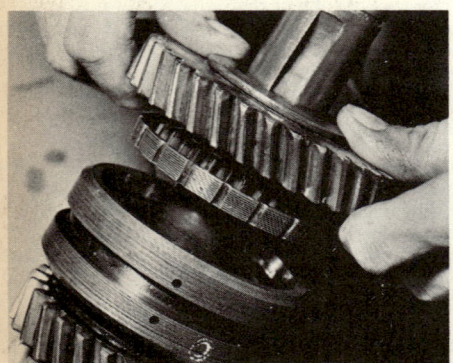
6.10 Sliding the 1st gear and bush into the shaft

7.1 Inserting the mainshaft assembly into the casing

7.2 Fitting the mainshaft rear bearing into the gear casing

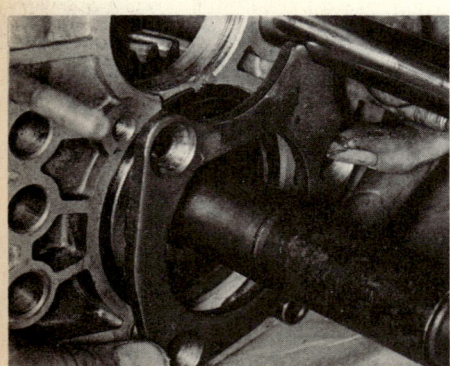
7.3 Fitting the reverse idler gear shaft and bearing retaining plate

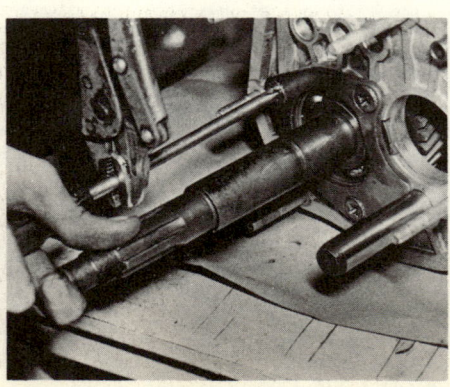
7.4 Tightening the bearing plate screws

7.5 Inserting the needle roller bearing into the end of the input shaft

7.6 Positioning the input shaft in the casing

7.7 Mainshaft and input shafts assembled in the casing

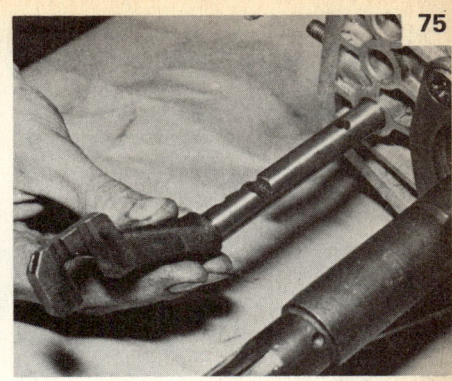
7.8 Fitting the 1st/2nd gear selector rail

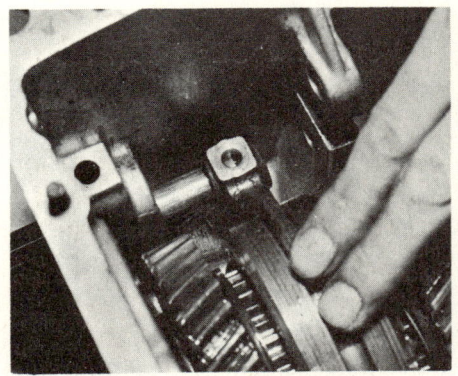
7.9 Positioning the 1st/2nd selector fork on the rail

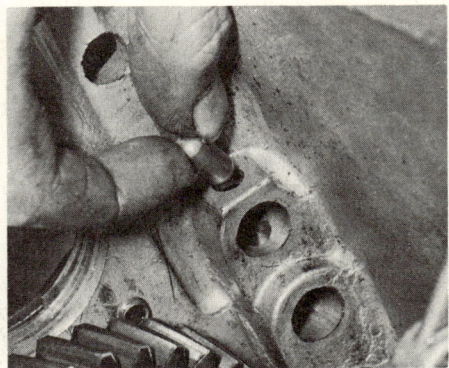
7.11 Inserting the first of the interlock plungers

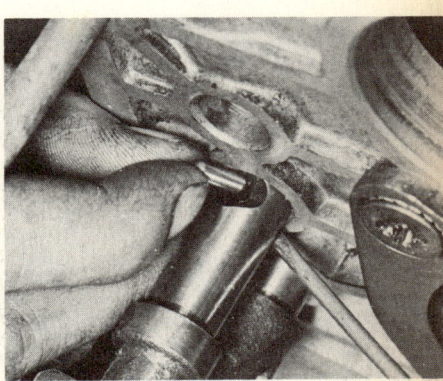
7.12 Fitting the interlock plunger in the 3rd/4th gear selector rail

7.13 Tightening the 3rd/4th gear selector fork retaining bolt

7.14 Inserting the laygear into the casing

7.15 Fitting the laygear front bearing

7.16 Fitting the bearing retaining bolt

7.17 Fitting the laygear rear bearing

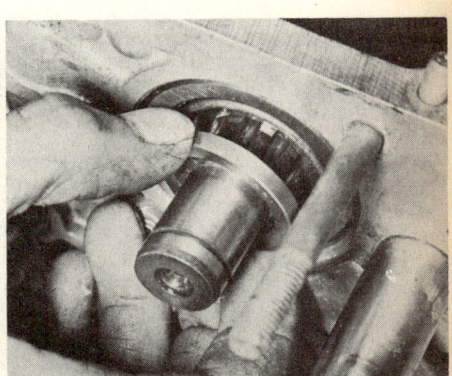
7.18 Replacing the laygear bearing spacer

Chapter 6 Gearbox

7.19 Fitting the laygear bearing convex washer

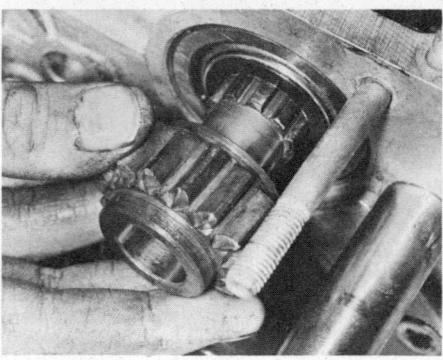
7.20a Fitting reverse gear to the layshaft

7.20b Fitting the circlip to the layshaft reverse gear

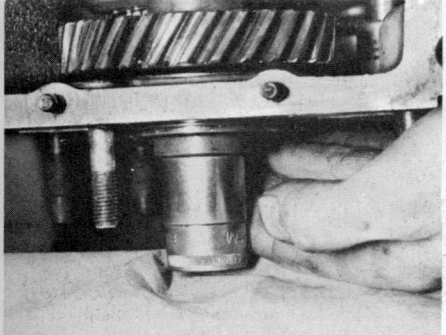
7.21 Supporting the front end of the laygear with a socket

7.22a Driving on the layshaft rear end circlip

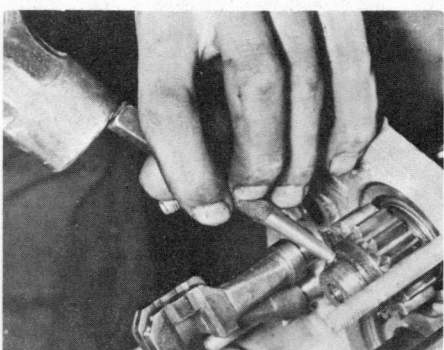
7.22b Tapping the laygear circlip fully into position

7.23 Fitting the Woodruff key into the mainshaft

7.24 Fitting reverse gear onto the mainshaft

7.25 Fitting the washer and circlip onto the mainshaft

21 To fit the circlip it has to be driven down against the pressure of the convex washer. Fit a socket on the bolt at the other end of the laygear and then stand the casing up on end on it (photo).
22 Using another tube or socket of suitable diameter drive the circlip down into position (photo). It may be necessary to finally tap the circlip ends in with a centre punch (photo).
23 Fit the large Woodruff key into the mainshaft (photo).
24 Fit the reverse gear on to the mainshaft with the hub extension rearwards and slide it over the key (photo).
25 Fit the convex washer over the shaft followed by the circlip. The convex side faces the circlip (photo).
26 To fit the circlip into the groove a suitable tube is once again needed. A piece of water pipe slotted across the end and tapped inwards to the correct diameter will serve the purpose (photo).
27 The bolt on the front of the layshaft should be tightened next to a torque of 69 lbf ft (9.5 kgf m) (photo). The shaft can be locked by engaging two gears together.
28 With the two forward selector rails at neutral insert the last of the interlock plungers (photo).
29 Fit the sleeve over the reverse gear selector rail and refit the fork also if it has been taken off. Fit the rail a little way into the hole in the casing (photo).
30 Fit the reverse sliding gear onto the shaft together with the selector fork (photo).
31 Push the selector rail right home until the fork lug lines up with the other two. Secure the fork retaining bolt.
32 Turn the casing on its side and refit the detent balls and springs (photos).
33 Fit a new gasket and refit the cover (photo).
34 Fit the locking ball in the tail end of the mainshaft and refit the speedo drive gear so that the cutout engages the ball (photo).
35 Using a suitable tube, drive on the mainshaft rear bearing behind the speedo gear (photo).
36 With the gear lever stop pin moved out to permit sideways movement, and having checked that the lever return spring is intact, prepare to replace the rear cover.
37 Make sure that the selector rails are all lined up in the neutral position (photo).

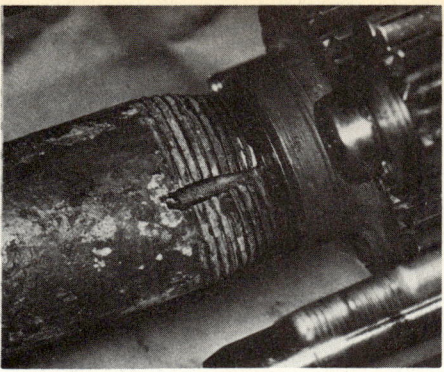
7.26 Driving on the mainshaft rear circlip

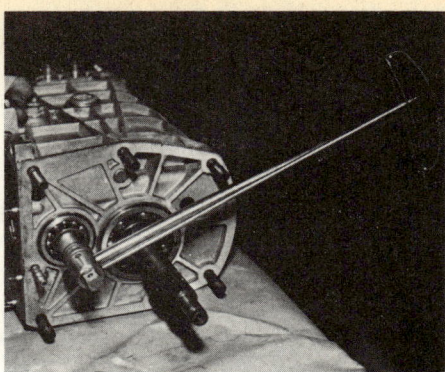
7.27 Torqueing the laygear front bearing bolt

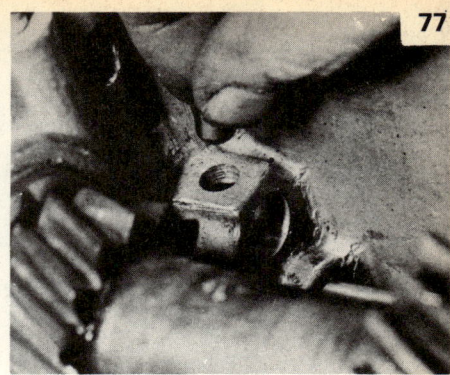
7.28 Inserting the last interlock plunger

7.29 Fitting the reverse gear selector rail complete with sleeve and fork

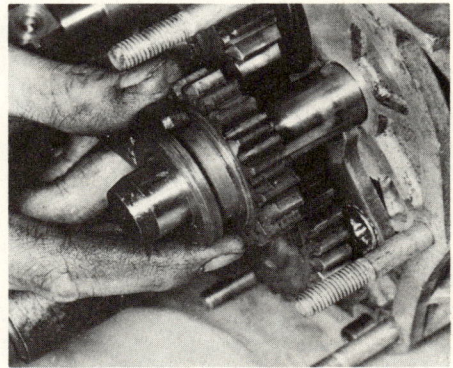
7.30 Fitting the reverse idler gear with selector fork to the idler shaft

7.32a Inserting the detent balls into the gearcase

7.32b Inserting the detent springs into the gearcase

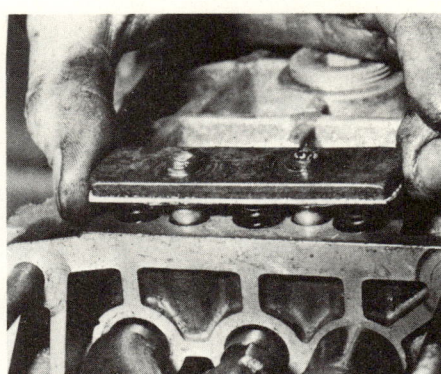
7.33 Detent balls and springs retained by cover and gasket

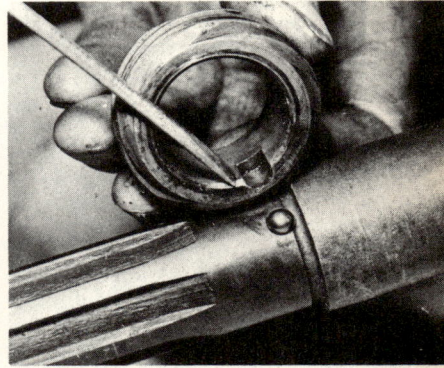
7.34 Speedometer drive gear and locking ball

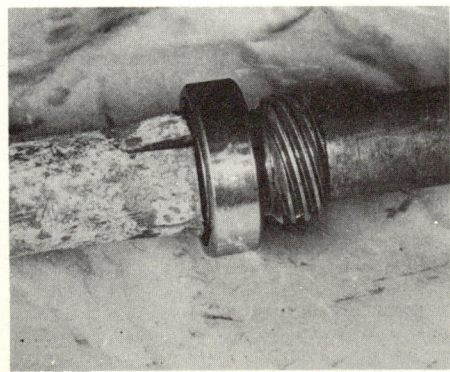
7.35 Driving on the mainshaft rear bearing

7.37 Selector rails correctly aligned in the neutral position

7.38 Fitting the rear cover

38 Offer up the rear cover having put a new gasket over the rear casing studs (photo).
39 Fit the nuts to the studs, not forgetting the exhaust pipe anchor bracket on the lower right-hand corner (photos). There is a special square-headed bolt for this, also one nut and washer goes on the stud before the bracket.
40 Fit the bottom cover plate using a new gasket (photo).
41 The bellhousing gasket is fitted over the studs on the front of the casing (photo).
42 Fit the convex washer in position in the front of the housing using grease to hold it in position (photo).
43 Offer up the bellhousing to the casing and refit the securing nuts.
44 Refit the speedometer cable drive unit (photo).
45 Fit the flexible coupling spider on to the rear of the mainshaft (photo) followed by the tab washer and nut. Tighten the nut to 58 lbf ft (8 kgf m) and bend over the lockwasher (photo).
46 Refit the rubber dust cover followed by the coupling centring ring and circlip (photo).
47 Fit the flexible coupling to the spider with the three bolts in readiness for replacement.

8 Speedometer drive gear

1 The speedometer gear drive unit cannot be removed from the gearbox when it is in position in the car, due to the proximity of the transmission tunnel. When the gearbox is out of the car, the speedometer drive gear can be removed without disturbing any other components. Its internal spindles and gears may be renewed as required (Fig. 6.10).
2 Having disconnected the knurled cable retaining collar, remove the nut from the single securing stud and draw the unit out of the casing. If the car is jacked up at the front there is every likelihood that some oil will come from the hole in the casing. Be ready to collect it.
3 If only the speedometer drive cable is to be removed it is simpler to take it out from the speedo head end after removing the instrument cluster (see Chapter 10), although if broken it may be necessary to undo the bottom end also in order to retrieve the lower broken part.
4 When refitting the drive gear make sure the gasket is in good condition. Top up the gearbox oil level if necessary.

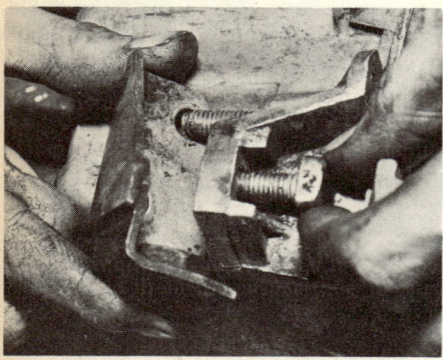
7.39a Fitting the captive bolt in the exhaust pipe mounting bracket

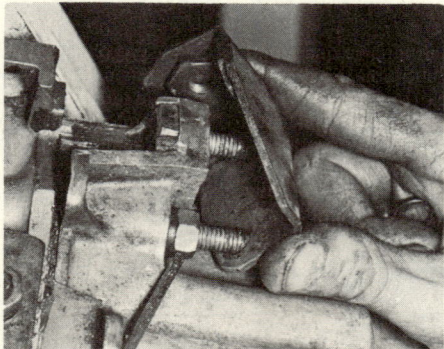
7.39b Fitting the exhaust pipe bracket

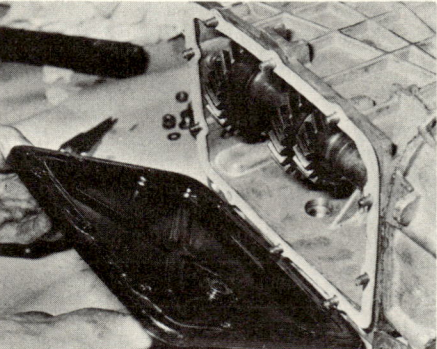
7.40 Fitting the bottom cover plate

7.41 Fitting the clutch housing gasket

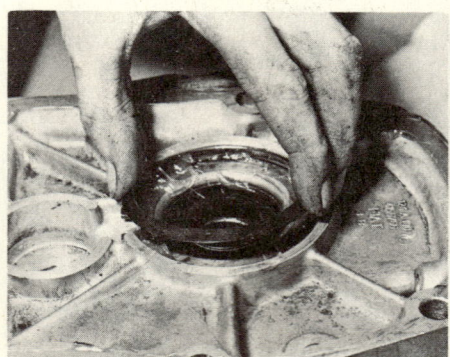
7.42 Positioning the washer in the clutch housing

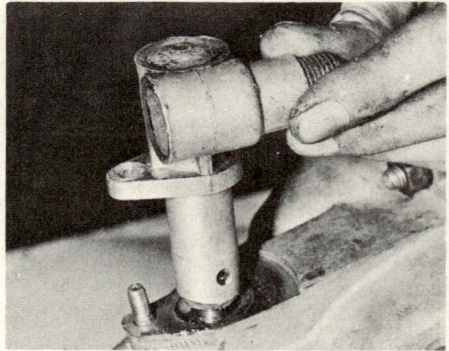
7.44 Replacing the speedometer drive gear unit

7.45a Sliding the coupling spider onto the mainshaft

7.45b Bending over the spider nut lock washer

7.46 Fitting the circlip onto the end of the mainshaft

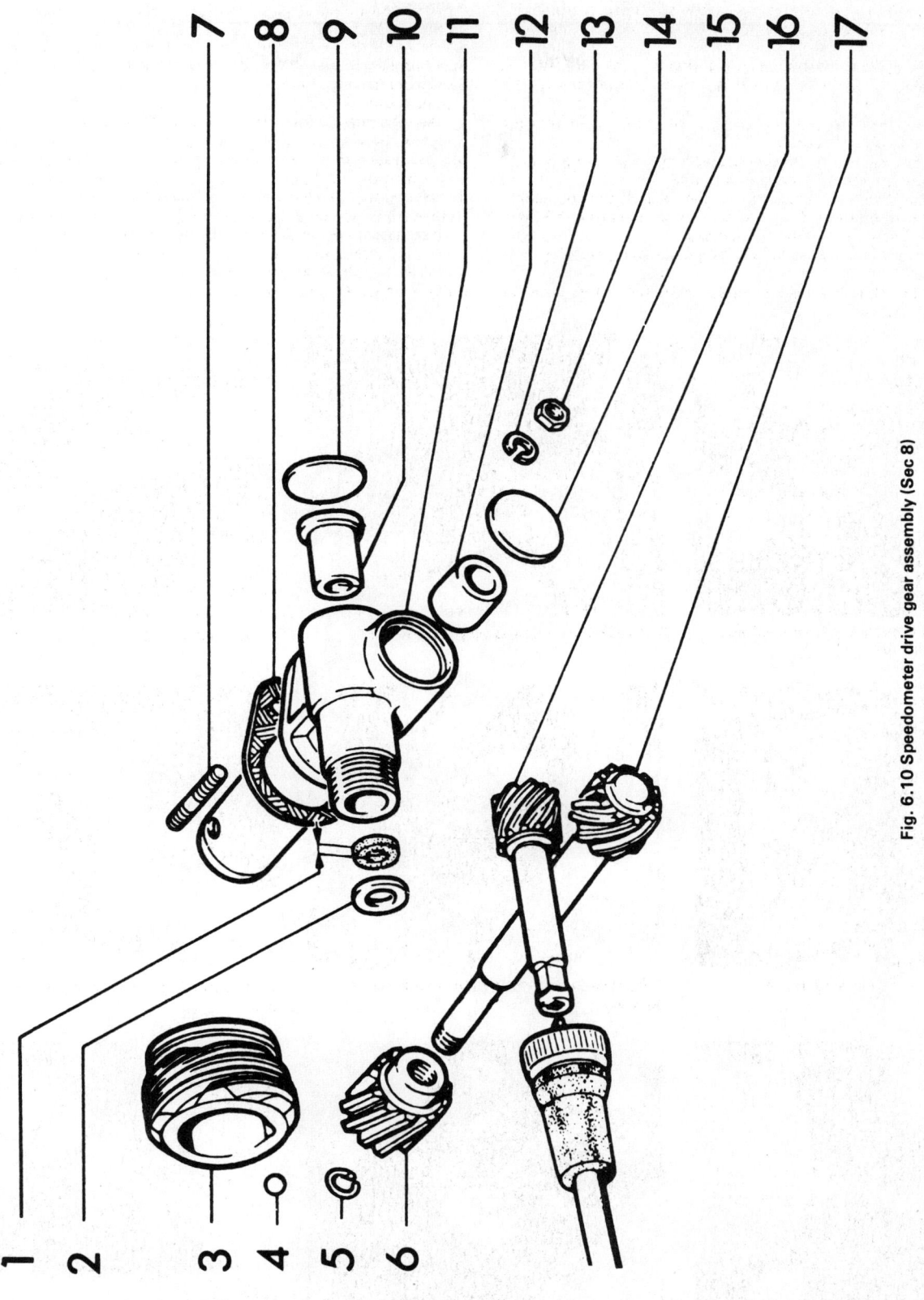

Fig. 6.10 Speedometer drive gear assembly (Sec 8)

1 Seal
2 Washer
3 Gearbox mainshaft gear
4 Locking ball
5 Circlip
6 Gear
7 Stud
8 Gasket
9 Blanking plug
10 Bush
11 Housing
12 Bush
13 Spring washer
14 Nut
15 Blanking plug
16 Shaft and gear
17 Shaft and gear

9 Fault diagnosis – gearbox

Symptom	Reason/s
Weak or ineffective synchromesh	Synchronising cones worn, split or damaged
	Synchromesh dogs worn, or damaged
Jumps out of gear	Broken gearchange fork rod spring
	Gearbox coupling dogs badly worn
	Selector fork rod groove badly worn
Excessive noise	Incorrect grade of oil in gearbox or oil level too low
	Gear teeth excessively worn or damaged
	Laygear thrust washers worn allowing excessive end play
Difficulty in engaging gears	Clutch pedal adjustment incorrect

Chapter 7 Propeller shaft

For modifications, and information applicable to later models, see Supplement at end of manual

Contents

Centre bearing – removal and refitting	4
Fault diagnosis – propeller shaft	7
Flexible coupling – removal and refitting	5
General description	1
Propeller shaft – balancing	6
Propeller shaft – removal and refitting	2
Universal joints – dismantling, inspection, repair and reassembly	3

Specifications

Type .... Two section with centre bearing. Flexible coupling and sliding joint at front end

Sliding joint lubricant
Type/specification .... Multi-purpose lithium-based grease with molybdenum disulphide, to NLGI No 2 (Duckhams LBM 10)

Torque wrench settings

	lbf ft	kgf m
Flexible coupling bolts	50	6·9
Rear axle pinion flange bolts	24	3·3
Centre bearing yoke nut	70	9·7

1 General description

The Lada has a split propeller shaft. This means that instead of one long tubular shaft with a universal joint at each end there are two short shafts with a universal joint in the middle as well. Advantages of this are the shallower angles at the joints, the reduced intrusion of a tunnel for the rear end of the shaft into the body floor and the greater rigidity obtained from 2 shorter shafts.

The front end of the forward shaft is attached to the gearbox mainshaft by means of a rubber 'doughnut' type flexible coupling. The rear end of the forward shaft runs in a bearing mounted in a flexibly cushioned support attached to the body shell. The rear shaft has a conventional universal joint at each end connecting it to the forward shaft and rear axle pinion flange.

The propeller shaft requires periodic lubrication of the splined joint of the forward shaft where it engages with the flexible coupling on the gearbox output shaft. Some makes of replacement universal joint spiders have provision for grease lubrication of the needle roller bearings as well.

It is essential to realise that the propeller shaft assembly is a finely balanced collection of components and that they are balanced only as they remain assembled in the alignment in which they were originally balanced. Therefore whenever dismantling the propeller shaft, pay particular attention to the reference alignment marks on the shaft components so that they are correctly reassembled. It is equally important to realise that if any single or small group of major components is renewed (except universal joint spider assemblies) the whole shaft assembly must be dynamically balanced professionally.

2 Propeller shaft – removal and refitting

This type of propeller shaft does not permit the separate removal of the forward and rear sections of shaft. Therefore proceed to remove the whole shaft assembly as follows:

1 Begin by raising the rear of the car onto chassis stands or car ramps. Make sure that the car is safely supported, because efforts involved in the removal and refitting of the shaft assembly could easily topple an inadequately supported vehicle.
2 Working underneath the car, remove the long bolts and nuts that hold the rubber 'doughnut' of the forward flexible coupling onto the spider fitting on the end of the forward shaft. To counteract the tension in these bolts it may be necessary to compress the coupling by tightening a clamp or worm drive clip around its outer periphery.
3 Next remove the handbrake return spring from the propeller shaft support crossmember and then remove the nuts and washers that hold the centre bearing support crossmember and the safety strap crossmember to the bodyshell.
4 Some light support should be offered to the shaft at this point to prevent it from being grazed on the ground.
5 Finally remove the four nuts and bolts that hold the rear propeller shift to the flange fitting on the bevel pinion shaft of the rear axle (photo).
6 The propeller shaft assembly may now be lifted from the car and transferred to a bench ready for examination and repair.
7 The refitting procedure for the shaft assembly is simply the reversal of the removal procedure, except that the following checks

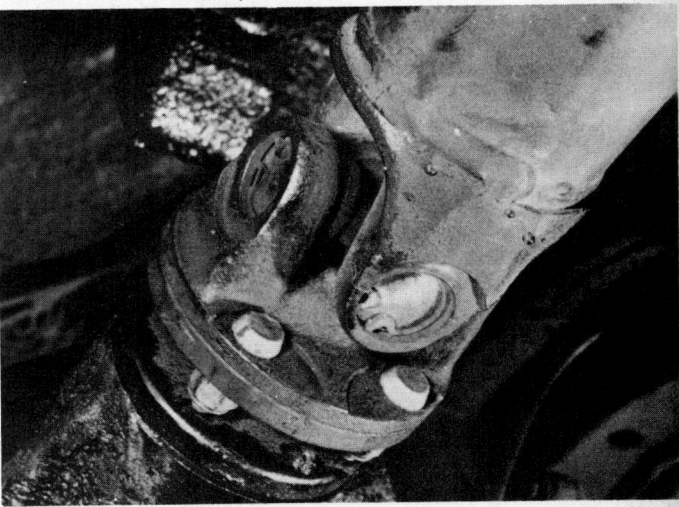

2.5 Propeller shaft rear end attachment

Chapter 7 Propeller shaft

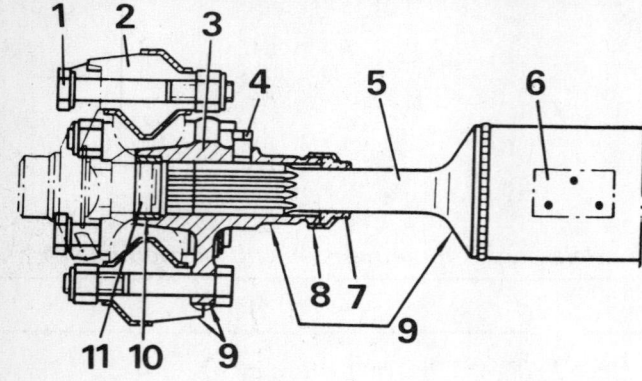

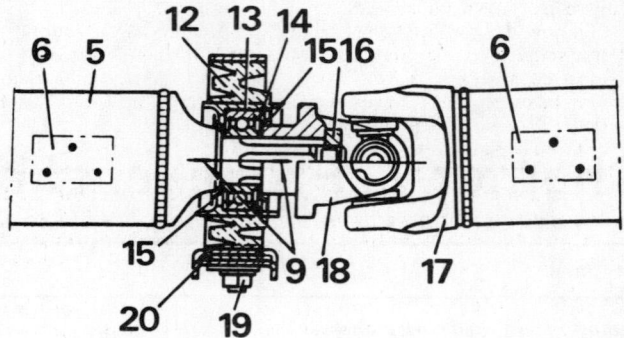

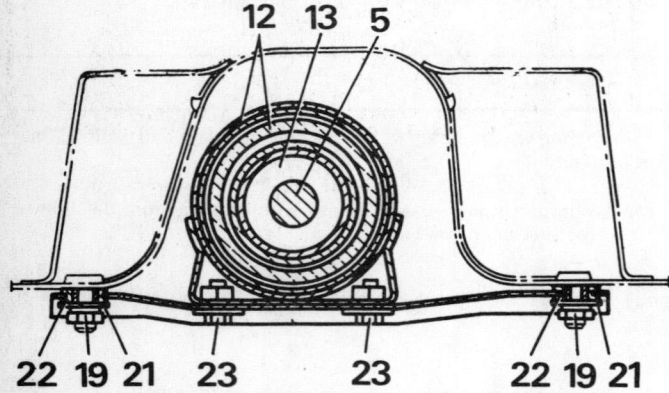

Fig. 7.1 Cut-away view of propeller shaft (Sec 1)

1	Flexible coupling to gearbox output flange bolts	12	Centre bearing housing
2	Flexible coupling	13	Centre bearing
3	Propeller shaft spider	14	Circlip
4	Lubrication plug	15	Dust excluders
5	Propeller shaft front end	16	Yoke retaining nut
6	Balance plates	17	Propeller shaft rear section
7	Grease seal	18	Yoke
8	Seal retainer	19	Support crossmember bolts
9	Alignment marks	20	Crossmember
10	Bush	21	Washer
11	Ring	22	Flexible bushes
		23	Centre bearing support bolts

must be made before the car is taken on the road:

(a) Make sure all shaft mating marks are in alignment
(b) Tighten all mounting nuts and bolts to the torque specified at the beginning of the Chapter
(c) Check that the handbrake operates properly

3 Universal joints – dismantling, inspection, repair and reassembly

1 Wear in the needle roller bearings in the rear shaft universal joints is characterised by vibration in the transmission, clonks on taking up drive and in extreme cases of lack of lubrication, metallic squeaking and ultimately grating and shrieking sounds as the bearings break up.
2 It is easy to check if the needle roller bearings are worn with the propeller shaft in position, by trying to turn the shaft with one hand, whilst holding the rear axle flange with the other hand. A second check is to try and lift the shaft whilst looking for any movement in the joints.
3 If wear is evident, a repair kit should be purchased from your Lada dealer; this contains a new spider unit and the associated bearings, seals and retainers. Alternatively an exchange rear shaft assembly complete with rear universal joints can be obtained.

Universal joints – dismantling

Note: *Before removing the rear shaft from the forward shaft/bearing assembly, mark the shafts and universal joint yokes to ensure that they are reassembled in their originally aligned positions in relation to each other.*

4 Clean away all traces of dirt and grease from the circlip located on the bearing cups in the yokes. Remove the clips by pressing their open ends together with a pair of circlip pliers and lift them out with a screwdriver. If they are difficult to remove, tap the bearing cup top with a mallet to ease the pressure on the circlip.
5 Hold one side of the joint, normally the tubular shaft side to begin with, and remove the bearing cups and needle rollers by tapping the yoke at each bearing with a copper or hide faced hammer. As soon as the bearing cups begin to emerge from their bores, they can be drawn out with either your fingers or a pair of pliers. If the bearing cups refuse to move then place a small drift against the inside of the bearing and tap it gently until the cup begins to move.
6 With all four cups removed together with their needle rollers, the spider can easily be extracted from the yokes. Once the spider is free the bearing faces may be wiped clean with a petrol damped cloth and the surfaces inspected. If any grazing scores or ridges are found the spider will need replacing. On some occasions when the universal joint has failed through lack of lubrication, the bores in the yokes in which the bearing cups fit can be worn. Again once the joint has been dismantled it is easy to check the condition of the yokes.

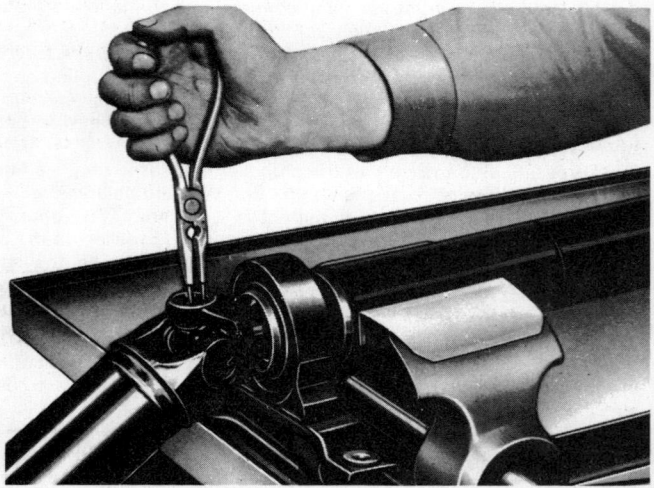

Fig. 7.2 Extracting a universal joint circlip (Sec 3)

Universal joints – reassembly

7 Thoroughly clean the yokes and bores. Remember to check that the circlip grooves are clear.
8 Fit new grease seals and retainers on the new spider journals and fit the spider into the shaft yoke. Assemble the needle rollers in the bearing cups and hold in place by smearing them with a medium lithium base grease. In new assemblies the needles should pack so well that they each retain the other in place.
9 Carefully ease the cups, packed with the needle rollers into the yoke bore and onto the appropriate spider journal. It is all too easy to hurry and a single roller might fall from place and prevent the cup from seating properly on the journal.
10 It may be necessary to tap the bearing cups home in the final stages and once the cup top face has passed the circlip groove in the yoke refit the circlips.
11 Once all the whole universal joint has been assembled there is provision on most makes of replacement universal joints for injecting extra grease into the spider bearings, before the propeller shaft is refitted to the car. A small grub-screw in the centre of the spider can be unscrewed and a grease nipple temporarily fitted to enable the joint to be greased.

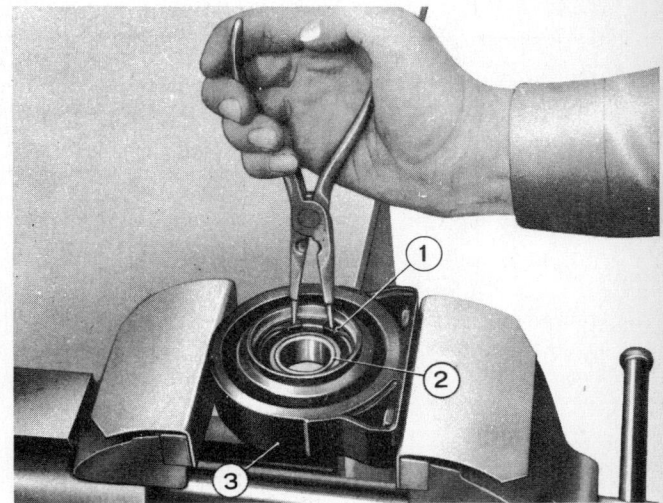

Fig. 7.3 Extracting the centre bearing circlip (Sec 4)

1 Circlip
2 Bearing
3 Housing

4 Centre bearing – removal and refitting

1 Remove the propeller shaft complete from the car as described in Section 2.
2 Grip the forward section of the propeller shaft in a vice.
3 Extract the universal joint bearing cup circlips and dismantle the joint by ejecting the bearing cups and needles as described in the preceding Section. Keep the cups and needles identified as to which yoke location they came from so that they can be returned to their original positions.
4 Unscrew the now accessible nut and separate the yoke from the front section of the shaft. To separate the yoke, use a suitable puller or screw the nut on two or three threads and supporting the yoke, tap the nut with a copper or soft-faced mallet.
5 Remove the centre bearing assembly from the front section of the propeller shaft by using one of the methods described in the preceding paragraph.
6 To dismantle the centre bearing, extract the circlip. Using a puller or a bolt, nut, washers and distance piece, draw the bearing from the support housing.
7 Reassemble the centre bearing (having first pressed the recommended lubricant into it) by driving it into its support housing. Fit a new circlip.
8 Fit the dust excluder to the rear end of the front section of the propeller shaft, drive the centre bearing assembly onto the shaft using a piece of tubing, fit the second dust excluder, the yoke and retaining nut. Tighten the nut to the specified torque.
9 Join the front and rear sections of the propeller shaft by reassembling the universal joint. Keep the yokes in their originally fitted positions and return the bearing cups and needles to their original seats.
10 Refit the propeller shaft to the vehicle as described in Section 2.

5.4 Propeller shaft front spider with shaft sliding section withdrawn

5 Flexible coupling – removal and refitting

1 Remove the propeller shaft from the vehicle (refer to Section 2) or alternatively disconnect the front of the shaft from the flexible coupling. Remove the centre bearing housing bolts and push the propeller shaft to one side.
2 The flexible coupling will remain attached to the flange on the gearbox output shaft. Leave the clamp or wormdrive clip in position and then unbolt and remove the coupling from this flange.
3 If the original flexible coupling is being refitted, align the mating marks on the coupling and the shaft spider. If a new coupling is being fitted, then after fitting of the coupling and propeller shaft, it may be necessary to balance the shaft as described in Section 6 but only if vibration is felt during road testing throughout the speed range.
4 The spider can be withdrawn from the shaft splines if the grease seal cap is released. Renew the seal if it is worn (photo).

6 Propeller shaft – balancing

1 Obviously the best way to cure vibration in a propeller shaft is to have it professionally weighted. A new shaft will have balance plates welded to it.
2 The home mechanic can rebalance a shaft to within acceptable limits if it has become unbalanced due to the fitting of new parts.
3 Place a wormdrive clip round the propeller shaft. Road test the car and adjust the position of the wormscrew by moving the clip round the shaft and also sliding it towards the front or rear of the shaft. Retest between each adjustment.
4 Where out of balance is severe, two or more clips can be used or a balance plate inserted under the clip band below the wormscrew.

Chapter 7 Propeller shaft

7 Fault diagnosis – propeller shaft

Symptom	Reason/s
Vibration	Wear in sliding sleeve splines
	Worn universal joint bearings
	Propeller shaft out of balance
	Distorted propeller shaft
Knock or clunk when taking up drive	Worn universal joint bearings
	Worn rear axle drive pinion splines
	Loose rear drive flange bolts
	Excessive backlash in rear axle gears
Noise on overrun	Splined coupling worn or unlubricated

Chapter 8 Rear axle

For modifications, and information applicable to later models, see Supplement at end of manual

Contents

Axleshafts and oil seals – removal and refitting	2
Axleshaft bearings – renewal	3
Differential carrier – overhaul	7
Differential carrier – removal and refitting	5
Fault diagnosis – rear axle	8
General description	1
Pinion oil seal – renewal	4
Rear axle assembly – removal and refitting	6

Specifications

Rear axle type .... Semi-floating, hypoid

Final drive ratio
- 1200, 1300 and 1500 saloon models .... 4·3 : 1
- 1200, 1300 and 1500 estate models .... 4·4 : 1
- 1600 saloon and estate models .... 4·1 : 1

Lubrication
- Oil type/specification .... Hypoid gear oil, viscosity SAE 90 or 75W/90, to API GL5 (Duckhams Hypoid 75W/90S)
- Capacity .... 2.3 pints (1.3 litres)

Torque wrench settings

	lbf ft	kgf m
Differential carrier to axle casing bolts	30	4·2
Differential bearing cap bolts	37	5·2
Crownwheel bolts	72	10·0
Propeller shaft rear flange bolts	24	3·3
Roadwheel bolts	50	7·0
Rear shock absorber mounting bolt	43	6·0
Rear suspension trailing arm bolt	58	8·0

1 General description

The live rear axle is conventional, with hypoid final drive gears and semi-floating axle shafts running on a single ball bearing at the outer end. The axle shaft inner ends are splined into the differential side gears.

The pinion runs in two taper roller bearings which are preloaded and also located by a collapsible spacer between them. The crownwheel to pinion mesh is set by the lateral movement of the differential cage. This lateral position is controlled by screwed retaining rings which both position and pre-load the differential side bearings.

Axle shafts can be removed without difficulty and the differential casing and assembly can be removed from the rear axle as a unit, once the propeller shaft has been detached and the axle shafts withdrawn a short distance.

2 Axle shafts and oil seals – removal and refitting

1 Loosen the wheel nuts of the wheel on the shaft to be removed and then jack up the car and support the axle casing on a stand. If both shafts are to be taken out it is important that the rear axle is supported firmly on two stands.
2 Remove the roadwheel and then undo the retaining screws and pull of the brake drum.
3 Remove the four nuts that secure the oil baffle and brake anchor plate to the axle casing flange, but do not remove the anchor plate or bolts.
4 The shaft and bearing have to be drawn out together and this normally involves some percussion to force the bearing from the housing. If a slide hammer is available there is no problem. Otherwise an alternative is to refit a wheel (or if possible an old rim without a tyre) and strike it from the inside to draw the axle out. Do not under any circumstances strike the axle shaft flange directly. It can be distorted and damaged too easily.
5 With the axle removed, the oil seal and O-ring behind the bearing may be taken out of the axle casing.
6 Refitting of the axle shaft is a straightforward reversal of the removal procedure. Do not forget the oil seal which should preferably be a new one, or the O-ring, which should certainly be new.
7 Check and top up the oil level if required.

3 Axle shaft bearings – renewal

1 If the ball bearing on the shaft is obviously worn and in need of renewal it is best done by someone with the proper equipment. The bearing is held on the shaft by a retaining ring which requires heating to 300°C and 5 tons weight under a press to get it on. Somewhat more than 5 tons is needed to get it off and the use of hammers, cold chisels, hacksaws and the like will probably have a net result of a ruined axle shaft. Take it along to a Lada franchise garage who will have the part and the appropriate tools and press to fit it without damage to either the axle or bearing.

4 Pinion oil seal – renewal

The pinion oil seal can be renewed with the axle in position in the car provided the following operations are carefully followed:
1 Jack up the rear of the car and support the axle on stands.

Chapter 8 Rear axle

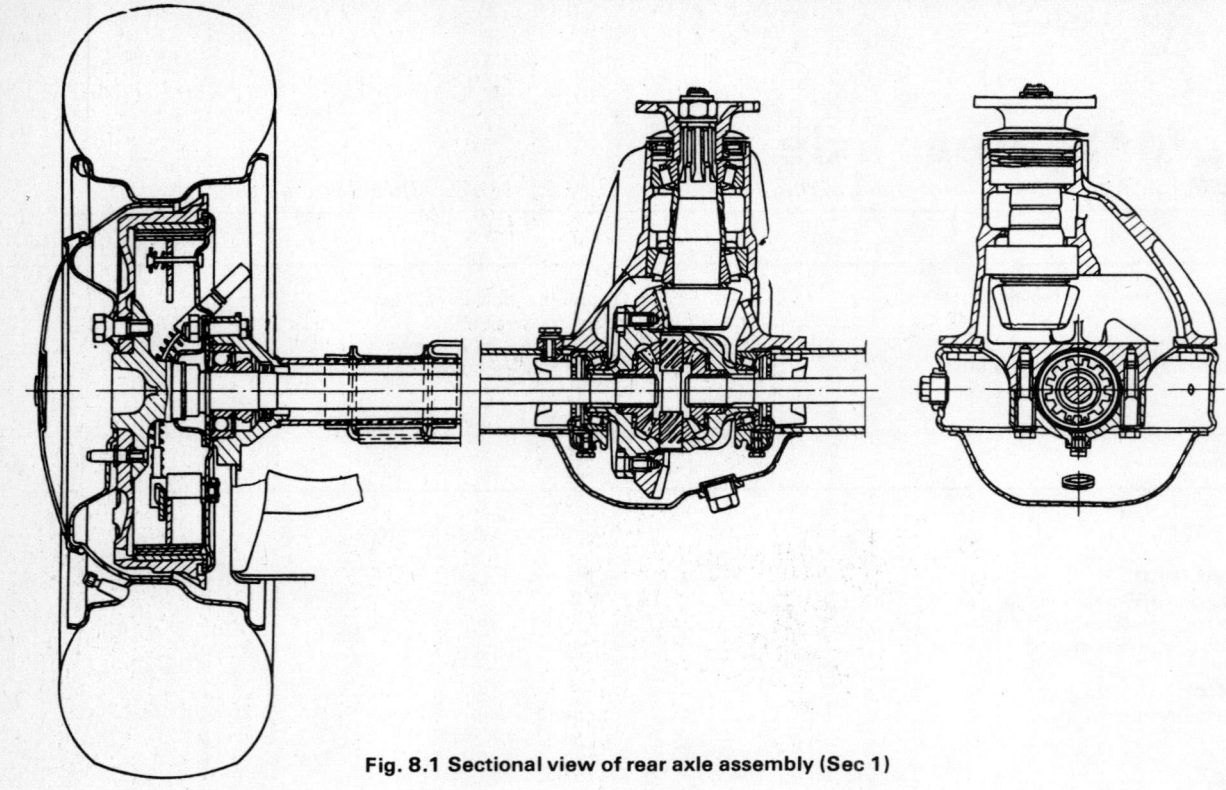

Fig. 8.1 Sectional view of rear axle assembly (Sec 1)

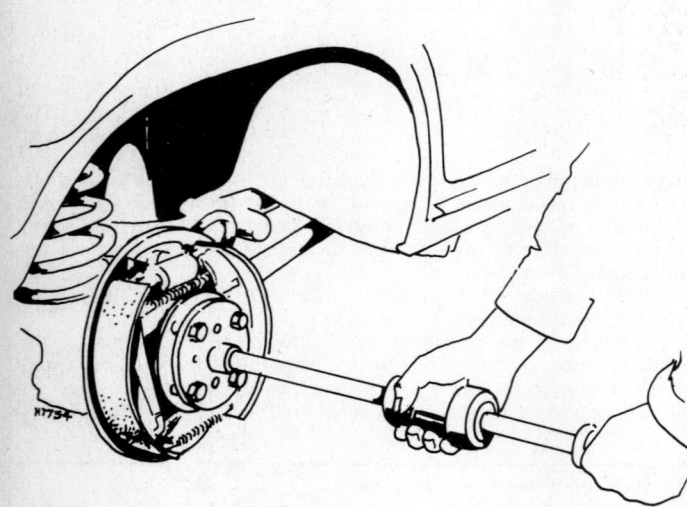

Fig. 8.2 Removing axleshaft using a slide hammer (Sec 2)

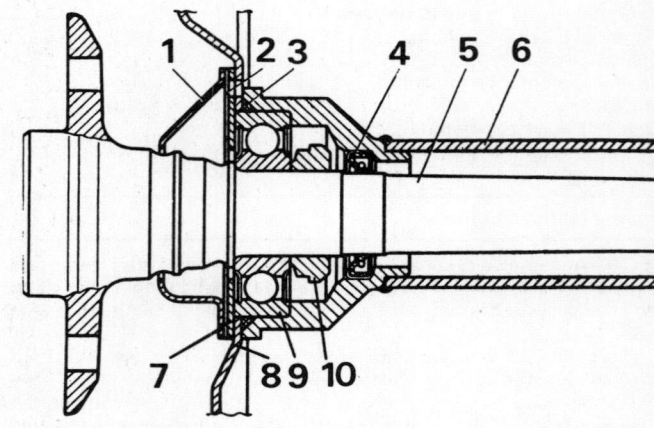

Fig. 8.3 Sectional view of rear axle hub (Sec 2)

1 Oil baffle
2 Gasket
3 Sealing ring
4 Oil seal
5 Axleshaft
6 Axle casing
7 Bearing lockplate
8 Brake backplate
9 Bearing
10 Bearing locking collar

2 Mark the edges of the propeller shaft and pinion coupling flanges to ensure exact refitting.
3 Remove the four coupling bolts, detach the propeller shaft at the axle pinion flange and tie the propeller shaft to one side.
4 Remove both rear roadwheels and brake drums to eliminate any drag.
5 Wind a cord round the pinion flange coupling and exerting a steady pull, note the reading on a spring balance, this should be between 2 and 5 lb (0·9 and 2·3 kg). The spring balance reading indicates the pinion bearing preload.
6 Mark the coupling in relation to the pinion splines for exact refitting.
7 Hold the pinion coupling flange by placing two 2 in long bolts though two opposite holes, bolting them up tight, undo the self-locking nut whilst holding a large rod or bar between the two bolts as a lever.
8 Remove the defective oil seal by drifting in one side of the seal as far as it will go to force the opposite side of the seal from the housing.
9 Refit the new oil seal, first having greased the mating surfaces of the seal and the axle housing. The flanges of the oil seal must face inwards. Using a piece of brass or copper tubing of suitable diameter, carefully drive the new oil seal into the axle housing recess until the face of the seal is flush with the housing. Make sure that the end of the pinion is not knocked during this operation.
10 Refit the coupling to its original position on the pinion splines after

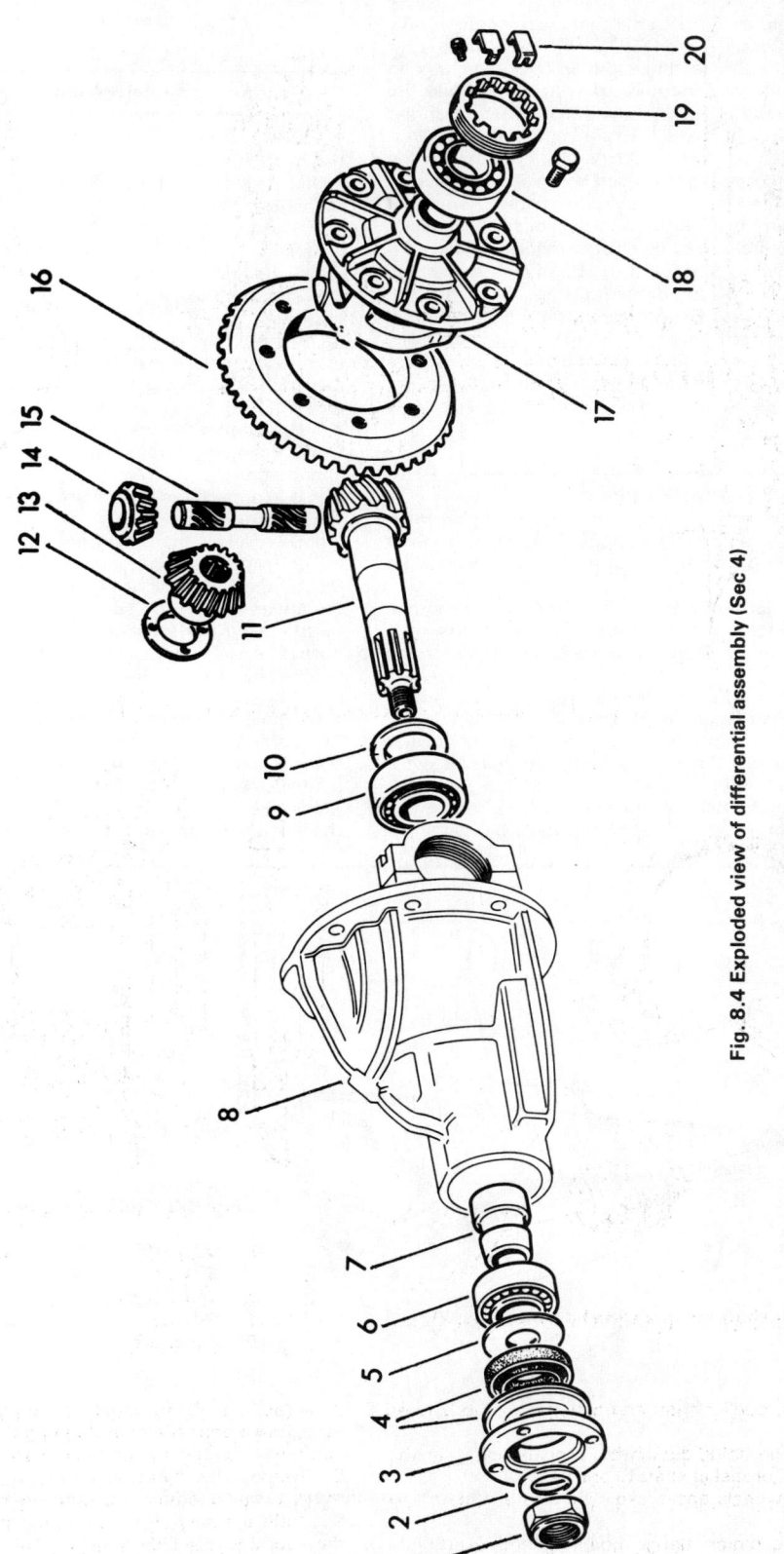

Fig. 8.4 Exploded view of differential assembly (Sec 4)

1 Pinion nut
2 Spring washer
3 Flange/coupling
4 Oil seal
5 Oil slinger
6 Smaller tapered roller bearing
7 Collapsible spacer
8 Differential carrier
9 Larger tapered roller bearing
10 Spacer shim
11 Pinion
12 Side gear shims
13 Side gear
14 Pinion gear
15 Pinion shaft
16 Crownwheel
17 Differential case
18 Differential case roller bearing
19 Adjuster rings
20 Adjuster ring locktabs

first having located the dust cover.

11 Fit a new pinion nut and holding the coupling still with the screwdriver or tyre lever, tighten the nut until the pinion endfloat only just disappears. Do not overtight.

12 Rotate the pinion to settle the bearing and then check the preload using the cord and spring balance method previously described. By slight adjustment of the nut and rotation of the pinion, obtain a spring balance preload figure to match that which applied before dismantling.

13 On no account overtighten the pinion nut as it cannot be slackened without introducing endfloat caused by over compressing the collapsible spacer shown in Fig. 8.4. Should this happen, withdraw the pinion nut, coupling, taper roller bearing and the collapsible spacer. Fit a new spacer and reassemble the other components. Tighten the pinion nut to a torque of 87 to 130 lbf ft (12 to 18 kgf m) if the spacer sleeve wall is 1·2 mm thick, or 130 to 188 lbf ft (18 to 26 kgf m) if the spacer sleeve wall is 1·7 mm thick. Tighten the nut only a fraction of a turn at a time once the lower specified torque setting has been reached and check the preload as previously described.

14 Remove the two holding bolts and refit the propeller shaft, making sure to align the mating marks. Refit the brake drums and roadwheels and lower the car.

5 Differential carrier – removal and refitting

1 Removal of the final drive and differential is possible without disturbing the crownwheel to pinion setting or without removal of the axle assembly.

2 Jack up the car, support it on stands and draw out both axle shafts about 3 in only to disengage them from the differential side gears.

3 Disconnect the propeller shaft from the drive pinion as described in Chapter 7.

4 Undo the drain plug in the bottom of the casing and let the oil out into a clean container. It may be re-used.

5 Support the differential casing and remove the eight bolts that secure it to the front of the axle. The whole unit may then be withdrawn.

6 Refitting of the differential assembly into the rear axle casing is a reversal of the removal procedure. Use a new gasket between the carrier and casing and tighten the retaining bolts to the specified torque.

7 Do not forget to refill the casing with oil.

6 Rear axle assembly – removal and refitting

1 Removal of the rear axle assembly is a rare enough job in view of the relative ease with which the sub-assemblies may be taken from it whilst in position. In certain circumstances however, such as external damage to the casing or rear suspension, it may be necessary.

2 Slacken the wheel bolts and jack up the rear of the car. Support it on stands under the body jacking points at each side. Then securely chock the front wheels in front and behind.

3 Remove the wheels and disconnect the propeller shaft from the rear axle.

4 Inside the engine compartment, remove the hydraulic fluid reservoir cap and seal the reservoir by putting plastic film under the cap before screwing it back on again. Disconnect the flexible hose where it joins the brake pipe on top of the rear axle casing.

5 Disconnect the brake cables from each lever at the rear wheel and then disengage the cable outer conduits from the adjacent anchorages.

6 Disconnect the two short suspension arms from the brackets on the axle casing.

7 Disconnect the brake regulator control rod from the axle housing by removing the bolt and nut.

8 Support the weight of the axle using a jack and blocks. Disconnect the lower ends of the telescopic shock absorbers from the axle casing by removing the retaining nuts and washers. Do not lose the rubber bushes.

9 Finally remove the bolts that secure the two main trailing arms and the transverse Panhard rod to their respective brackets on the axle. The axle may then be lowered and drawn back from the car.

10 Refitting the rear axle assembly is a reversal of the removal procedure. Make sure that the springs and their seats are properly located and the necessary bushes fitted to the upper ends of the dampers. Move the assembly into position under the rear of the car and then lift the axle on the jack until the ends of the trailing arms and the Panhard

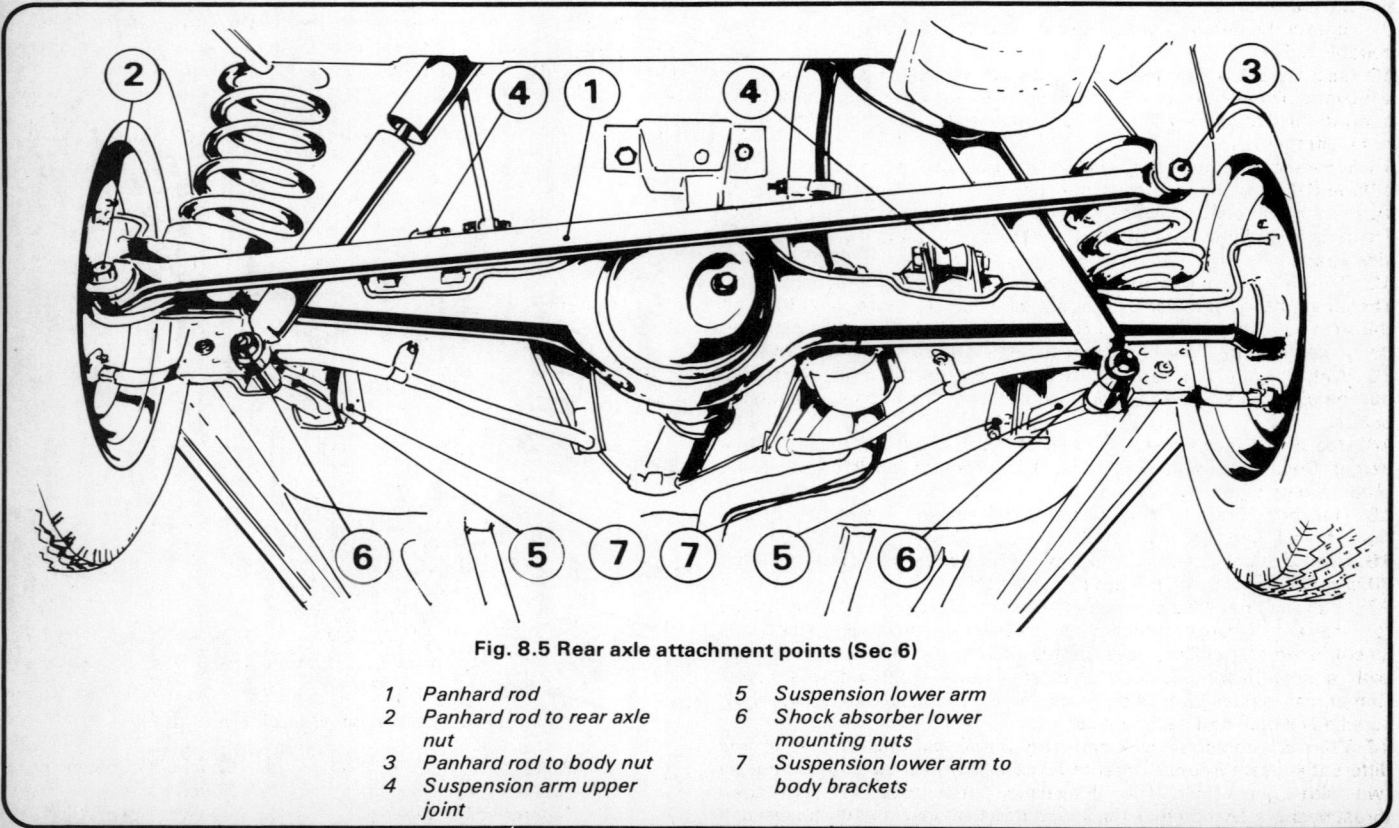

Fig. 8.5 Rear axle attachment points (Sec 6)

1 Panhard rod
2 Panhard rod to rear axle nut
3 Panhard rod to body nut
4 Suspension arm upper joint
5 Suspension lower arm
6 Shock absorber lower mounting nuts
7 Suspension lower arm to body brackets

rod can be reconnected to their brackets. Do not tighten the nuts and bolts more than finger tight.

11 Pull the shock absorbers downwards and refit the lower ends to the axle casing brackets. Do not forget the rubber bushes.

12 Refit the brake regulator rod and tighten the nut and bolt. Reconnect the two short longitudinal rods but do not fully tighten the nuts and bolts at this stage.

13 Reconnect the handbrake cables and the hydraulic brake hose. Bleed the hydraulics and adjust the handbrake as described in Chapter 9.

14 Refit the propeller shaft, refit the wheels and lower the car to the ground. Check the rear axle oil level.

15 Finally with the car standing on level ground, bounce the rear end of the car several times to settle the suspension. Tighten the bolts that secure the suspension arms, Panhard rod and shock absorber mountings, to the specified torque wrench setting.

7 Differential carrier – overhaul

1 The differential carrier houses the pinion, crownwheel and differential assembly. The pinion and crownwhheel are a matched pair and must never be renewed individually.

2 Complete overhaul requires skill and the use of special tools. With a severely worn assembly it will probably be more economical to purchase a new unit.

3 With the differential carrier removed from the rear axle casing, clean away all external dirt and place the casing on the bench.

4 Unscrew and remove the bolts that hold the differential bearing adjuster locking tabs to the bearing caps. Remove the locking tabs and unscrew the bearing adjusting rings. Record the number of turns required to remove the ring as this information will prove useful at reassembly.

5 Lock the differential by inserting a metal rod through the apertures in the case, then loosen the pinion nut.

6 Unscrew the bearing cap bolts, mark the positions of the caps and then remove them.

7 With the caps removed, withdraw the differential case complete with crownwheel from the differential carrier.

8 Unbolt the crownwheel from the differential case. Mark their relative positions for reassembly.

9 Extract the differential case outer roller bearings and tracks using a suitable puller.

10 Using a brass or copper drift, tap out the small pinion shaft on which the pinion gears run. Turn the side bevel gears simultaneously to bring the pinion gears into alignment in the apertures of the differential case. Lift them out.

11 Now remove the side gears by moving them into the centre of the differential case then out through the apertures. Retrieve any thrust washers from behind the side gears, noting their number and position. These shims control the backlash between the pinion gears and the side gears.

12 The pinion should now be removed from the differential carrier. The pinion flange nut was previously loosened (see paragraph 5) but if this was overlooked, it can be unscrewed now if the pinion companion flange is bolted to a length of steel angle or flat to be used as a lever.

13 With the nut removed, slide the flange from the pinion and prise out the oil seal using two screwdrivers as levers positioned at opposite points.

14 Use a soft-faced mallet to drive the pinion out of the differential carrier. Collect the pinion bearings. Discard the collapsible spacer and obtain a new one.

15 The shims fitted behind the gear on the pinion should be retained ready for refitting.

16 Drive out any bearing tracks (which are being renewed) from the differential carrier using a brass or copper drift.

17 With the final drive completely dismantled, inspect all components for wear and damage and renew as necessary. Always renew bearings as complete assemblies; never be tempted to use an old bearing track with a new bearing just because the track is difficult to remove. Remember, pinion gears, side gears and the crownwheel/pinion are all supplied as matched pairs or sets.

18 With all components clean and ready for reassembly, fit the differential gears into the differential case. Commence by inserting the two side gears into the differential case together with their thrustwashers (A nominal thickness thrustwasher should be used if new side gears are being fitted). Now fit the pinion bevel gears into the differential case. To do this, turn the side gears to align the pinion gears with the pinion shaft holes.

19 Insert the pinion shaft and tap it home with a brass drift.

20 Using either a dial gauge or feeler gauge, measure the endfloat of the side gears. If it is outside the tolerance specified (0·020 in to 0·508 mm) select new thrust washers and substitute them for those previously fitted.

21 Bolt the crownwheel into position on the differential case, making sure that any marks are in alignment. Tighten the bolts to the specified torque.

22 Fit the tapered roller bearings to the differential case, making sure that they are fitted the correct way round (smallest diameter further from the differential case).

23 The differential case can now be put aside until the pinion has been fitted to the differential carrier.

24 Tap the two pinion bearing outer tracks into the differential carrier. Check that these are the right way round with their smaller diameters towards each other.

25 Special tools are available from Lada to set the pinion but the following method will enable the job to be done without them.

26 Fit the pinion (complete with spacer shim behind the pinion gear) together with its front and rear bearings.

27 The pinion will now have taken up approximately its original position. Fit the flange to the pinion and tighten the nut so that the bearings are just pulled together enough to eliminate any pinion endfloat, no tighter!

28 Mount the differential case onto the carrier, fit the bearing caps and tighten the cap bolts just enough to nip up the case bearings.

29 Screw on the bearing adjusting rings the same number of turns they were unscrewed.

30 Check for any endfloat and then tighten the bearing cap bolts to the specified torque.

31 Rotate the differential case several times to settle the bearings and then, using a dial gauge or feeler blades, calculate the backlash between the crownwheel and pinion. If a dial gauge is being used, place its stylus on one of the crownwheel teeth. Hold the pinion absolutely still and rock the crownwheel backwards and forwards. The backlash should be between 0·004 to 0·006 in (0·101 to 0·152 mm).

32 Using a brass drift, tap the bearing adjusting rings round in whichever direction is necessary to move the differential case and so adjust the backlash. If the adjusting rings are hard to move, slightly release the bearing cap bolts. After each adjustment and *before* checking the backlash, tighten the cap bolts to the specified torque.

33 Clean the crownwheel and pinion teeth and apply engineer's blue

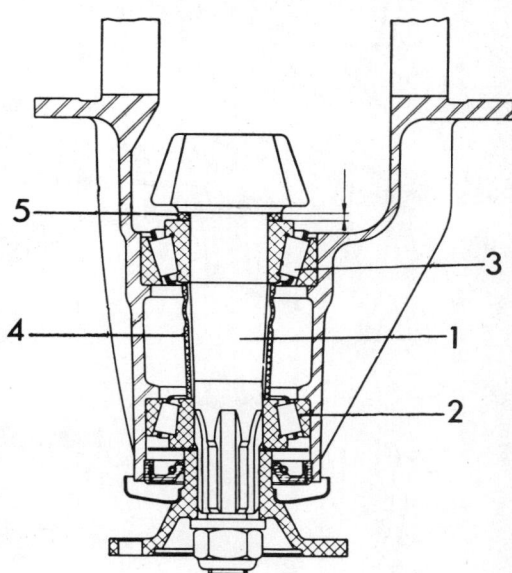

Fig. 8.6 Cutaway view of pinion (Sec 7)

1 Pinion
2 Tapered roller bearing
3 Tapered roller bearing
4 Collapsible spacer
5 Shim/spacer

to several teeth of both gears. Turn the pinion and then assess the pattern marked on the gears with the examples shown in Fig. 8.7. Correct as indicated. Move the crownwheel by turning the adjusting rings. Move the pinion by varying the thickness of the washer located between the pinion gear and the bearing inner race. These washers are available in various thicknesses in increments of 0·002 in (0·051 mm).

34 As soon as the pinion/crownwheel adjustment is satisfactory, withdraw the differential case from the carrier.

35 Unscrew the pinion nut, pull off the flange and withdraw the pinion.

36 Remove the smaller of the pinion bearings complete with track.

37 Re-insert the pinion into the differential carrier, passing it through the larger rear bearing track.

38 A new collapsible spacer must now be slid onto the pinion.

39 Fit the smaller pinion bearing and track. Fit the oilslinger and tap a new oil seal into the recess in the pinion nose of the differential carrier.

40 Smear grease onto the oil seal lips. Fit the flange, washer and nut (finger tight).

41 Tighten the pinion nut only enough to eliminate any endfloat of the pinion.

42 Refit the differential case assembly to the differential carrier but without the adjusting rings. Mark their positions before removing them.

43 Pass a bar through the differential case to hold it while the pinion nut is tightened to 87 lbf ft (12 kgf m). From now on the nut must only be tightened in small increments and the pinion turning torque checked between each adjustment. With a cord wound round the pinion flange and a steady pull exerted, the force required to turn the

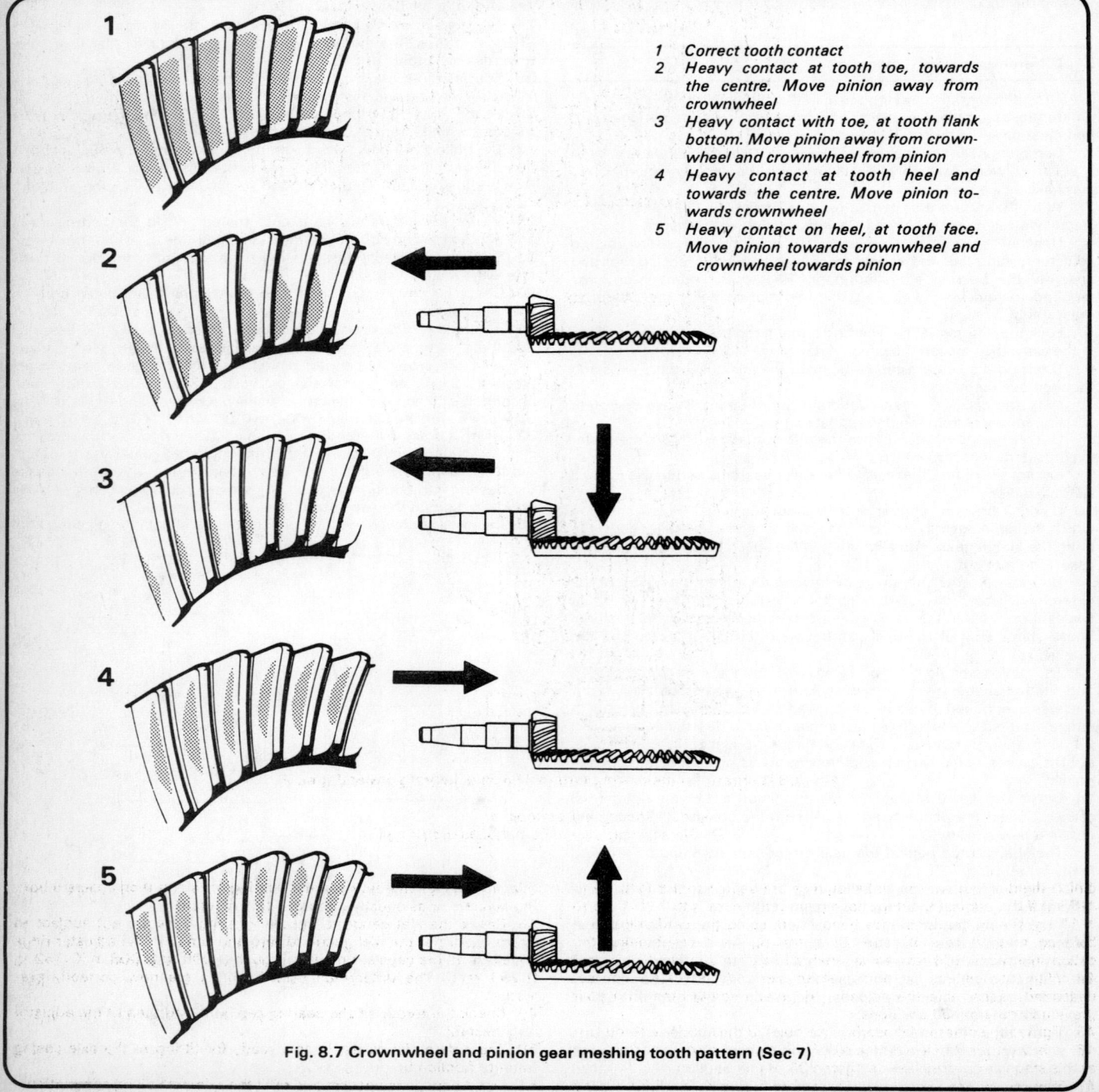

1. Correct tooth contact
2. Heavy contact at tooth toe, towards the centre. Move pinion away from crownwheel
3. Heavy contact with toe, at tooth flank bottom. Move pinion away from crownwheel and crownwheel from pinion
4. Heavy contact at tooth heel and towards the centre. Move pinion towards crownwheel
5. Heavy contact on heel, at tooth face. Move pinion towards crownwheel and crownwheel towards pinion

Fig. 8.7 Crownwheel and pinion gear meshing tooth pattern (Sec 7)

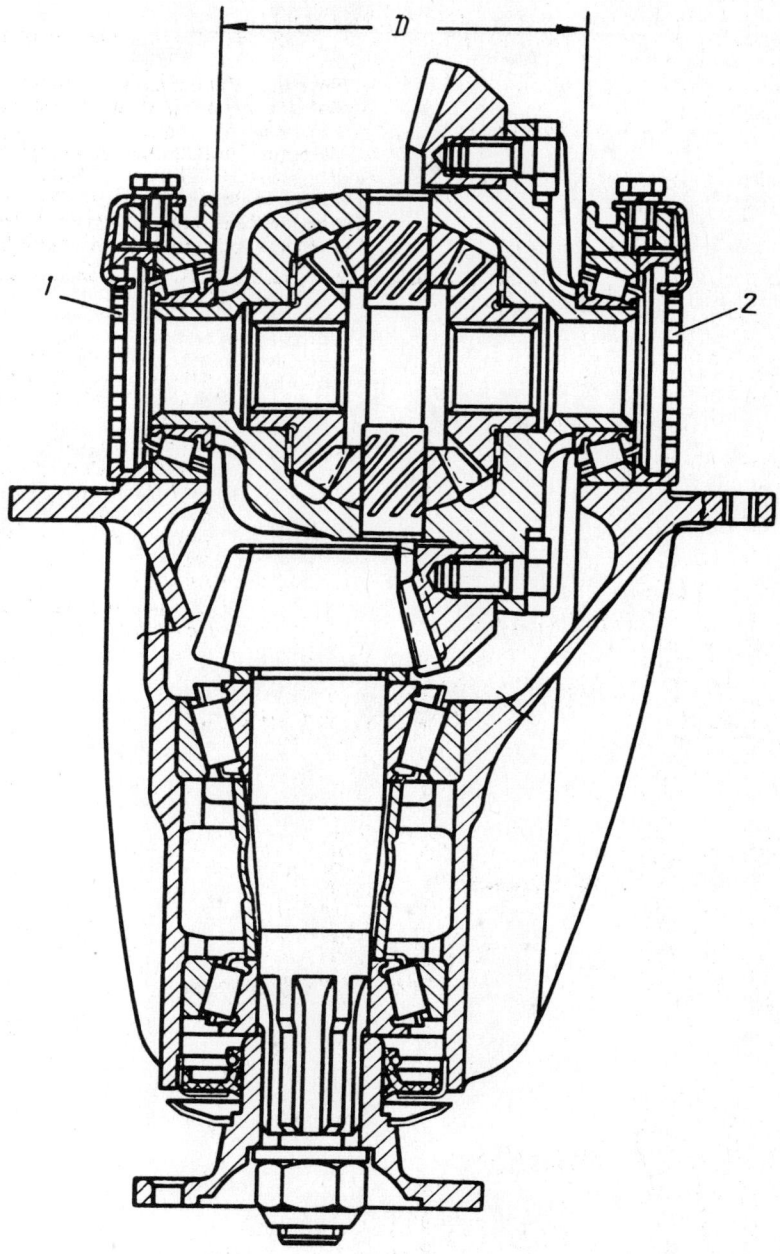

Fig. 8.8 Diagram for measuring differential case bearing preload (Sec 7)

1 and 2 Bearing adjusting nuts
D Distance between differential bearing caps

pinion (holding bar withdrawn) should be between 1 to 3·5 lb (0·45 to 1·6 kg) if the original bearings have been refitted, or 3 to 7 lb (1·36 to 3·17 kg) if new bearings have been fitted, using the cord and spring balance method (see Section 4). If the nut is overtightened, the collapsible spacer will have to be renewed (refer to Section 4).

44 With the pinion bearing preload correctly adjusted as just described, reassemble the adjusting rings and screw them into their previously determined positions.

45 Tighten the differential bearing cap bolts to the specified torque.

46 It is worth rechecking the pinion/crownwheel tooth pattern again at this stage as described in paragraph 33 of this Section.

47 Now check the preload on the differential case bearings. To do this, fit a dial gauge between the bearing caps and then unscrew both the adjuster rings equally through $\frac{1}{2}$ to $\frac{3}{4}$ of a turn.

48 Check the dial gauge to ensure that the caps are not subject to lateral load. Set the dial gauge to zero and screw in the adjuster rings equally until the caps are moved through 0·006 to 0·008 in (0·152 to 0·203 mm). The differential case bearings are now correctly preloaded.

49 Check the torque of the bearing cap bolts and then fit the adjuster ring locktabs.

50 The differential carrier is now ready for fitting to the axle casing (refer to Section 5).

51 Use a new flange gasket and refill the axle with oil on completion.

Chapter 8 Rear axle

8 Fault diagnosis – rear axle

Symptom	Reason/s
Vibration	Worn axleshaft bearings Loose driveflange bolts Out of balance propeller shaft Wheels require balancing
Noise	Insufficient lubrication Worn gears and differential components generally Propeller shaft splines worn or unlubricated
Clunk on acceleration or deceleration	Incorrect crownwheel and pinion mesh Excessive backlash due to wear in crownwheel and pinion teeth Worn axleshaft or differential side gear splines Loose driveflange bolts Worn drive pinion flange splines
Oil leakage	Faulty pinion or axleshaft oil seals May be caused by blocked axle housing breather

Chapter 9 Braking system

For modifications, and information applicable to later models, see Supplement at end of manual

Contents

Brake and clutch pedal assembly – removal, renovation and refitting	18
Brake disc and drum – examination and renovation	21
Brake servo unit – description	19
Brake servo unit – removal and refitting	20
Caliper unit – dismantling and reassembly	10
Caliper unit – removal and refitting	9
Fault diagnosis – braking system	22
Flexible brake hoses – inspection, removal and refitting	6
Front disc pads – renewal	5
General description	1
Handbrake – adjustment	15
Handbrake cable – renewal	16
Handbrake lever assembly – removal and refitting	17
Hydraulic system – bleeding	14
Maintenance	2
Master cylinder – dismantling and reassembly	12
Master cylinder – removal and refitting	11
Rear brake pressure compensator valve	13
Rear brake shoes – renewal	4
Rear brake wheel cylinder – removal, overhaul and refitting	8
Rear drum brakes – adjustment	3
Rigid brake lines – inspection, removal and refitting	7

Specifications

System type .... Four wheel hydraulic, dual circuit, disc front, drum rear. Vacuum servo assistance with pressure regulator in rear circuit. Handbrake mechanical to rear wheels

Hydraulic fluid
Type/specification .... Hydraulic fluid to SAE J1703, or FMVSS 116 DOT 3 or DOT 4 (Duckhams Universal Brake and Clutch Fluid)

Front disc brakes
Disc diameter	10 in (254 mm)
Disc thickness	0·39 in (10·0 mm)
Disc minimum thickness after refinishing	0·37 in (9·5 mm)
Maximum disc run-out	0·005 in (0·15 mm)
Minimum pad (friction lining) thickness	0·059 in (1·5 mm)
Caliper piston diameter	1·89 in (48·0 mm)

Rear drum brakes
Drum diameter	9·8 in (250 mm)
Maximum internal diameter after regrinding	9·9 in (251 mm)
Minimum shoe lining thickness	0·078 in (2·0 mm)

Torque wrench settings
	lbf ft	kgf m
Caliper mounting bolts	25	3.5
Disc to hub bolts	40	5·5
Rear brake backplate bolts	15	2·1
Foot pedal pivot bolt	20	2·8

1 General description

The braking system is four wheel, hydraulically operated with discs on the front and drums on the rear.

The hydraulic circuit is of the dual type, using a tandem master cylinder to ensure that in the event of a break or fault developing in one circuit, the remaining circuit remains fully operational.

A vacuum servo brake booster and fluid leakage indicator are fitted to all later model vehicles, as is a pressure regulator valve to prevent the rear wheel locking under heavy braking applications.

A mechanically operated handbrake is incorporated which operates on the rear wheels only.

2 Maintenance

1 Every 250 miles (400 km) or weekly, whichever occurs first, check the fluid level in both the master cylinder reservoirs. If necessary, top-up with fluid of the specified type which has been stored in an airtight container and has remained unshaken for the previous 24 hours.
2 Check that the reservoir cap breather holes are clear.
3 If topping-up is required frequently in one reservoir, inspect the hydraulic pipes of that particular circuit for leaks.
4 At the intervals specified in Routine maintenance, or more frequently if pedal travel becomes excessive, adjust the rear brakes as described in Section 3.

Chapter 9 Braking system

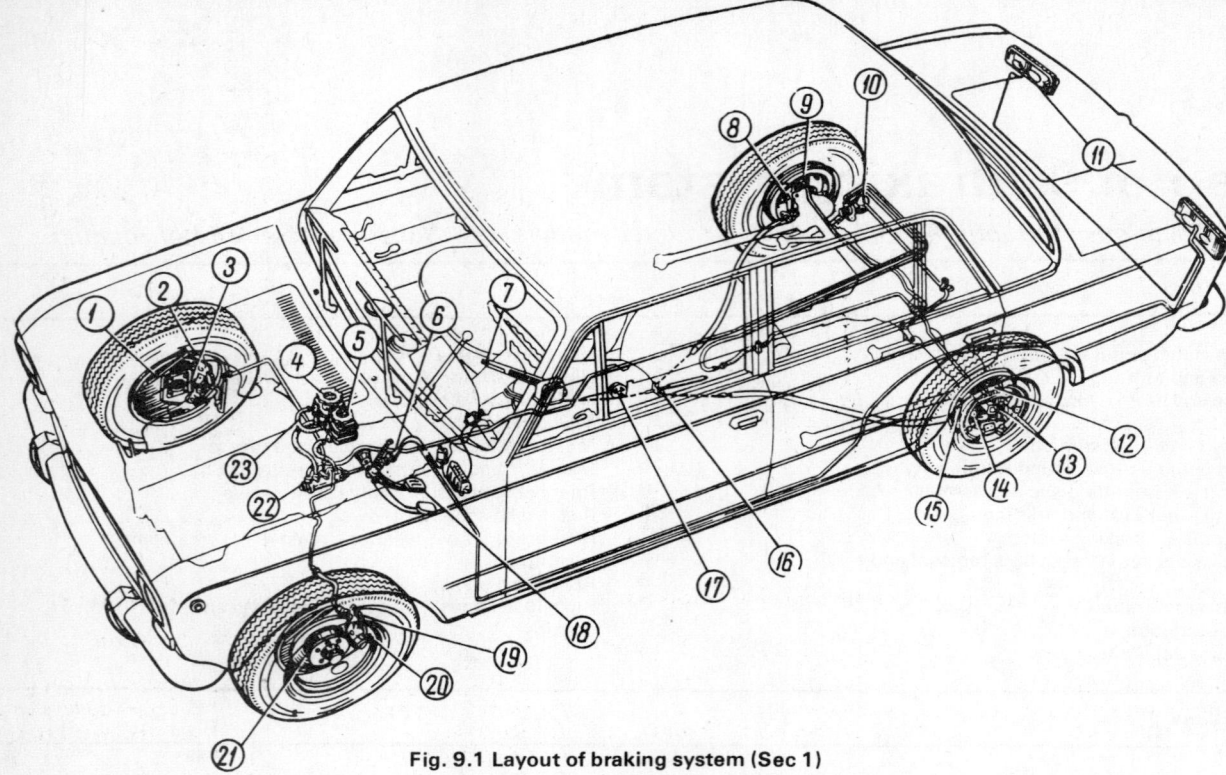

Fig. 9.1 Layout of braking system (Sec 1)

1 Disc shield
2 Interconnecting pipe
3 Caliper
4 Fluid reservoir (rear brake circuit)
5 Fluid reservoir (front brake circuit)
6 Stop lamp switch
7 Handbrake lever
8 Rear shoe adjuster
9 Bleed nipple
10 Pressure regulator valve
11 Stoplamps
12 Rear wheel cylinder
13 Handbrake shoe operating lever
14 Shoe adjuster cam
15 Rear brake shoes
16 Handbrake cable equalizer
17 Handbrake cable pulley
18 Brake pedal
19 Wheel cylinder interconnecting pipe
20 Caliper bleed nipple
21 Disc
22 Brake master cylinder
23 Fluid feed hose

5 At similar mileage intervals, remove the rear brake drums, inspect the linings and renew them if they are worn to $\frac{1}{32}$ in (0.8 mm) with bonded type; or with riveted linings, down to the rivet heads. Where the linings are in good condition, brush out any dust from the rivet head recesses and the interior of the drums before refitting the drums.
6 Examine the thickness of the friction lining material of the front disc brake pads. If it is worn down to $\frac{1}{16}$ in (1.6 mm) then all the disc pads should be renewed at the same time as a set. No adjustment is required to disc brakes.
7 At the intervals specified in Routine maintenance, bleed the hydraulic system of old fluid and refill with fresh.
8 At the intervals specified in Routine maintenance, renew all flexible hoses and rubber seals within the hydraulic components.

3 Rear drum brakes – adjustment

1 Release the handbrake fully and depress the footbrake pedal several times to position the shoes.
2 Chock the front wheels and then raise the rear roadwheels from the ground by placing a jack under the differential housing.
3 Turn the two hexagonal adjusters on each brake backplate in the direction shown in Fig. 9.2 until by turning the wheel, the brakes can be felt to bind. Now back off the adjusters until the shoes are no longer in contact with the drums.
4 Do not confuse transmission drag with rubbing shoe linings.
5 Apply the footbrake pedal hard and re-check the adjustment.
6 Note that on later estate car models the rear brakes are self-adjusting and no manual adjustment is necessary or provided for.

4 Rear brake shoes – renewal

1 Jack-up the rear axle differential carrier and support the axle

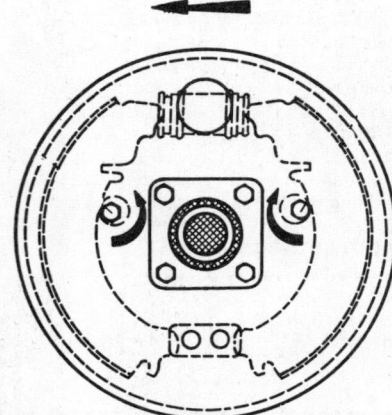

Fig. 9.2 Rear brake shoe adjusters (arrows indicate direction of travel to take up wear) (Sec 3)

casing on axle-stands.
2 Remove the roadwheel, undo the two retaining bolts and pull off the brake drum (photos). If the drum will not pull off, try screwing two bolts into the tapped holes provided. If this fails, try tapping them off using a block of hardwood and a hammer. Do not apply too much

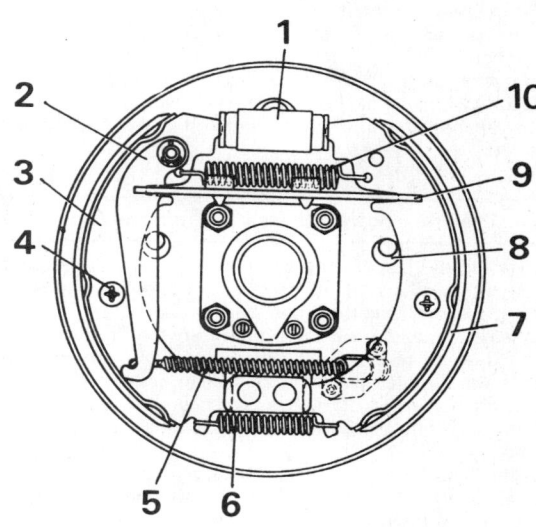

Fig. 9.3 Identification of rear brake assembly components (Sec 4)

1 Wheel cylinder
2 Handbrake operating lever
3 Brake shoe
4 Shoe steady spring and cup
5 Handbrake cable
6 Lower shoe return spring
7 Friction lining
8 Adjusting cam
9 Expander strut
10 Shoe upper return spring

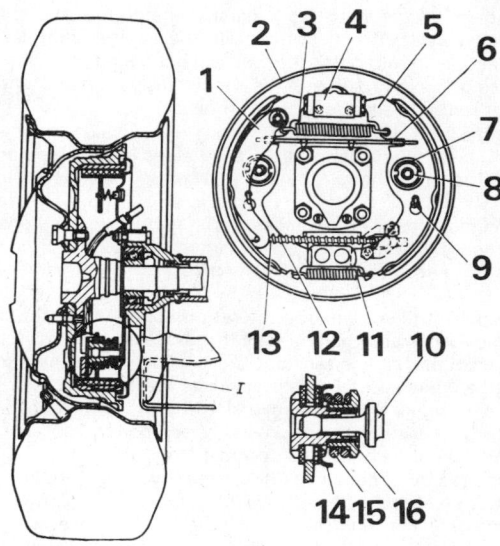

Fig. 9.4 Self-adjusting type rear brake (Sec 4)

1 Handbrake operating lever
2 Backplate
3 Return spring
4 Wheel cylinder
5 Shoe
6 Expander strut
7 Friction disc
8 Pin sleeve
9 Guide spring
10 Pin
11 Shoe return spring
12 Handbrake cable spring
13 Handbrake cable end fitting
14 Seat
15 Coil spring
16 Sleeve

4.2a Removing the brake drum retaining bolts

4.2b Pulling off a brake drum

4.4 Brake shoe steady spring and cup

4.5 View of a rear brake assembly

4.6 Handbrake cable and lever

4.7 Top brake shoe return spring

force or apply the hammer directly to the drum as they are made from aluminium with iron liners and can easily fracture. If the drum will still not release, back off the adjusters fully, start the engine and apply the handbrake fully. Select first gear and with the engine idling let in the clutch. This action should break the rust seal between the drum and the axleshaft. Using this method, with the car still jacked up, it is important that the front roadwheels are securely chocked.

3   Release the backplate adjuster fully.
4   Using a pair of pliers, depress the shoe steady spring cup and turn it through 90°, then withdraw the cup and spring from the steady post (photo).
5   Note and mark (or sketch if necessary), the relative positions of the shoes with regard to leading and trailing ends (photo).
6   Unhook the end of the handbrake cable from the operating lever (photo).
7   Using a pair of pliers, unhook the return springs from the holes in one of the shoe webs and then remove the shoes and springs (photo).
8   On no account depress the foot brake pedal while the shoes are removed or the wheel cylinder pistons will be ejected.
9   Fit the new brake shoes by reversing the removal procedure but apply a trace of grease to the backplate shoe engagement slots. Refit the drum and adjust the brakes as described in Section 3.
10 Check the reservoir fluid level for the rear hydraulic circuit.
Note: *On models fitted with self-adjusting brakes, each shoe is fitted with an adjusting mechanism comprising of a sleeve, spring and friction discs. If the new shoes are not supplied complete with adjusters, it will be necessary to unscrew the adjuster sleeves from the old shoes using Lada tool No A72259 and fit the sleeves and associated components to the new shoes. Refer to Fig. 9.4 to ensure that the adjuster spring and friction discs are fitted to the shoes in the correct order. The adjuster spring height should be at least 0.787 in (20 mm). If they are below this they should be renewed.*

5   Front disc pads – renewal

1   Remove the front wheels and inspect the amount of friction material left on the friction pads. The pads must be renewed when the thickness of the material has worn down to $\frac{1}{16}$ in (1.5 mm).
2   With a pair of pliers pull out the small wire clip that holds the main retaining pins in place (photo).
3   Remove the main retaining pins which run through the caliper and lift out the pad retaining clips (photo).
4   The friction pads can now be removed from the caliper. If they prove difficult to move by hand a pair of long nosed pliers can be used (photo).
5   Carefully clean the recesses on the caliper in which the friction pads and shims lie, and the exposed faces of each piston from all traces of dirt and rust.
6   Check the thickness of the friction pad (not including its metal backing plate). If the thickness is $\frac{1}{16}$ in (1.5 mm) or less, renew the pads. Pads must be renewed as complete front wheel sets of four to maintain even braking.
7   If new pads are to be fitted, the caliper unit pistons must be pushed back into their cylinders to accommodate the thicker pads. Use a flat lever to do this, but ensure that each piston is pressed squarely in and only sufficiently to enable the pad to enter the caliper unit opening. During this operation, the fluid reservoir level will rise and it may be necessary to syphon some off.
8   Fit the pads into the caliper unit recess.
9   Fit the two pad retaining clips and slide the pins through the holes in the caliper assembly and pads. Secure the pins with the wire clips and springs.
10 Depress the footbrake several times to settle the pads and then check the hydraulic fluid level in both reservoirs.
11 Lower the vehicle to the ground and check the brakes for correct operation.

6   Flexible brake hoses – inspection, removal and refitting

1   Periodically, inspect the condition of the flexible brake hoses. If they appear swollen, chafed, or when bent double with the fingers tiny cracks are visible, then they must be renewed.
2   Always uncouple the rigid pipe from the flexible hose first, then release the end of the flexible hose from the support bracket. Now unscrew the flexible hose from the caliper or connector. If this method is followed, no kinking of the hose will occur.
3   When refitting the hose, always use a new copper sealing washer (if fitted).
4   When refitting is complete, check that the flexible hose does not rub against the tyre or other adjacent components. Its attitude may be altered to overcome this by releasing its bracket support locknut and twisting the hose in the required direction by not more than one quarter turn.
5   Bleed the hydraulic system (Section 14).

7   Rigid brake lines – inspection, removal and refitting

1   At regular intervals, wipe the steel brake pipes clean and examine them for signs of rust or denting caused by flying stones.
2   Examine the fit of the pipes in their insulated securing clips and bend the tongues of the clips if necessary to ensure a positive fit.
3   Check that the pipes are not touching any adjacent component or rubbing against any part of the vehicle. Where this is observed, bend the pipe gently away to clear.
4   Any section of pipe which is rusty or chafed should be renewed. Brake pipes are available to the correct length and fitted with end unions from most Lada dealers and can be made to pattern by many accessory suppliers. When fitting the new pipes, use the old pipes as a guide to bending and do not make any bends sharper than is necessary.
5   The system will of course have to be bled when the circuit has been reconnected.

8   Rear brake wheel cylinder – removal, overhaul and refitting

1   The wheel cylinder should be removed when the seals need renewal. It is not wise to try to overhaul the cylinder in situ.
2   Remove the brake drums and brake shoes as directed in Section 4 of this Chapter.
3   Next, wipe the top of the brake fluid reservoir and unscrew the

5.2 Disc pad pins and retaining clips

5.3 Withdrawing the disc pad retaining pins

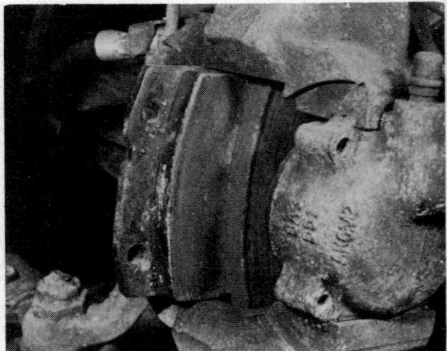
5.4 Removing the disc pads

Chapter 9 Braking system

cap. Place a thin sheet of polythene over the top and replace the cap; alternatively tape over the vent hole in the cap. The vacuum created will prevent loss of fluid from the system when the pipe/hose connections are dismantled on the rear brake backplate.

4 Using an open-ended spanner, unscrew the union fitting retaining the brake pipe in the wheel cylinder.

5 Once the brake pipe is detached, the two bolts that retain the cylinder to the backplate can be undone and the cylinder removed.

6 Once the cylinder is removed and on a clean bench, proceed to pull out the pistons, seals, spring and relay plungers which act between the pistons and shoes.

7 Examine the piston and cylinder for signs of wear or scoring; if there are any the whole assembly must be renewed. If the metal surfaces of the pistons and cylinder are in good condition only the seals need to be renewed (Fig. 9.5).

8 Clean all the parts with hydraulic fluid or methylated spirit. Do not use any other solvent.

9 Immerse the cleaned parts and new seals in new hydraulic fluid before beginning to reassemble the slave cylinder.

10 Fit the new seals to the spring seats so that the lips face away from the centre of the piston. Make sure that the lip enters the cylinder cleanly.

11 Once the two pistons and the dust seals are in place, the pistons should be retained with a rubber band around the cylinder.

12 Refit the cylinder back to the backplate, reassemble the brake shoes and drum. Then bleed the brake system. **Note**: *Wheel cylinder diameters vary with the date of production. Always have two matching cylinders on the rear wheel if renewing.*

9 Caliper unit – removal and refitting

1 Apply the handbrake, chock the rear wheels and jack-up the front of the car. Support the front of the car securely on axle-stands and then remove the appropriate roadwheel.

2 Wipe the top of the brake fluid reservoir, unscrew the cap and place a thin sheet of polythene over the top. Refit the cap. The vacuum created will stop the leakage of fluid from the system when subsequently dismantled. Alternatively tape over the vent in the reservoir cap.

3 Wipe clean the area around the caliper flexible hose to metal pipe connection and the hose to caliper connection. Detach the metal pipe from the hose. Hold the fixed end of the hose with a spanner whilst the union fitting on the pipe is unscrewed from the pipe. Then remove the nut which retains the hose in the support bracket.

4 Once the inside end of the hose is detached, proceed to remove the brake pads and store them safely so that the surfaces will not be contaminated with oil or dirt.

5 Finally undo the two bolts that secure the caliper to the wheel support (Fig. 9.6) and lift off the caliper assembly.

6 Refit the caliper amd brake pads using the reverse of the removal procedure. Make sure that the caliper retaining bolts are tightened to the specified torque wrench setting and then lock them by bending over the tab washers.

7 Finally bleed the front brake system as described in Section 14.

10 Caliper unit – dismantling and reassembly

1 Remove the caliper from the wheel support as detailed in Section 4 of this Chapter.

2 Remove both brake pads and their retaining pins and clips.

3 Temporarily, reconnect the caliper to the hydraulic system and

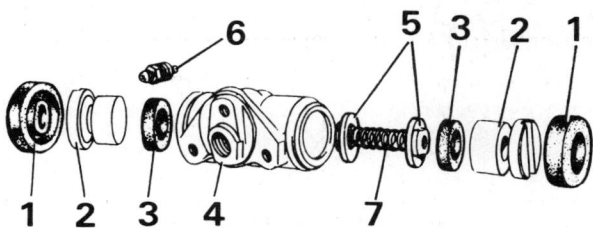

Fig. 9.5 Exploded view of a rear wheel cylinder (Sec 8)

1 Dust excluding boot
2 Piston
3 Seal
4 Cylinder body
5 Spring seats
6 Bleed nipple
7 Spring

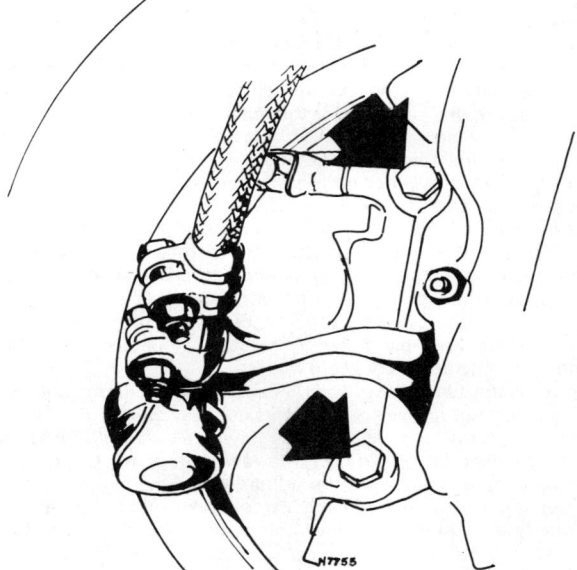

Fig. 9.6 Caliper mounting bolts (Sec 9)

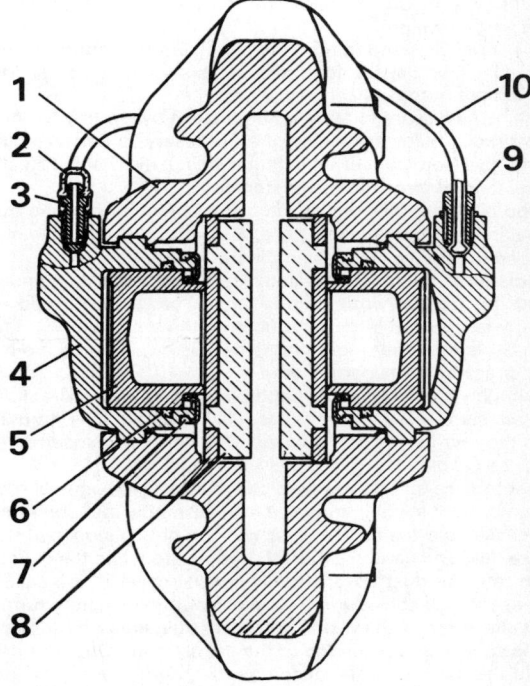

Fig. 9.7 Sectional view of caliper (Sec 10)

1 Caliper body
2 Bleed nipple cap
3 Bleed nipple
4 Cylinder body
5 Piston
6 Seal
7 Dust excluder retaining ring
8 Disc pad
9 Pipe union
10 Pipe

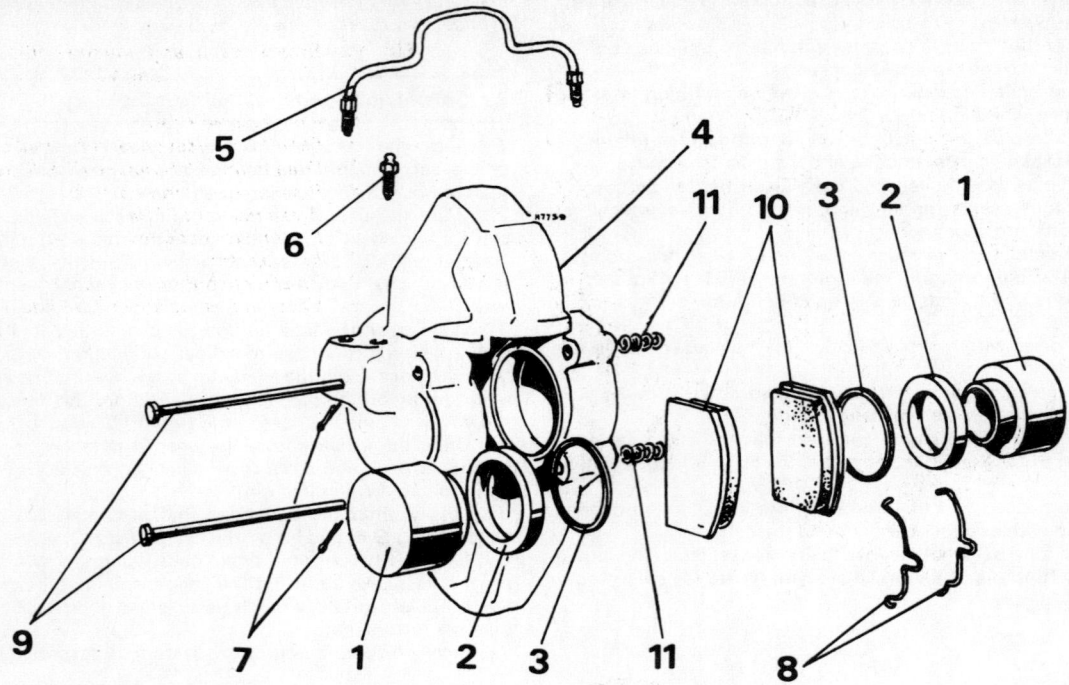

Fig. 9.8 Exploded view of caliper (Sec 10)

1 Piston
2 Dust excluding boot
3 Boot retaining ring
4 Caliper body
5 Cylinder interconnecting pipe
6 Bleed nipple
7 Spring clips
8 Pad retaining clips
9 Pad retaining pins
10 Disc pads
11 Coil springs

support its weight. Do not allow the caliper to hang on the flexible hose; support its weight.

4 Use a small G-clamp to hold the piston in the mounting half of the caliper in place, whilst the footbrake is gently depressed to force the other piston out of the caliper.

5 It will be necessary to remove the pistons over a tray to collect the spilled hydraulic fluid. It may also be necessary to partially bleed the system to the front wheel installation being overhauled in order to enable the fluid to force the free piston out.

6 As the piston comes out of the caliper block extract the dust seal and once it is completely out, use a plastic knitting needle or thin wooden rod to remove the seal ring in the caliper bore.

7 The piston in the mounting side is removed in the same manner as described in the previous paragraphs. The piston freed first is temporarily re-inserted into its old bore (without seals) and retained by a small G-clamp. The mounting side piston is forced out under hydraulic pressure as its partner was.

8 Thoroughly clean the caliper and pistons in clean hydraulic fluid or methylated spirit. Any other fluids will damage the internal seals between the two halves of the caliper. Note: *Do not separate the two halves of the caliper.*

9 Inspect the caliper bores and piston surfaces for signs of scoring or corrosion which, if evident, means a new assembly must be fitted.

10 To reassemble the caliper, first wet the new piston seal ring with new brake fluid and carefully insert it into its groove in the caliper bore.

11 Then refit the dust cover into its special cover in the caliper bore rim. Release the bleed valve in the caliper by one complete turn.

12 Coat the sides of the piston with hydraulic fluid and then position the piston squarely over the top of the caliper bore. Once in position at the top of the bore, ease the piston gently into the bore until just over $\frac{1}{4}$ in (6 mm) is left protruding from the caliper. Engage the dust seal onto the groove on the rim of the piston and then push the piston as far as it can go into the caliper.

13 Repeat the operations described in paragraphs 10, 11, and 12 to refit the other piston into the caliper.

14 The pads may be fitted to the caliper assembly before it is refitted to the car. In any event bleed the brakes when the complete caliper has been refitted to the car.

11 Master cylinder – removal and refitting

1 Disconnect all the fluid pipes from the master cylinder body and push a cap over the open ends of the pipes to prevent dirt entering the system (Figs. 9.9 and 9.10).

2 Unscrew and remove the two master cylinder flange securing nuts and withdraw the unit from the front of the brake vacuum servo unit.

3 Refitting is a reversal of removal but the hydraulic system must be bled as described in Section 14.

12 Master cylinder – dismantling and reassembly

1 Clean all dirt from the external surfaces of the master cylinder body, taking care that none enters the fluid outlet holes.

2 Remove the reservoir caps and filters and tip out the brake fluid.

3 Extract the circlip from the end of the cylinder body.

4 Lightly grip the cylinder body in the protected jaws of a vice and remove the large screwed plug from the front end of the cylinder. Next remove the two stop bolts from the underside of the cylinder.

5 Refer to Fig. 9.11 and push out the pistons, seals and springs. Ensure the components are kept in the correct order of removal.

6 Clean all the components with clean hydraulic fluid or methylated spirit and inspect the cylinder bore and piston sides for scores or wear. If scores are found replace the whole master cylinder assembly. Do not try to polish out scratches or score marks.

7 Once all the parts have been cleaned, gather them together for reassembly. Do not re-use seals, always use new ones.

8 Begin reassembly by soaking the new seals in brake fluid before slipping the secondary seals onto the rear ends of each piston. Then slip the primary seal cups and the primary seals themselves onto their seating on the forward end of each piston. Make sure the primary seals are correctly aligned with the fine lips towards the forward end of the piston.

9 Insert the rear piston (which is recessed to accept the pushrod from the brake pedal) with the groove downwards, making sure that both primary and secondary seals are not damaged. Insert the first

Chapter 9 Braking system

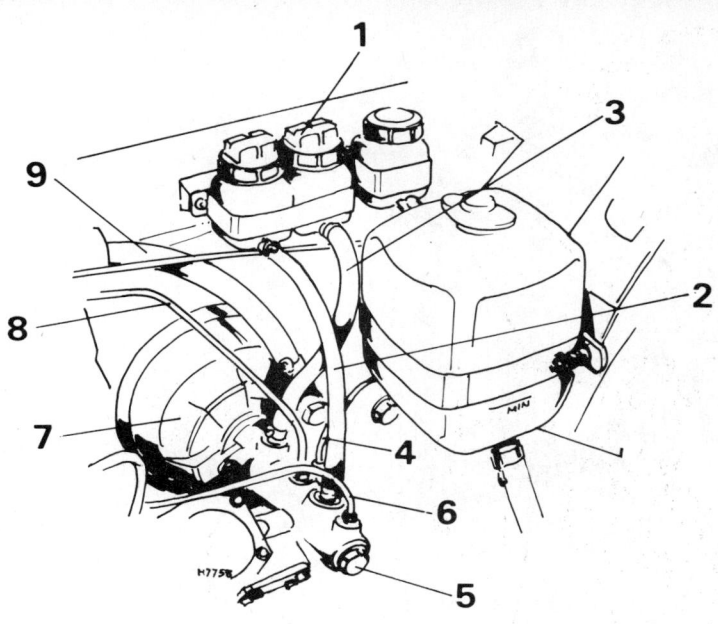

Fig. 9.9 Brake servo and master cylinder connections (Sec 11)

1 Fluid reservoir cap
2 Feed to rear (secondary) hydraulic circuit
3 Feed to front (primary) hydraulic circuit
4 Pipeline to left front caliper
5 Brake master cylinder
6 Pipeline to rear circuit
7 Vacuum servo
8 Pipeline to right front caliper
9 Vacuum hose to servo from intake manifold

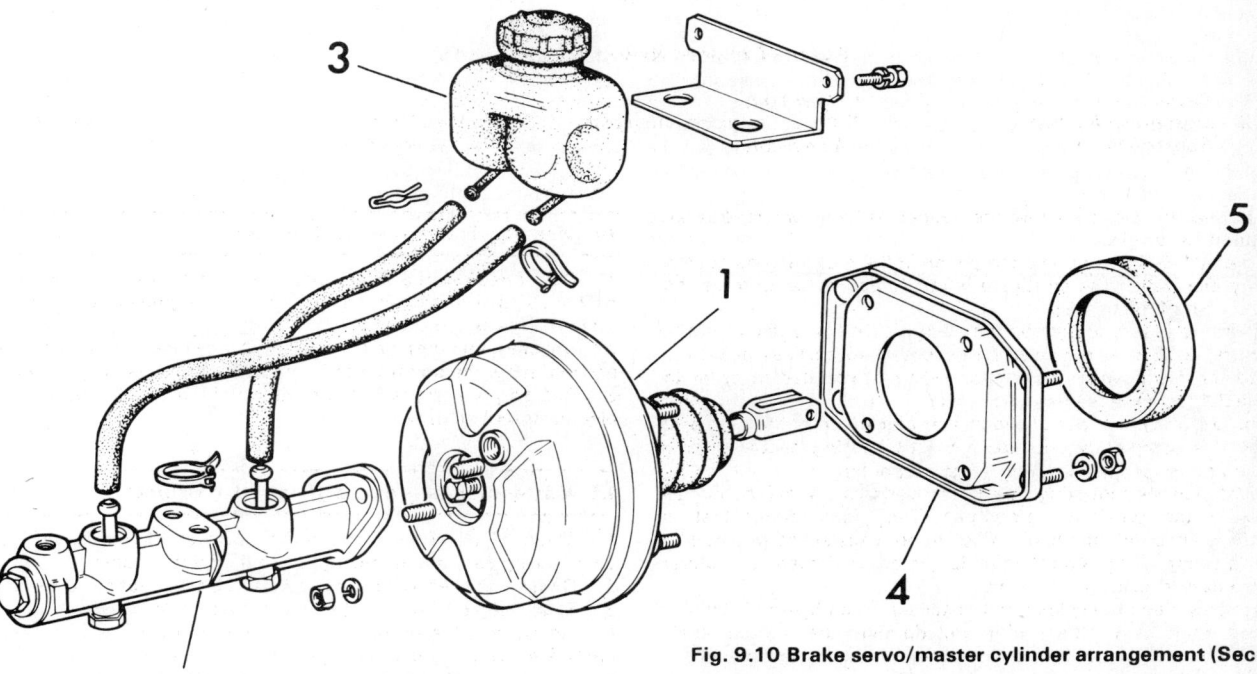

Fig. 9.10 Brake servo/master cylinder arrangement (Sec 11)

1 Servo
2 Master cylinder
3 Reservoir
4 Mounting plate
5 Sealing ring

stop bolt to register into the groove in the piston. The bolt serves to limit the rearward movement of the piston.
10 Drop the buffer and seal spreading springs down the cylinder bore onto the first piston. Follow the springs with the second piston, groove downwards, complete with both its secondary and primary seals. Screw in the second stop bolt into the cylinder. The bolt registers in the groove in the second (forward most) piston and limits the rearward movement of that piston.
11 Finish the assembly by dropping the second buffer and seal spreader springs and then the special end bolt. Screw in the end bolt tightly to seal the threads.
12 If you have removed the inlet unions, they should be refitted now using new star washers and seal.
13 The cylinder is now ready to be refitted to the car.

13 Rear brake pressure compensator valve

1 As mentioned in the introduction, a special valve is incorporated in the hydraulic fluid line to the rear wheels. This is mounted on the underbody. It consists basically of a plunger in a housing which when released reduces the pressure on the outlet side of the valve. The plunger is held in the housing by a rod and link bar attached to the rear axle. When the distance between the body and axle increases, as it would under sharp braking causing a nose down tail up attitude, the link/rod lets the plunger out and the braking pressure to the rear wheels is reduced. This prevents the wheels from locking and consequent skidding.

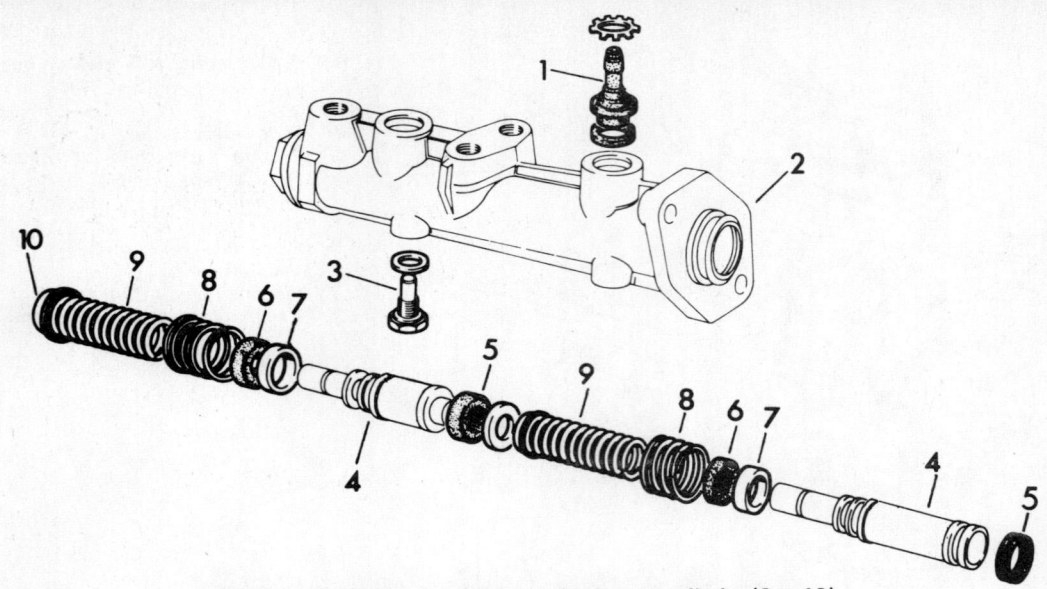

Fig. 9.11 Exploded view of the brake master cylinder (Sec 12)

1 Fluid inlet union	4 Piston	7 Seal retainers	9 Spreader spring
2 Cylinder body	5 Secondary seals	8 Buffer spring	10 End plug
3 Piston stop bolts	6 Primary seals		

2 If the pressure regulator is suspected of malfunction, first check that the operating bar is correctly set. The end of the operating rod which connects to the top end of the link shackle, should be set 7.87 in (200 mm) from the underside of the body. This will involve jacking up the body in relation to the axle and disconnecting it. In this position the other end of the operating bar should just touch the end of the plunger under the rubber dust cover.

3 The plunger position may be altered by slackening the two regulator mounting screws. One hole is slotted so the unit may be tilted as required and tightened in the correct position.

4 The unit itself may be removed from the car after undoing the two hydraulic pipe unions and removing the mounting bolts. It is difficult to test the working properties without proper equipment. Seals can be renewed if the cover plug is taken off but the usual practice is to renew the whole unit (Fig. 9.14).

5 When refitting the brake lines note that the lower port on the regulator is for the line from the master cylinder.

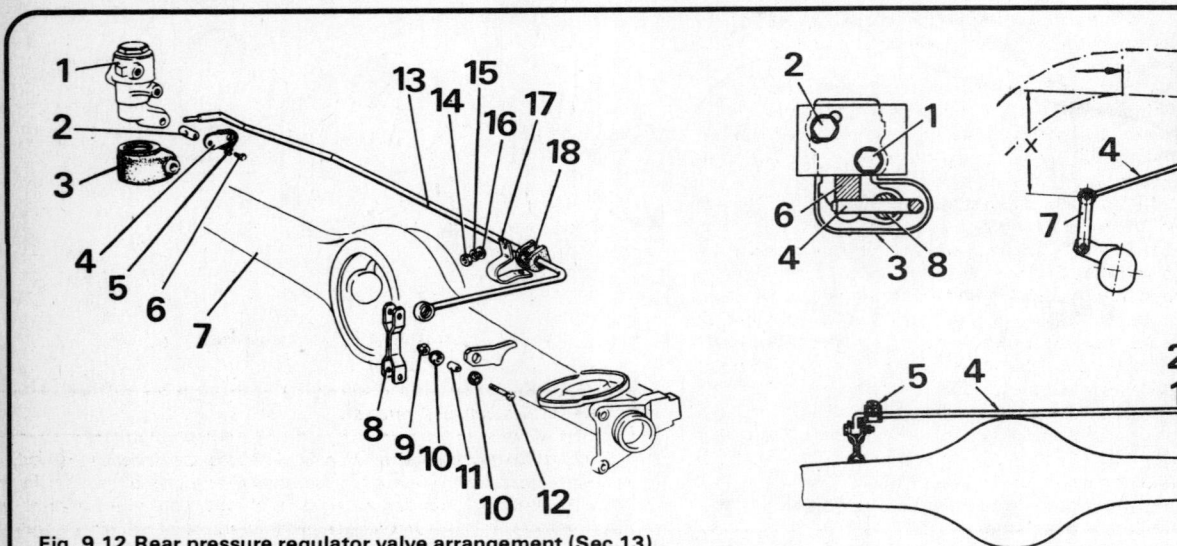

Fig. 9.12 Rear pressure regulator valve arrangement (Sec 13)

1 Regulator valve	10 Bush
2 Shaft	11 Sleeve
3 Boot	12 Link bolt
4 Lockplate	13 Operating rod
5 Spring washer	14 Clamp nut
6 Bolt	15 Spring washer
7 Rear axle casing	16 Plain washer
8 Link	17 Clamp
9 Nut	18 Flexible bush

Fig. 9.13 Rear pressure regulator valve fitting and adjustment diagram (Sec 13)

1 and 2 Mounting bolts	8 Pin
3 Boot	9 Fluid inlet from master cylinder
4 Operating rod	
5 Clamp	10 Fluid outlet to rear wheel cylinders X = 7.87 in (200 mm)
6 Piston	
7 Link	

Chapter 9 Braking system

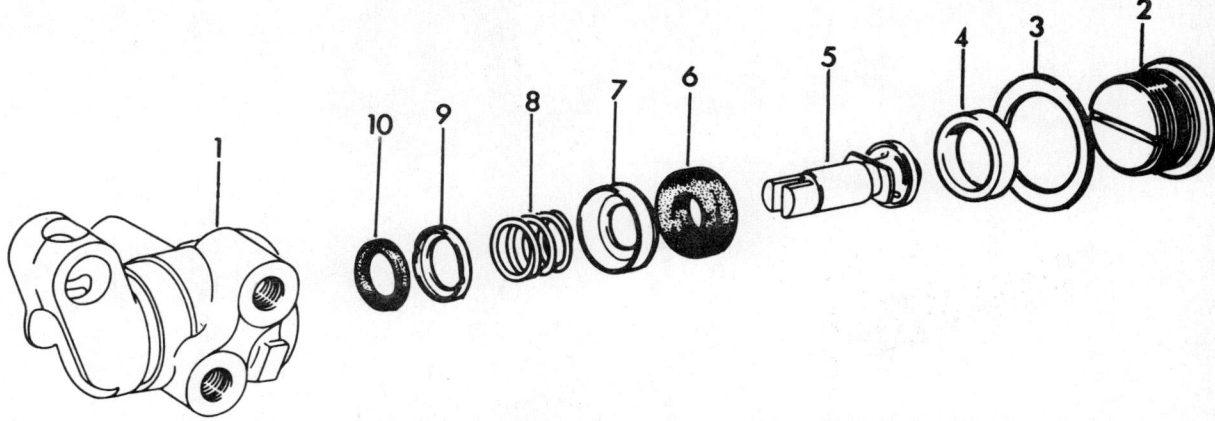

Fig. 9.14 Exploded view of rear pressure regulator valve (Sec 13)

1 Valve body	4 Spacer ring	7 Cup	9 Cup
2 End plug	5 Plunger	8 Spring	10 Secondary seal
3 Sealing washer	6 Primary seal		

14 Hydraulic system – bleeding

1 Removal of air from the hydraulic system is essential to the correct operation of the brakes. Whenever either of the hydraulic circuits has been 'broken' or a component removed and refitted the system must be bled.
2 If the master cylinder has been removed and refitted, initial bleeding should be carried out using the nipples on the master cylinder body.
3 An indication of air in the system is a 'spongy' pedal or when the pedal travel is reduced by repeated applications of the brakes. In the latter case, the trouble may be due to a worn or faulty master cylinder and this should be rectified immediately.
4 If there is any possibility of incorrect fluid having been put into the system, drain all the fluid out and flush through with methylated spirit. Renew all piston seals and cups since these will be affected and could possibly fail under pressure.
5 Gather together a clean jam jar, a length of tubing which fits tightly over the bleed nipples, and a tin of the correct brake fluid.
6 To bleed the system, clean the areas around the bleed valves. Start on the front brakes first by removing the rubber cap over the bleed valve, and fitting a rubber tube in position.
7 Place the end of the tube in a clean glass jar containing sufficient fluid to keep the end of the tube submerged during the operation.
8 Open the bleed valve with a spanner and quickly press down the brake pedal. After slowly releasing the pedal, pause for a moment to allow the fluid to recoup in the master cylinder and then depress again. This will force air from the system. Continue until no more bubbles can be seen coming from the tube. At frequent intervals make certain that the reservoir is kept topped up, otherwise air will enter at this point again. Tighten the bleed valve when the pedal is fully depressed.
9 Continue the operations on the rear brakes.
10 Always discard fluid which has been bled from the system and top-up the system with fluid which has been stored in an airtight container and has remained unshaken for the preceding 24 hrs.

15 Handbrake – adjustment

1 The handbrake is adjusted automatically whenever the rear brake shoes are adjusted. However, due to cable stretch, additional adjustment may be required when the handbrake lever can be pulled more than three notches (clicks) to the full-on position.
2 The adjustment mechanism can be found underneath the car just to the rear of the propeller shaft centre bearing support member.
3 Simply loosen the locknut and turn the adjusting nut next to the cable harness to take up the slack in the cable, until the handbrake lever travel is as desired. Secure the adjusting nut in position by tightening the locknut against the adjusting nut.

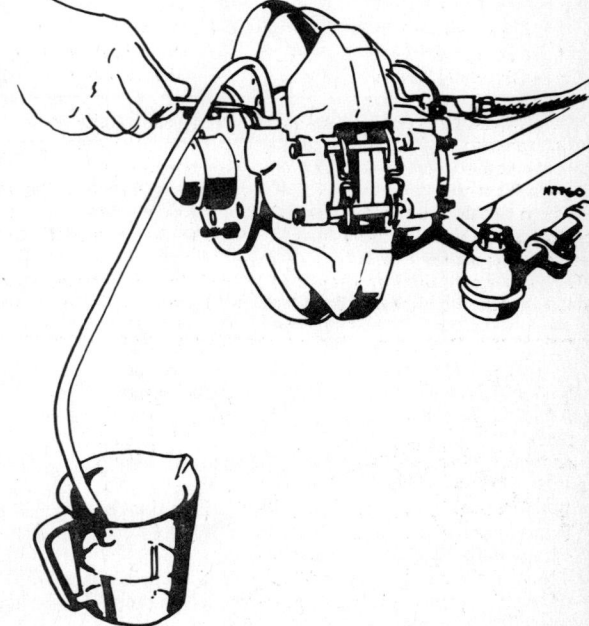

Fig. 9.15 Bleeding a front caliper (Sec 14)

16 Handbrake cable – renewal

1 The cable will need renewal when it has been necessary to adjust it regularly to accommodate an increasing amount of stretch in the cable. Normally the handbrake linkage should only need adjustment once a year but if it becomes necessary to adjust at monthly intervals, then the cable merits renewal.
2 Begin the removal task by raising the rear of the car onto car ramps or chassis stands.
3 Undo the cable adjusting nut and its locknut and unscrew them off the threaded rod from the handbrake lever.
4 Remove the cable equalizer from the threaded rod and lift the cable from it. Undo the nuts that retain the cable sleeve support brackets to the bodyshell and remove the brackets.
5 Remove the rear wheels and brake drums and disconnect the ends of the cables from the shoe operating levers. Detach the cables from the back of the brake anchor plate (photo). Withdraw the cable assembly from beneath the vehicle.

Chapter 9 Braking system

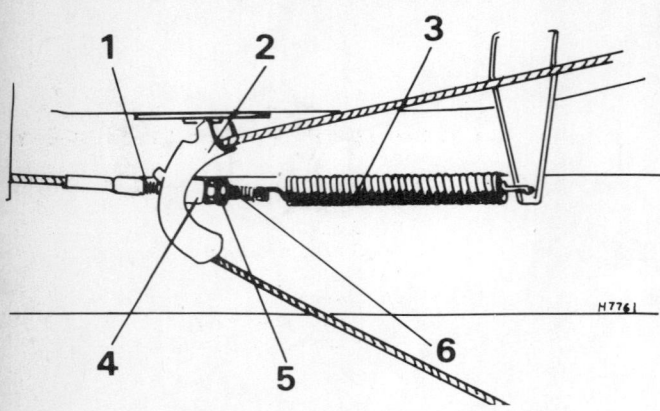

Fig. 9.16 Handbrake cable adjuster (Sec 15)

1 Threaded end fitting on primary cable
2 Cable equalizer
3 Return spring
4 Spacer
5 Locknut
6 Adjuster nut

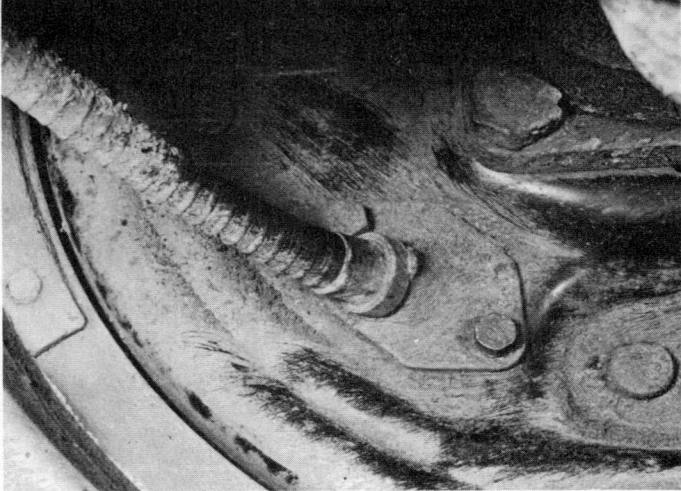

16.5 Entry of handbrake cable on brake backplate

6 Fit the new cable using the reverse of the removal procedure and adjust it as described in Section 15. Grease the cable groove in the equalizer. Lower the car to the ground.

17 Handbrake lever assembly – removal and refitting

1 The handbrake lever assembly exists as three sub-assemblies. The lever and ratchet sub-assembly is the largest, the link cable sub-assembly which joins the lever to the rear cable harness is next and finally there is a pulley sub-assembly round which the link rod runs.
2 Remove the centre trim which covers the lever pivot assembly.
3 Pull out the split pin which retains the link rod to the output lever from the handbrake lever pivot assembly. Remove the link rod.
4 Undo and remove the four bolts that secure the lever pivot assembly to the bodyshell, and lift away the lever assembly.
5 The link cable can now be removed once the adjusting lock nut and return spring are removed from the threaded end of the rod.
6 The link cable pulley runs on a bush on a bolt screwed into a bracket in the transmission tunnel.
7 Once the pulley bolt is removed the bush and pulley can be separated.
8 There is not any scope for repair work on the handbrake lever and its associated linkage; quite simply if it is worn or the cable is stretched, the parts should be renewed.
9 Refitting the handbrake lever assemblies follows the reversal of the removal procedure: remember to adjust the cable linkage as described in Section 15 before returning the car to the road.

18 Brake and clutch pedal assembly – removal, renovation and refitting

1 The brake and clutch pedals are mounted on a substantial channel member which at its lower end is bolted to the engine compartment bulkhead, and at the top end to the dashboard to support the upper section of the steering column.
2 The pedals themselves pivot on bushes running on a common long bolt which passes from one side of the channel to the other.
3 It is not necessary to disturb this channel member when removing either clutch or brake pedal.
4 Begin removal of either pedal by removing the return springs from the pedal assemblies.
5 Next undo and remove the nut on the end of the long bolt and then draw the bolt from the channel.
6 Remove the clutch pedal from the channel and withdraw it complete with push-rod from the clutch master cylinder.
7 The brake pedal can be removed once the clip that retains the servo push-rod pin in place has been removed. The push-rod can then be detached from the pedal assembly.
8 Once the pedals are free, an inspection can be made of the bolt, bushes and spacers. If the bushes are oval and worn, renew them.
9 The brake pushrod and pivot are available as individual spares and therefore both should be inspected for wear.

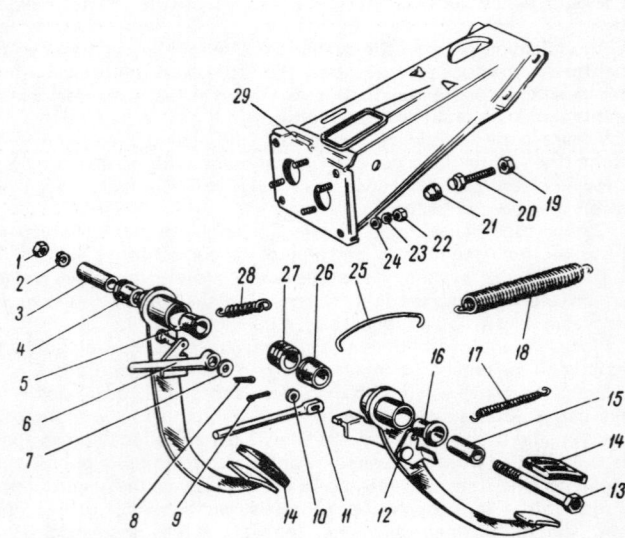

Fig. 9.17 Brake and clutch pedal components (Sec 18)

1 Nut
2 Spring washer
3 Brake pedal spacer sleeve
4 Brake pedal bush
5 Brake pedal bush
6 Pushrod
7 Washer
8 Split pin
9 Split pin
10 Plain washer
11 Clutch pushrod
12 Clutch pedal
13 Brake and clutch pedal shaft
14 Pedal rubber
15 Clutch pedal spacer sleeve
16 Clutch pedal bush
17 Clutch pedal return spring
18 Clutch pedal booster spring
19 Nut
20 Clutch pedal stop bolt
21 Rubber stop
22 Nut
23 Spring washer
24 Plain washer
25 Booster spring staple
26 Clutch pedal bush
27 Spacer sleeve
28 Brake pedal return spring
29 Master cylinder/pedal bracket

Chapter 9 Braking system

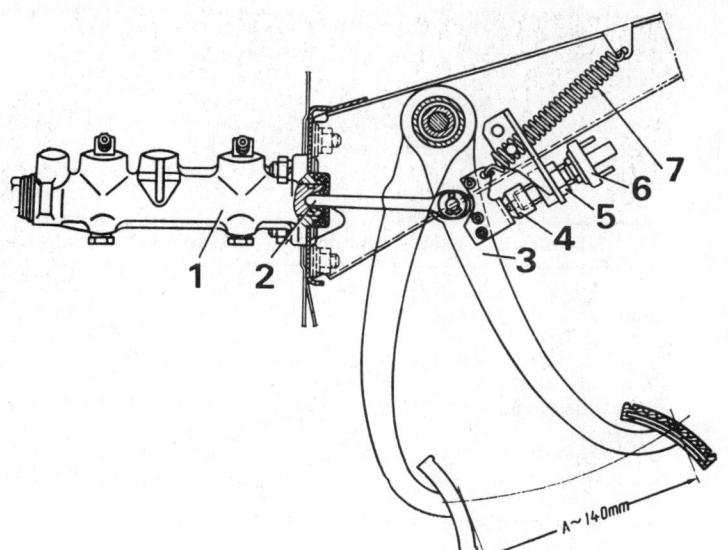

Fig. 9.18 Brake pedal adjustment diagram (Sec 18)

1. Master cylinder
2. Pushrod
3. Brake pedal
4. Stop lamp switch buffer
5. Locknut
6. Switch
7. Pedal return spring
A = Pedal height from floor

10 The reassembly procedure is the reversal of the dismantling procedure, except that it is as well to check the brake pedal free movement which should be between 0.12 to 0.20 in (3.0 to 5.0 mm). Adjust by screwing the stop lamp switch in or out. Check the operation of the brake light switch with the ignition on. **Note:** *Faulty brake stop lamps can sometimes be due to the stop lamp switch wiring being trapped under the seat runner.*

19 Brake servo unit – description

The vacuum servo unit is fitted into the brake system in series with the master cylinder and brake pedal to provide power assistance to the driver when the brake pedal is depressed. The unit operates by vacuum obtained from the induction manifold, and comprises basically a booster diaphragm and a non-return valve.

The servo unit and hydraulic master cylinder are connected together so that the servo unit pushrod acts as the master cylinder pushrod. The driver's braking effort is transmitted through another pushrod to the servo unit piston and its built in control system.

The servo unit piston does not fit tightly into the cylinder but has a strong diaphragm to keep its periphery in contact with the cylinder wall so assuring an air-tight seal between the two parts. The forward chamber is held under vacuum conditions created in the inlet manifold of the engine and during periods when the engine is not in use the controls open a passage to the rear chamber so placing it under vacuum. When the brake pedal is depressed, the vacuum passage to the rear chamber is cut off and the chamber is opened to atmospheric pressure. The consequent rush of air into the rear chamber pushes the servo piston forward into the vacuum chamber and operates the pushrod to

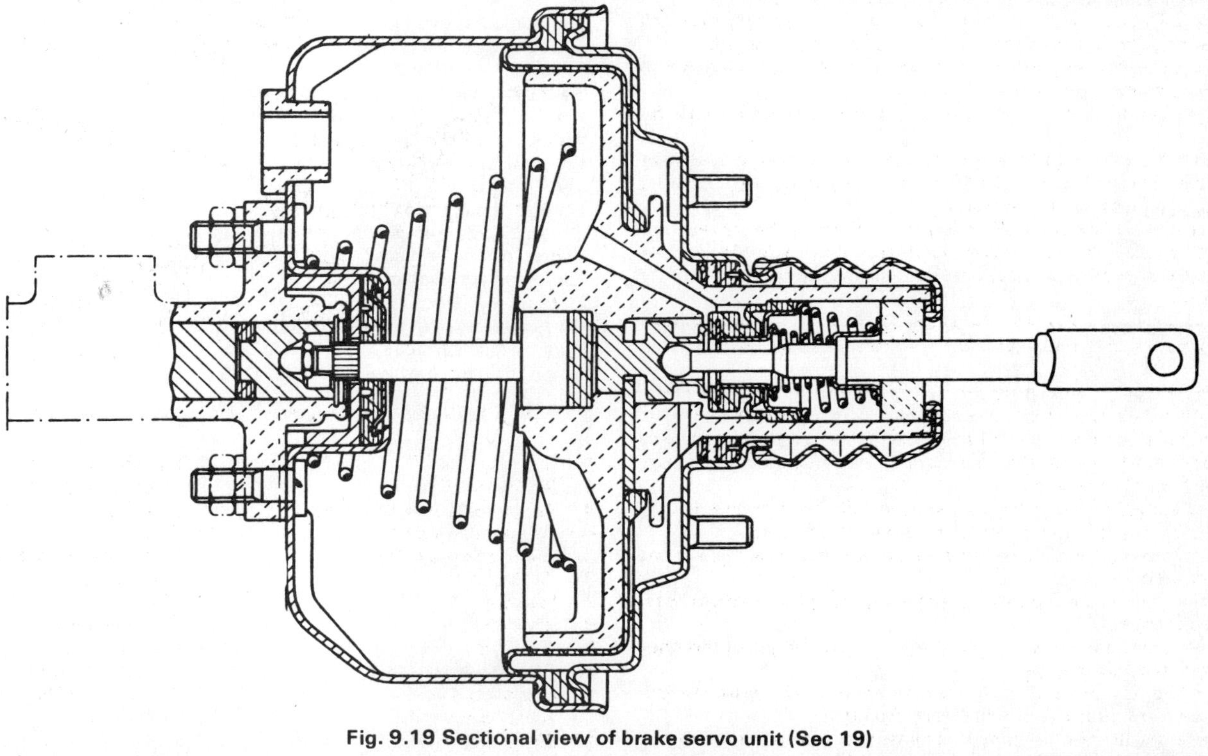

Fig. 9.19 Sectional view of brake servo unit (Sec 19)

the master cylinder. The controls are designed so that assistance is given under all conditions. When the brakes are not required, vacuum is re-established in the rear chamber when the brake pedal is released.

Air from the atmosphere passes through a small filter before entering the control valves and rear chamber and it is only this filter that will require periodic attention. To renew the filter, pull off the boot from the rear of the servo unit and slide it along the pushrod. Extract the circlip and stop plate and cut the air filter diagonally to remove it. Cut the new filter in a similar manner, push it into position and refit the stop plate, circlip and boot.

20 Brake servo unit – removal and refitting

1 Refer to Section 9 of this chapter and remove the brake master cylinder.
2 Slacken the hose clip and remove the vacuum hose from the inlet manifold from the union on the forward face of the servo unit.
3 Next remove the clip that retains the small pin that joins the servo pushrod to the pedal lever. Remove the pin and the pedal return spring and separate the pushrod and pedal.
4 The servo unit is attached to the mounting plate by four nuts on studs in the servo unit. The mounting plate has another four bolts which pass through the rear engine compartment bulkhead and servo to join not only the servo unit to the bulkhead but also the pedal support channel on the rear side of the bulkhead.
5 Undo and remove the four nuts on those studs which join the pedal support channel to the compartment bulkhead. The servo unit and mounting plate can then be lifted away from the forward engine side of the bulkhead. Retrieve the thick seal sandwiched between the mounting plate and bulkhead.
6 Once the unit is free, transfer it to a clean bench and separate the mounting plate.
7 Refitting the brake servo unit follows the reversal of the removal procedure. Remember to use new spring lockwashers and tighten nuts and bolts to their appropriate torques.

21 Brake disc and drum – examination and renovation

1 After a considerable mileage the internal diameter of the rear drums may become out of round, worn beyond the permissible limit, or tapered.
2 Dependent upon whether the brake shoes have been renewed before the rivets have scored the internal surface of the drum; so the drums may require renewal or regrinding particularly if deep scoring is visible.
3 Where any of the foregoing conditions are evident, remove the drums and either renew them or have them prefessionally ground, always provided that the new dimensions do not exceed the tolerances given in the Specifications Section.
4 The appearance of the front discs will show even light scoring which is normal. Any deep grooves will indicate the need for renewal or grinding as will excessive run-out, measured with a dial gauge.
5 Always check the tolerances specified in the Specifications Section before having an original disc ground or refaced. A disc can be detached from the hub after unscrewing the securing bolts.

22 Fault diagnosis – braking system

Symptom	Reason/s
Brake grab	Out of round drums Excessive run-out of discs Rust on drum or disc Oil stained linings or pads
Brake drag	Faulty master cylinder Foot pedal return impeded Reservoir breather blocked Seized caliper or wheel cylinder piston Incorrect adjustment of handbrake Weak or broken shoe return springs Crushed, blocked or swollen pipe lines
Excessive pedal effort required	Linings or pads not yet bedded-in Drum, disc or linings contaminated with oil or grease Scored drums or discs Faulty vacuum servo unit
Brake pedal feels hard	Glazed surfaces of friction material Rust on disc surfaces Seized caliper or wheel cylinder piston

Chapter 10 Electrical system

For modifications, and information applicable to later models, see Supplement at end of manual

Contents

Alternator brushes – renewal	11
Alternator – fault diagnosis and repair	10
Alternator – general description	6
Alternator – removal and refitting	9
Alternator – routine maintenance	7
Alternator – special procedures	8
Battery – charging	5
Battery – electrolyte replenishment and testing	4
Battery – maintenance and inspection	3
Battery – removal and refitting	2
Direction indicators – fault diagnosis	25
Direction indicator switch – removal and refitting	24
Fault diagnosis – electrical system	39
Flasher side repeater lights	31
Front side and flasher light assemblies	29
Fuses	19
General description	1
Headlights and bulbs – adjustment, removal and refitting	28
Headlight switch	27
Horns and horn switch	26
Ignition switch (with steering lock) – removal and refitting	35
Instrument cluster – removal, bulb renewal, and refitting	33
Interior lights (boot, glove compartment, engine compartment, passenger compartment)	32
Radios and tape players – fitting (general)	37
Radios and tape players – suppression of interference	38
Rear light assemblies	30
Relays	36
Speedometer cable	34
Starter motor circuit – testing	13
Starter motor – dismantling, repair and reassembly	15
Starter motor drive pinion – inspection and repair	16
Starter motor – general description	12
Starter motor – removal and refitting	14
Voltage regulator – general description	17
Voltage regulator – maintenance and renewal	18
Windscreen wiper motor and linkage – dismantling and reassembly	22
Windscreen wiper motor and linkage – removal and refitting	21
Windscreen wipers – fault diagnosis	20
Windscreen wiper switches – removal and refitting	23

Specifications

System .... 12V negative earth

Battery .... 55 Ah at 20 hr rate

Alternator
Type .... AC T221
Output .... 42 A at 5000 rpm

Voltage regulator
Type .... PP 380
Voltage setting .... 14·2 ± 0·3 volts

Starter motor
Type .... CT 221 pre-engaged
Rated power .... 1·3 kW

Wiper blades and arms
Arm .... Champion CCA2
Blade .... Champion X-3303

Fuses

Fuse No	Rating	Circuit protected
1 (at left-hand end of fuse block)	16A	Interior lamps, inspection lamp, cigar lighter, horns, stoplamp (not 1500 models), front door open warning lamp
2	8A	Windscreen wiper, heater fan, windscreen washer pump
3	8A	LH headlamp (main beam), main beam pilot lamp
4	8A	RH headlamp (main beam)
5	8A	LH headlamp (dipped beam)
6	8A	RH headlamp (dipped beam)
7	8A	LH front side lamp and pilot lamp, RH tail lamp, LH number plate lamp (except 1600), RH number plate lamp (1600 only), boot lamp, instrument cluster lamps
8	8A	RH front side lamp, LH tail lamp, RH number plate lamp (except 1600), LH number plate lamp (1600 only), under bonnet lamp, cigar lighter lamp

Chapter 10 Electrical system

Fuse No	Rating	Circuit protected
9	8A	Oil pressure warning lamp or gauge, water temperature gauge, fuel level and low level gauge and lamp, handbrake 'ON' lamp, brake fluid level warning lamp, carburettor electromagnetic valve (1500 and 1600), tachometer (1500, 1600 models), glove compartment lamp, choke out warning lamp (1500, 1600), direction indicator circuit, reversing lamps, ignition warning lamp, brake stoplamp (1500 model)
10	8A	Voltage regulator and alternator field winding
11 (1600 early models only)	8A	Spare
12 (1600 early models only)	8A	Spare
13 (1600 early models only)	8A	Spare
14 (1600 early models only)	16A	Spare
15 (1600 early models only)	16A	Radiator fan (in-line fuse on later models)
16 (1600 early models only)	8A	Hazard warning lamps (No 1 fuse on later models)

Bulbs

	Wattage
Headlamp	45/40
Front parking/flasher	21/5
Rear stop/tail	21/5
Rear flasher	21
Rear number plate lamp	5
Under bonnet lamp	5
Interior lamps	5
Side repeater lamp	4
Luggage boot lamp	4
Glove compartment lamp	4
Cigar lighter lamp	4
Instrument illumination and warning lamps	3
Reversing lamp	21
Front door open edge lamps	5

1 General description

The major components of the 12 volt negative earth system comprise a 12 volt battery, an alternator (driven from the crankshaft pulley), and a starter motor.

The battery supplies a steady amount of current for the ignition, lighting and other electrical circuits and provides a reserve of power when the current consumed by the electrical equipment exceeds that being produced by the alternator.

The alternator has its own regulator which ensures a high output if the battery is in a low state of charge and the demand from the electrical equipment is high, and a low output if the battery is fully charged and there is little demand from the electrical equipment.

When fitting electrical accessories to cars with a negative earth system it is important, if they contain silicone diodes or transistors, that they are connected correctly; otherwise serious damage may result to the components concerned. Items such as radios, tape players, electronic ignition systems, electronic tachometer, automatic dipping etc, should all be checked for correct polarity.

It is important that both battery leads are always disconnected if the battery is to be charged while it is in the car. If body repairs are to be carried out using electric welding equipment, the alternator must be disconnected otherwise serious damage can be caused. Make sure that the battery terminals are connected to their correct leads. Whenever working on electrical equipment, it is a good idea to disconnect the battery earth lead.

2 Battery – removal and refitting

1 The battery is situated on the right-hand side of the engine compartment, and is held in place by two tie rods and a pressed steel plate.
2 To remove the battery begin by disconnecting the negative earth lead from the battery and bodyshell. Then disconnect the positive lead from the battery.
3 Once the leads have been removed, the two nuts which tension the tie rods onto the battery retaining plate, may be loosened and the plate moved aside.
4 Lift the battery from its seating in the bodyshell, taking great care not to spill any of the highly corrosive electrolyte.
5 Refitting is the reversal of this procedure. Connect the positive lead first and smear the clean terminal posts and lead clamp assembly beforehand with petroleum jelly in order to prevent corrosion. DO NOT USE ORDINARY GREASE.

3 Battery – maintenance and inspection

1 Normal weekly battery maintenance consists of checking the electrolyte level of each cell to ensure that the separators are covered by $\frac{1}{4}$ inch (6 mm) of electrolyte. If the level has fallen top-up the battery using distilled water only. Do not overfill. If a battery is overfilled or any electrolyte spilled, immediately wipe away and neutralize as electrolyte attacks and corrodes any metal it comes into contact with very rapidly.
2 If the battery has the Auto-fil device fitted, a special topping-up sequence is required. The white balls in the Auto-fit battery are part of the automatic topping up device which ensures correct electrolyte level. The vent chamber should remain in position at all times except when topping-up or taking specific gravity readings. If the electrolyte level in any of the cells is below the bottom of the filling tube top-up as follows:

(a) Lift off the vent chamber cover
(b) With the battery level, pour distilled water into the trough until all the filling tubes are full
(c) Immediately refit the cover to allow the water in the trough and tubes to flow into the cells. Each cell will automatically receive the correct amount of water

3 As well as keeping the terminals clean and covered with petroleum jelly, the top of the battery, and especially the top of the cells, should be kept clean and dry. This helps prevent corrosion and ensures that the battery does not become partially discharged by leakage through dampness and dirt.
4 Once every three months remove the battery and inspect the battery securing bolts, the battery clamp plate, tray, and battery leads for corrosion (white fluffy deposits on the metal which are brittle to touch). If any corrosion is found, clean off the deposits with ammonia and paint over the clean metal with an anti-rust/anti-acid paint.
5 If topping-up of the batery becomes excessive and the case has been inspected for cracks that could cause leakage, but none are found, the battery is being overcharged and the voltage regulator will have to be checked by an automobile electrician.
6 With the battery on the bench at the three monthly interval check, measure the specific gravity with a hydrometer to determine the state of charge and condition of the electrolyte. There should be very little variation between the different cells and if variation in excess of 0·025 is present it will be due to either:

(a) Loss of electrolyte from the battery at some time caused by spillage or a leak resulting in a drop in the specific gravity of

the electrolyte, when the deficiency was replaced with distilled water instead of fresh electrolyte

(b) An internal short circuit caused by buckling of the plates or a similar malady pointing to the likelihood of total battery failure in the near future

7   The specific gravity of the electrolyte for fully charged conditions at the electrolyte temperature indicated, is listed in Table A. The specific gravity of a fully discharged battery at different temperatures of the electrolyte is given in Table B.

Table A
Specific Gravity – Battery Fully Charged
1·268 at 100°F or 38°C electrolyte temperature
1·272 at 90°F or 32°C electrolyte temperature
1·276 at 80°F or 27°C electrolyte temperature
1·280 at 70°F or 21°C electrolyte temperature
1·284 at 60°F or 16°C electrolyte temperature
1·288 at 50°F or 10°C electrolyte temperature
1·292 at 40°F or 4°C electrolyte temperature
1·296 at 30°F or –1·5°C electrolyte temperature

Table B
Specific Gravity – Battery Fully Discharged
1·098 at 100°F or 38°C electrolyte temperature
1·102 at 90°F or 32°C electrolyte temperature
1·106 at 80°F or 27°C electrolyte temperature
1·110 at 70°F or 21°C electrolyte temperature
1·114 at 60°F or 16°C electrolyte temperature
1·118 at 50°F or 10°C electrolyte temperature
1·122 at 40°F or 4°C electrolyte temperature
1·126 at 30°F or –1·5°C electrolyte temperature

4   Battery – electrolyte replenishment and testing

1   If the battery is in a fully charged state and one of the cells maintains a specific gravity reading which is 0·025 or more lower than the others, and where possible a check of each cell has been made with a voltmeter to check for short circuits (a four to seven second test should give a steady reading of between 1·2 to 1·8 volts), then it is likely that electrolyte has been lost from the cell with the low reading, or the battery is nearing the end of its useful life.
2   Adding electrolyte or testing the battery condition for remaining life should be left to your service station.

5   Battery – charging

1   In winter time when heavy demand is placed upon the battery, such as starting from cold, and most electrical equipment is continually in use, it is a good idea to occasionally have the battery fully charged from an external source at the rate of 3·5 to 4 amps.
2   Continue to charge the battery at this rate until no further rise in specific gravity is noted over a four hour period.
3   Alternatively, a trickle charger charging at the rate of 1·5 amps can be safely used overnight.
4   Specialy rapid 'boost' charges which are claimed to restore the power of the battery in 1 to 2 hours are not recommended as they can cause serious damage to the battery plates through over-heating.
5   While charging the battery, note that the temperature of the electrolyte should never exceed 100°F (37·8°C).

6   Alternator – general description

An alternator is fitted as standard equipment on all Lada models. The main advantage of the alternator lies in its ability to provide a relatively high power output at low revolutions. Driving slowly in traffic with a dynamo fitted invariably means a very small or even no charge at all reaching the battery. In similar conditions even with the wipers, heater, lights and perhaps radio switched on the alternator will still ensure a charge reaches the battery. The alternator is of the rotating field ventilated design and comprises principally a laminated stator, on which is wound a 3-phase output winding, and a twelve pole rotor carrying the field windings. Each end of the rotor shaft runs in ball race bearings which are lubricated for life. Aluminium end brackets hold the bearings and incorporate the alternator mounting lugs. The rear bracket supports the silicone diode rectifier pack which converts the AC output of the machine to DC for battery charging and output to the voltage regulator.

The rotor is belt-driven from the engine through a pulley keyed to the rotor shaft. A special centrifugal action fan adjacent to the pulley draws air through the machine. This fan forms an integral part of the alternator specification. It has been designed to provide adequate flow of air with the minimum of noise and to withstand the stresses associated with the high rotational speeds of the rotor. Rotation is clockwise when viewed from the drive end.

The rectifier pack of silicone diodes is mounted on the inside of the rear end casing, the same mounting is used by the brushes which contact the slip rings on the rotor to supply the field current. The slip rings are carried on a small diameter moulded drum attached to the rotor. By keeping the circumference of the slip rings to a minimum, the contact speed and therefore the brush wear is minimised.

7   Alternator – routine maintenance

1   The equipment has been designed for the minimum amount of maintenance in service, the only items subject to wear being the brushes and bearings.
2   Brushes should be examined after about 75 000 miles (120 000 km) and renewed if necessary. The bearings are prepacked with grease for life, and should not require further attention.
3   Check the drivebelt every 6000 miles (9600 km) for correct adjustment which should be 0·5 inch (13 mm) total movement at the centre of the longest run between pulleys.

8   Alternator – special procedures

Whenever the electrical system of the car is being attended to, or external means of starting the engine are used, there are certain precautions that must be taken otherwise serious and expensive damage can result.
1   Always make sure that the negative terminal of the battery is earthed. If the terminal connections are accidentally reversed or if the battery has been reverse charged the alternator diodes will be damaged.
2   The output terminal on the alternator must never be earthed but should always be connected directly to the positive terminal of the battery.
3   Whenever the alternator is to be removed or when disconnecting the terminals of the alternator circuit, always disconnect the battery terminal earth first.
4   The alternator must never be operated without the battery to alternator cable connected.
5   If the battery is to be charged by external means always disconnect both battery cables before the external charger is connected.
6   Should it be necessary to use a booster charger or booster battery to start the engine always double check that the negative cable is connected to negative terminal and the positive cable to positive terminal.

9   Alternator – removal and refitting

1   Disconnect the battery leads.
2   Note the terminal connections at the rear of the alternator and disconnect the wires (photo).
3   Remove the undertray (sump guard) from under the car. Remove the bolt which secures the alternator to the adjusting link (photo), slacken the mounting bolt and push the alternator towards the engine. Remove the drivebelt. There is no need to remove the adjusting link.
4   Remove the lower mounting bolt and lift the alternator away from under the car.
5   Take care not to knock or drop the alternator as it is easily damaged.
6   Refitting the alternator is the reverse sequence to removal.
7   Adjust the drivebelt so that it has 0·5 inch (13 mm) total movement at the centre of the longest run between pulleys.

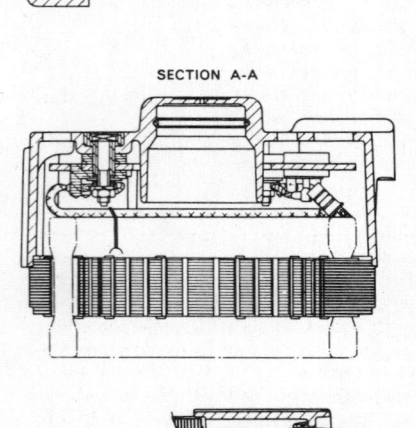

Fig. 10.1 Sectional views through the alternator (Sec 6)

9.2 Electrical connections on rear of alternator

9.3 Alternator adjusting link

Chapter 10 Electrical system

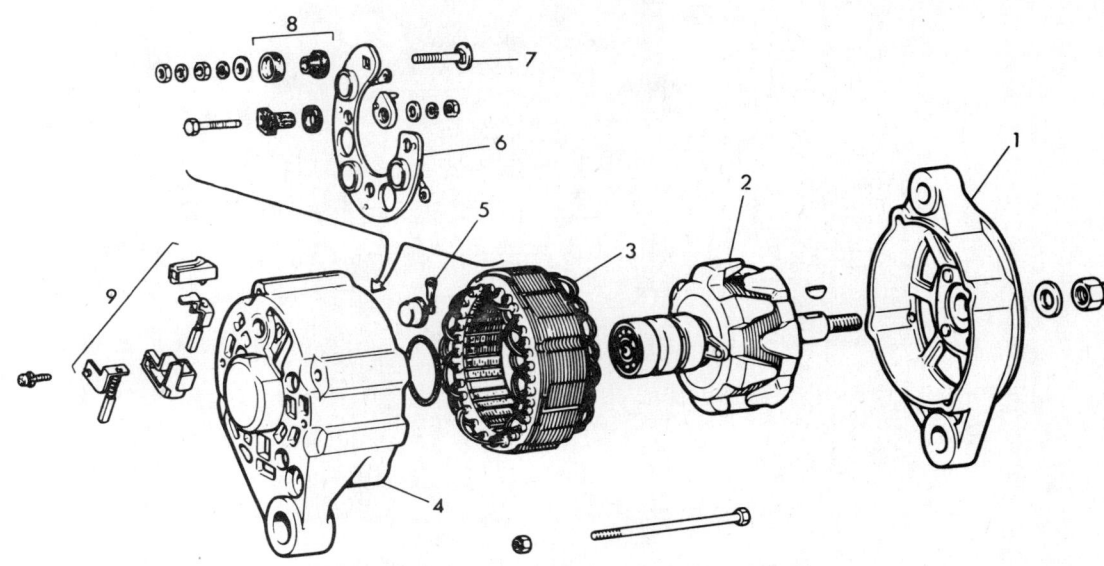

Fig. 10.2 Exploded view of the alternator (Sec 11)

1	Drive end bracket	4	Slip ring end bracket	7	Terminal screw
2	Rotor	5	Silicone diode	8	Insulators
3	Stator	6	Rectifier frame	9	Brushes

10 Alternator – fault diagnosis and repair

Due to the specialist knowledge and equipment required to test or service an alternator it is recommended that if the performance is suspect the car be taken to an automobile electrician who will have the facilities for such work. Because of this recommendation, information is limited to the inspection and renewal of the brushes. Should the alternator not charge or the system be suspect the following points may be checked before seeking further assistance:

(a) Check the drivebelt tension, as described in Section 7
(b) Check the battery, as described in Section 3
(c) Check all electrical cable connections for cleanliness and security

11 Alternator brushes – renewal

1 Remove the alternator from the car as described in Section 9.
2 Remove the crosshead retaining screw then withdraw the brush box from the rear end of the alternator.
3 Check that the carbon brushes are able to slide smoothly in their guides without any sign of binding.
4 Measure the amount by which the brushes protrude from the brush box. If this is less than 0·3 inch (7 mm), obtain and fit new brushes.
5 Refitting the brush box is a straightforward reversal of the removal procedure.

12 Starter motor – general description

1 The starter motor is the pre-engaged type, in which the switch solenoid is also employed to move the drive pinion along the starter motor shaft, into contact with the ring gear on the flywheel, before power is supplied to the motor for turning the engine.
2 There is a spring between the pinion and the actuating lever from the solenoid, so that in the event of an exact abutment of gearteeth as the pinion is impelled to engage with the flywheel ring gear, the solenoid switch will still continue and make power contact. The pinion will fall into engagement as soon as the motor shaft turns.
3 The starter motor is located on the right-hand side of the engine, and it is bolted to the clutch bellhousing. The motor/pinion shaft projects into the bellhousing.

13 Starter motor circuit – testing

1 If the starter motor fails to turn the engine when the switch is operated there are four possible reasons

(a) The battery is discharged
(b) The electrical connections between switch solenoid, battery and starter motor are somewhere failing to pass the necessary current from the battery through the starter to earth
(c) The solenoid switch is faulty
(d) The starter motor is either jammed or electrically defective

2 To check the battery, switch on the headlights. If they go dim after a few seconds the battery is discharged or defective. If the lamps glow brightly, next operate the starter switch and see what happens to the lights. If they go dim then you know that power is reaching the starter motor but failing to turn it. The starter will have to come out for examination. If the starter should turn very slowly go on to the next check.
3 If, when the starter switch is operated, the lights stay bright, then the power is not reaching the starter. Check all connections from battery to solenoid switch and starter for perfect cleanliness and tightness. With a good battery installed this is the most usual cause of starter motor problems. Check that the earth link cable between the clutch housing and frame is also intact and cleanly connected. This can sometimes be overlooked when the engine has been taken out.
4 If no results have yet been achieved turn off the headlights, otherwise the battery will go flat. You will possibly have heard a clicking noise each time the starter switch was operated. This is the solenoid switch operating but it does not necessarily follow that the main contact is closing properly. (**Note**: *if no clicking has been heard from the solenoid it is certainly defective*).The solenoid contact can be checked by putting a voltmeter or bulb across the main cable connection on the starter side of the solenoid and earth. When the switch is operated, there should be a reading or lighted bulb. If not, the solenoid switch is faulty. (Do not put a bulb across the two solenoid terminals. If the motor is not faulty the bulb will blow). If, finally it is established that the solenoid is not faulty and 12 volts are getting to the starter then the starter motor must be defective.

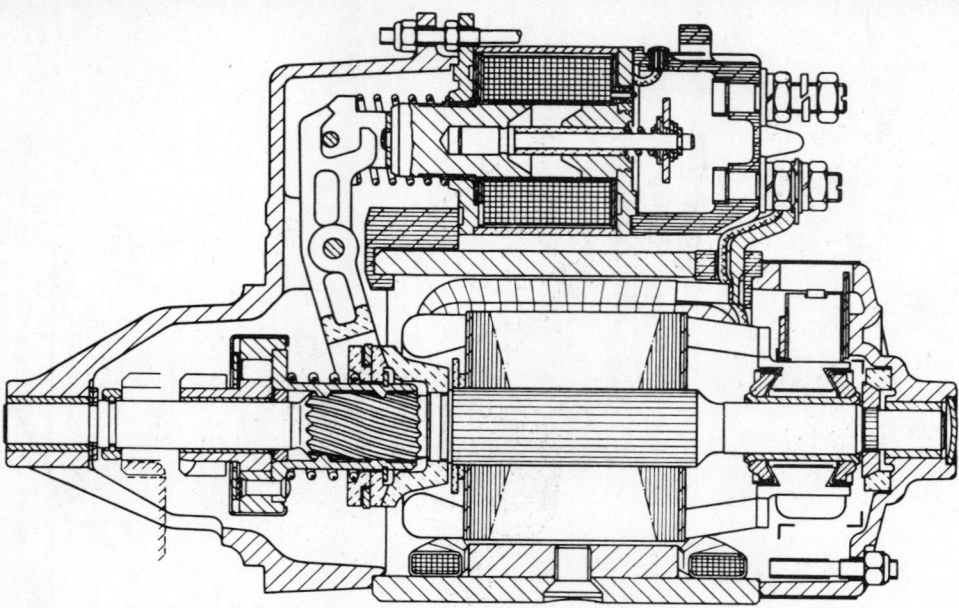

Fig. 10.3 Sectional view of starter motor (Sec 12)

14 Starter motor – removal and refitting

1 Removal of the starter motor is a difficult job and takes a lot of time because of all the other things that have to be taken away first. If you are removing it because it does not work (and if it does not work you can do nothing except remove it first) make quite sure first that it is not just a disconnected lead. One of the leads to the starter is a straight connector and it may be loose or dirty. You will have to feel round behind the exhaust shield plate to get to it. It cannot be seen.
2 Remove the bolts holding the heat shield over the starter motor. Then disconnect the leads from the starter solenoid.
3 Separate the exhaust pipe from the manifold (photo). This will also require the disconnection of the exhaust pipe underneath the car. A new exhaust pipe to manifold flange gasket should also be obtained.
4 The three starter mounting bolts have to be undone. A socket with a long extension and universal joint adaptor will be required. The bolts are accessible from under the car.
5 Withdraw the starter motor upwards or downwards from the engine compartment.
6 Refit the starter motor using the reverse of the removal procedure. As the exhaust pipe was removed, a new gasket should be used when refitting it to the manifold flange.

14.3 Moving the exhaust downpipe to gain access to the starter motor

15 Starter motor – dismantling, repair and reassembly

1 The starter motor assembly comprises three sub-assemblies: The motor itself, the solenoid switch and the pinion actuator housing. The actuator housing forms the mechanical link between the motor and the solenoid switch.
2 Such is the inherent reliability and strength of starter motors that it is very unlikely that a motor will ever need dismantling until it is totally worn out and in need of replacement as a whole.
3 The solenoid which is usually available individually as a spare is attached to the actuator housing by three nuts on three long bolts passing the length of the solenoid. Undo and remove these three end nuts and lift the solenoid from the starter motor assembly.
4 There is no possibility of repairing the solenoid and therefore if after reconnecting across the battery with two stout leads the unit remains lifeless or the switch part fails to work, the whole solenoid must be renewed.
5 *Starter motor brushes:* On the forward end of the motor there is a wide strap, with a single screw to tighten it in position, and it covers the aperture which allows access to the motor brushes.
6 The procedure for inspection and renewal of the brushes is straightforward. With the starter motor removed from the car and on a bench, slacken the single screw which clamps the end strap in position and slip the strap along the motor casing to uncover the brush access apertures.
7 The brushes are retained in their mountings by spiral springs. Move the ends of the spiral spring to allow the carbon brushes to be extracted from their mountings. Undo the small screw which secures the small lead from the brush to its terminal on the forward end fitting and remove the brush from the motor assembly. If the brush is worn to the extent when the spiral spring applies little force, then the brush should be renewed.
8 *Motor dismantling:* Having already removed the solenoid, the front end of the motor is the next unit to be separated from the motor assembly. The motor is held together by tie rods screwed into the actuator housing at the rear end and projecting through the front end

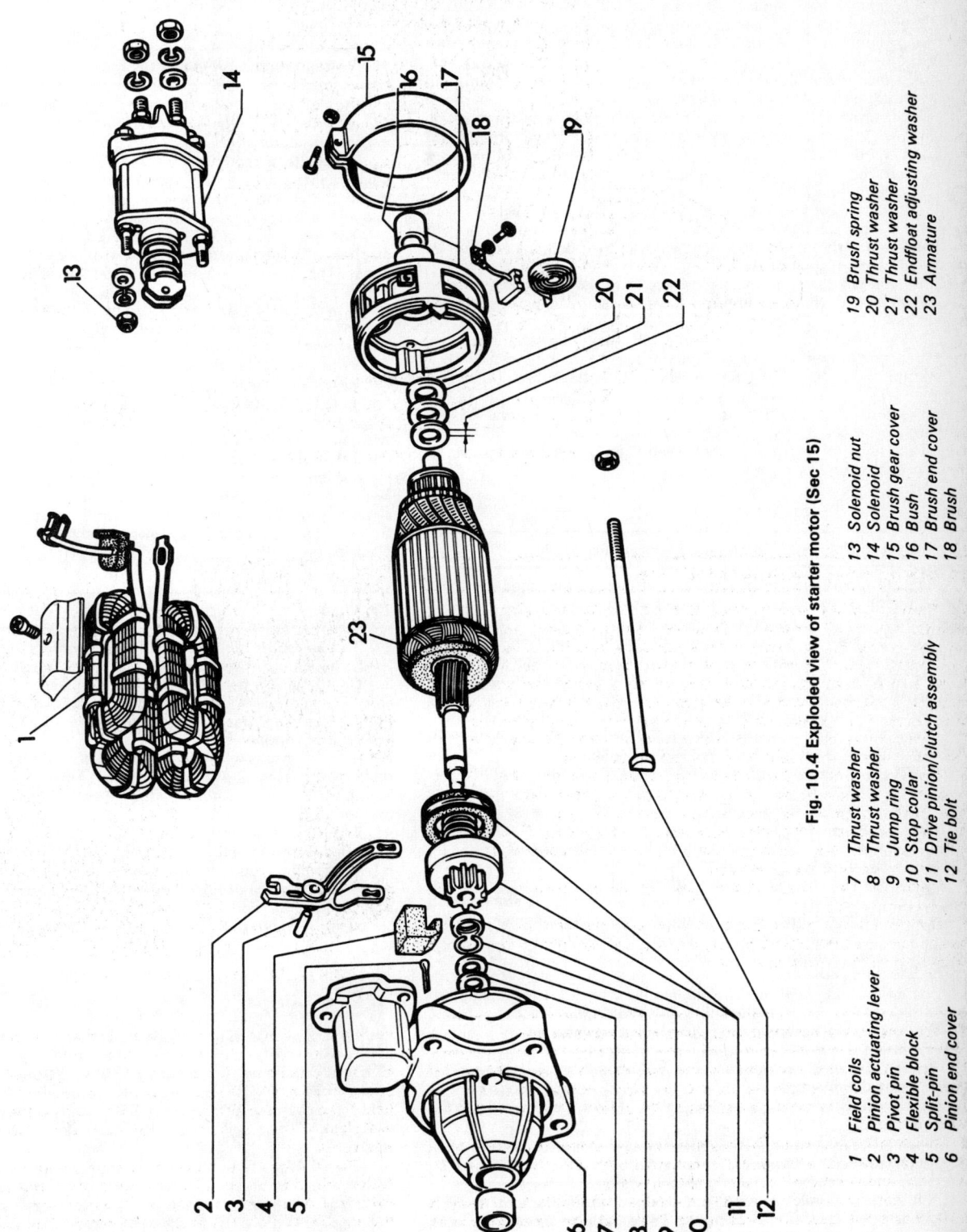

Fig. 10.4 Exploded view of starter motor (Sec 15)

1 Field coils
2 Pinion actuating lever
3 Pivot pin
4 Flexible block
5 Split-pin
6 Pinion end cover
7 Thrust washer
8 Thrust washer
9 Jump ring
10 Stop collar
11 Drive pinion/clutch assembly
12 Tie bolt
13 Solenoid nut
14 Solenoid
15 Brush gear cover
16 Bush
17 Brush end cover
18 Brush
19 Brush spring
20 Thrust washer
21 Thrust washer
22 Endfloat adjusting washer
23 Armature

fitting to accommodate nuts at the front.

9 Before proceeding to separate the motor sub-assemblies it is necessary to disconnect the electrical connections between them. In particular the forward end cover strap and the brushes should be removed. The electrical leads running from the field windings in the motor, to terminals in the forward end cover, should be detached from those terminals.

10 Once the nuts on the tie rods have been removed, the forward end fitting, motor casing and the pinion actuator housing can be separated. The motor armature and drive pinion are mounted on a shaft which runs in bush bearings housed in the actuator housing and forward end fitting. It will be necessary to drive the pinion actuating pivot pin from the actuator housing to permit the separation of the actuator housing and motor armature assembly.

11 Be careful to retrieve the spacer shims and thrust bearings on each end of the motor shaft when the motor shaft is freed. You should refit the spacers and thrust components exactly in the positions which they occupied before dismantling.

12 The field windings are held to the inside of the motor main casing by special blocks which are in turn secured by screws passing through the casing.

13 The armature and pinion assembly will usually be separated from the motor components in order to gain access to the pinion assembly. Inspection and repair of the pinion assembly is described in Section 16.

14 Reassembly of the starter motor follows the reversal of the dismantling procedure. Fortunately there is little in the way of adjustments to make on the motor, the assemblies usually fit together to their correct relative positions.

16 Starter motor drive pinion – inspection and repair

1 Persistent jamming or reluctance to disengage may mean that the starter pinion needs attention. The starter motor should be removed from the car first of all for general inspection.

2 With the starter motor removed, thoroughly clean all the grime and grease off with a petrol soaked rag. Take care to avoid any liquid running into the motor itself. If there is a lot of dirt, particularly on the pinion itself, this could be the trouble. The pinion should move freely along a spiral which is machined on the motor shaft. If the pinion motion is not smooth and easy against the springs which are fitted to the solenoid and pinion carriage to return it to its disengaged position, the motor should be dismantled and the armature/pinion assembly inspected and cleaned as follows.

3 Having removed the armature/pinion assembly the commutator may be cleaned with a petrol dampened rag. The pinion is retained on the motor shaft by a spring ring and sleeve. The sleeve should be driven off the end of the shaft exposing the spring ring which can now be slipped out of its groove seat and off the shaft.

4 Slide the pinion off the rotor shaft and then clean the spiral which is exposed. Wipe the internal spiral in the pinion clean. You should not dismantle the pinion clutch assembly. Individual parts are not available and if the pinion teeth are damaged then the pinion or the whole starter should be renewed.

5 The spiral splines should be lubricated with light grease before reassembly of the pinion to the shaft. The intermediate disc that forms the thrust bearing between the actuating lever ring and the pinion sleeve, should also be lubricated with a light grease.

6 Reassembly of the pinion onto the motor shaft follows the reversal of the removal procedure.

17 Voltage regulator – general description

1 The voltage regulator is specially designed to be compatible with the rectified output from the alternator. The alternator is self limiting as far as the current output is concerned and the regulator's function is to control the output voltage of the alternator. It is of the dual vibrating contact type.

2 The regulator does not incorporate a cut-out to prevent the discharge of the battery through the alternator because the main terminal on the alternator is connected into the diode bridge which apart from rectifying the alternator output also prevents current from flowing into the device from the battery.

3 There is no point in acquiring the equipment necessary to check the regulator, so if it is suspect it will be a job for a qualified auto electrician.

18 Voltage regulator – maintenance and renewal

The voltage regulator is located on the left-hand wing valance and the ignition light relay on the right-hand wing valance within the engine compartment.

1 The voltage regulator is a sealed unit without provision for easy adjustment, and the only maintenance that need and can be done whilst the regulator is operating satisfactorily is to check the wiring contacts to the regulator periodically.

2 The leads should be disconnected one at a time and their ends cleaned with methylated spirit. They may be rubbed clean with emery paper. The contacts on the regulator should receive the same treatment.

3 It will only be necessary to check that the regulator is working properly when checks have been completed on the alternator, cables, battery and terminals and no fault found.

4 When you are satisfied that the regulator must be at fault, take the car to an auto-electrician and he will have the equipment to enable him to tell you very quickly where the malfunction is.

5 It is as well to remember that the regulator is used in conjunction with a relay which is used to switch on or off the red ignition warning light. If your only indication that something is faulty is the glowing of the red ignition light, then the fault might be in the special relay which switches this light. Should the actuating coil in the relay fail, the relay switch will remain closed and the ignition light will glow.

6 The ignition light relay switch is readily checked by an auto electrician.

19 Fuses

1 The various electrical circuits are protected by fuses located in a holder beneath the instrument panel adjacent to the steering column.

2 On all models except early 1600 versions, nine 8 amp and one 16 amp fuses are used. On early 1600 models, thirteen 8 amp and three 16 amp fuses are used. A separate in-line 16 amp fuse is used for the heated rear window.

3 The symptom of fuse failure is the simultaneous failure of a number of electrical systems. The fuse which has blown can then be identified by the combination of electrical systems which do not operate.

4 The Specifications at the beginning of this Chapter detail the identification of each fuse and the circuits protected by them.

5 Never think you can leave fuses out or by-pass them, or substitute a fuse with a piece of tin foil or similar. A fuse blows for a reason and if the fault is not corrected before the fuse is renewed, you will do serious damage to the wiring on the circuit involved.

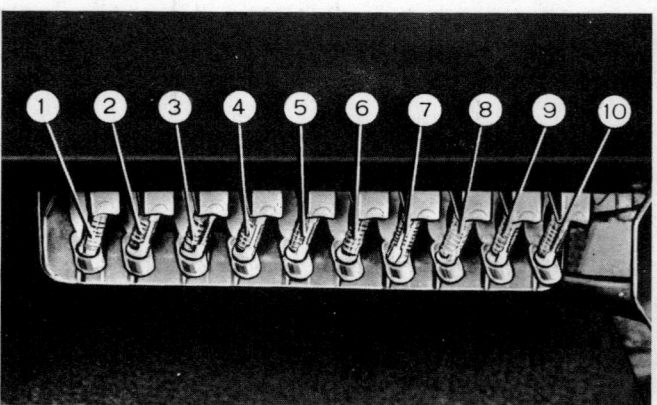

Fig. 10.5 Typical fuse holder (Sec 19)

For key see Specifications

Chapter 10 Electrical system

20 Windscreen wipers – fault diagnosis

1 On earlier Lada models the wiper motor is controlled by an off/intermittent/on switch on the dash panel. A plunger type washer pump is adjacent to the wiper switch. Later models have a combined windscreen wiper and washer stalk switch on the steering column.
2 Switch and lead failure will be indicated when only a particular mode of operation of the wiper is malfunctioning. The duplication of wiper motor activating circuits helps with fault diagnosis. If none of the circuits manage to operate the motor, it is unlikely that all circuits have failed simultaneously; the motor is suspect and should be removed.
3 When the wipers work in only one mode then the suspect circuit should be tested. Check the continuity of the leads in that circuit and the operation of the switch with a small battery, a bulb and some leads. Note the intermittent operation control unit is a separate piece of equipment, mounted adjacent to the wiper motor.
4 If the wipers run too slowly it will be due to something restricting the free operation of the linkage or a fault in the motor. In such cases it is well to check the current used by connecting an ammeter in the circuit. If it exceeds three amps something is restricting free movement. If it is less than three amps then the commutator and brush gear in the motor are suspect.
5 The wiper motor and gearbox are mounted behind the dashboard and operate the individual wipers through a crank linkage which is sufficiently exposed for a foreign object to interfere with the linkage movement (photo).
6 On some models, a relay is included in the wiper motor wiring circuit. It is worth removing the cover and checking that the twin contacts are correctly aligned. If necessary, bend the contacts or release the contact mounting bracket screw.

20.5 Location of windscreen wiper motor

21 Windscreen wiper motor and linkage – removal and refitting

1 Although it is possible to separate the motor from the wiper linkage in situ, it is advisable to remove the linkage and motor together. It is difficult to disconnect the motor from the mounting and linkage whilst it is still in position in the vehicle.
2 Begin by removing the windscreen wiper arms from the splined hub. Then undo and remove the nut and chromed trim with gasket which retains the wiper spindle housing on the bodyshell (photos).
3 Once the wiper spindle housings have been freed, the bolts which secure the motor mounting to the bodyshell bracket can be removed to allow the motor/wiper assembly to be lifted clear of the car.
4 The electrical plug connector should be separated as the wiper assembly is removed.
5 Refit the motor and linkage using the reversal of the removal procedure.

21.2A Removing the retaining bolt from wiper arm spindle

22 Windscreen wiper motor and linkage – dismantling and reassembly

1 The motor and gearbox unit is held to its mounting plate by three bolts which pass through rubber grommets to provide a flexible mounting for the motor.
2 If the motor is known to be faulty, there is no point in trying to effect a repair, a new motor/gearbox assembly should be purchased.
3 If the problem was one of sloppy operation with a great deal of slack in the operating mechanism, the bearing bushes at each end of both link rods should be closely inspected. The bearings are not renewable individually and it will be a matter of renewing the appropriate link rods. It should be remembered that if the bearings were worn, then so the pins on which the bearings run could also be worn.
4 Still with the investigation into sloppy operation, it is worth removing the top of the gearbox so that the condition of the worm driven gearwheel can be inspected. This main gearwheel should be available as a spare and may be renewed if found to be worn.
5 The wiper spindles are not renewable individually and if the spindles are a loose fit in their housings, it will be necessary to renew the whole main framework. This framework comprises both spindles, the housings and spacing member and the motor mounting.
6 Where unsatisfactory automatic parking is encountered, one of the following adjustments may be carried out with the motor and linkages in position on the vehicle.

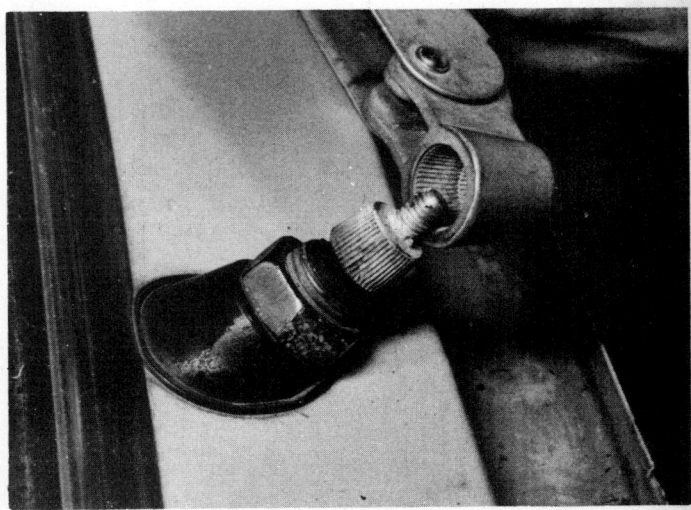
21.2B Lifting off the wiper arm

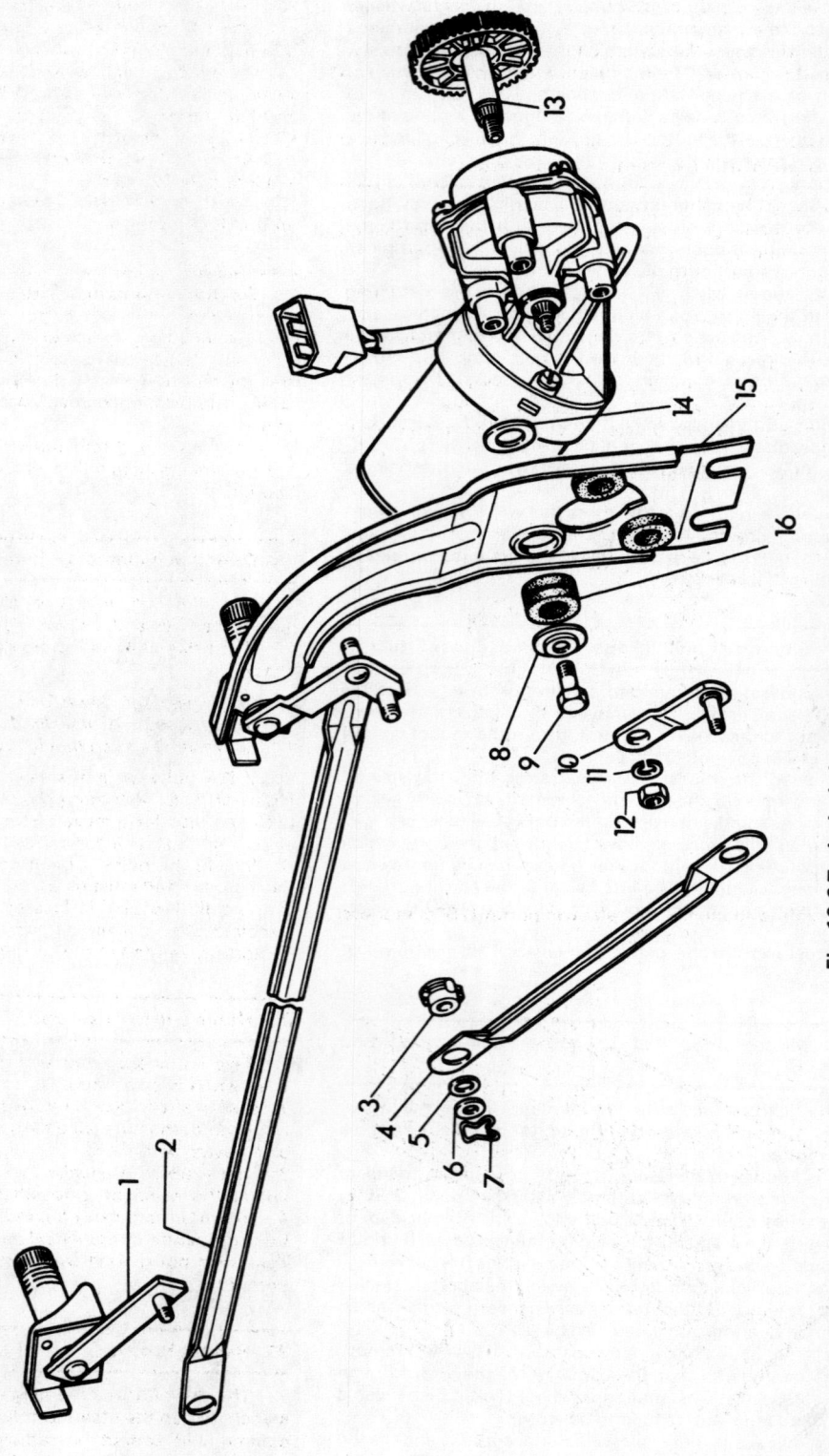

Fig. 10.6 Typical windscreen wiper assembly (Sec 22)

1 Drive spindle and crankarm
2 Link arm
3 Bush
4 Short link
5 Spacer
6 Washer
7 Clip
8 Washer
9 Screw
10 Crankarm
11 Spring washer
12 Nut
13 Drivegear
14 Washer
15 Mounting plate
16 Flexible bush

Chapter 10 Electrical system

7   *Larger motor with circular armature casing:* Remove the plastic cover and black plastic plate holding screws and rotate the plate until the correct parking angle is achieved when the wipers are switched off using the wiper switch.
8   *Small motor with cup type armature screwed to baseplate with two screws:* Remove the connecting link (2 in – 50 mm) by unscrewing the nut from the drive spindle located under the bulkhead. Position the wiper arms in the correctly parked position and refit the link and the nut.

23 Windscreen wiper switches – removal and refitting

1   Several types of switch have been employed to operate the wiper system. Currently the wiper switch is mounted on the steering column and is stalk operated. In order to remove switch units mounted on the steering column proceed as follows.
2   Disconnect the negative terminal from the battery.
3   Remove the steering wheel as directed in Chapter 11.
4   Undo and remove the screws securing the column switch half covers to the switch frame, and remove the covers.
5   Slacken the switch unit retaining strap which secures the lower end of the unit to the upper steering column support bracket.
6   Disconnect the block cable connectors that join the switch units to the vehicle electrical loom.
7   Slide the switch unit off the steering column.

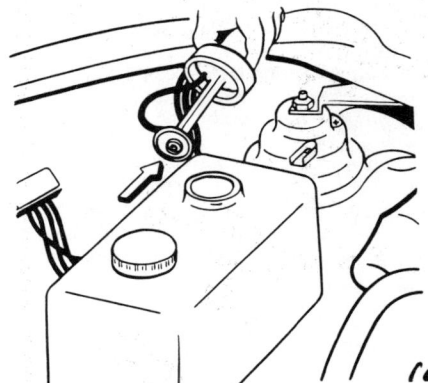

Fig. 10.7 Washer fluid reservoir and electric pump (1600 models) (Sec 23)

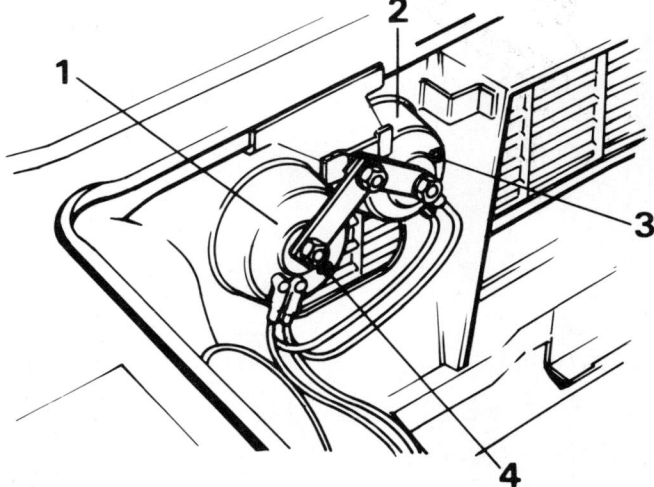

Fig. 10.8 Typical horn arrangement (Sec 28)

1	High tone	3	Adjusting screw
2	Low tone	4	Adjusting screw

8   The steering column switch unit is renewable only as a complete assembly.
9   Refitting the switch unit follows the reversal of the removal procedure.
10 *Dashboard mounted switches and controls:* Removal of switches from the dashboard is quite straightforward since they are a push-in spring fit in the apertures in the dashboard.
11 Testing is straightforward again, requiring the switch to be connected in series with a bulb and small battery. The light should come on when the switch is turned on.
12 *On 1600 models,* the wiper switch also operates the electric washer pump. Pull the switch control lever towards the steering wheel to actuate the washer.
13 The washer pump is located in the pick-up tube/filler cap assembly on the fluid reservoir.

24 Direction indicator switch – removal and refitting

1   The direction indicator switch is incorporated in the steering column group of switches. It is stalk operated. The various switches are not renewable individually and it will be necessary therefore to renew the whole column switch assembly if any one switch has broken.
2   To remove and refit the steering column switch assembly follow the operations detailed in paragraphs two to nine of the preceding Section.

25 Direction indicators – fault diagnosis

1   The direction indicator circuit comprises a flasher unit, the indicator switch and indicator lights connected in series in that order.
2   If the direction indicators do not operate, carry out the following checks:

  (a) Inspect for blown fuse
  (b) Test security of all leads and connections
  (c) Inspect switch mechanism

3   If the indicator lamps flash too slowly or too quickly or the fascia indicator lamp does not go out, check the lamp units for a burnt out bulb and also for a loose connection. If the flashing cycle is irregular, check for a bulb of incorrect wattage.
4   Where the bulbs, switch and wiring are found to be in order, the flasher unit itself must be at fault.
5   The flasher unit is located beneath the dash panel and can be removed after pulling off the wire connectors. Make sure the connectors are refitted to the correct terminals on the new flasher unit.

26 Horns and horn switch

1   Two horns are fitted, one giving a high and the other a low note. The switch is operated by a ring or pad on the steering column and is accessible after the ring has been removed.
2   Both horns have adjusting screws in the back and the volume can be adjusted with these.
3   Horn failure is usually due to a fault in the spring-loaded contact ring on the switch or corrosion on the switch contact surfaces.
4   The horn ring can be removed by first prising off the plastic cover from the centre of the wheel and undoing the retaining screws (photo). Take care not to lose the springs on each screw when replacing the horn ring.

27 Headlight switch

1   The main/dipped headlight selector switch is mounted in the switch unit on the steering column. Individual switches in this steering column unit cannot be removed and refitted easily and it will be necessary to remove the whole switch unit as described in Section 23 of this Chapter.
2   The supply to the headlight switch comes from the main light switch (for sidelights as well) mounted on the dashboard.
3   The rocker type switch is a spring fit in the dash panel and can be prised out for checking or renewal (photo).

Chapter 10 Electrical system

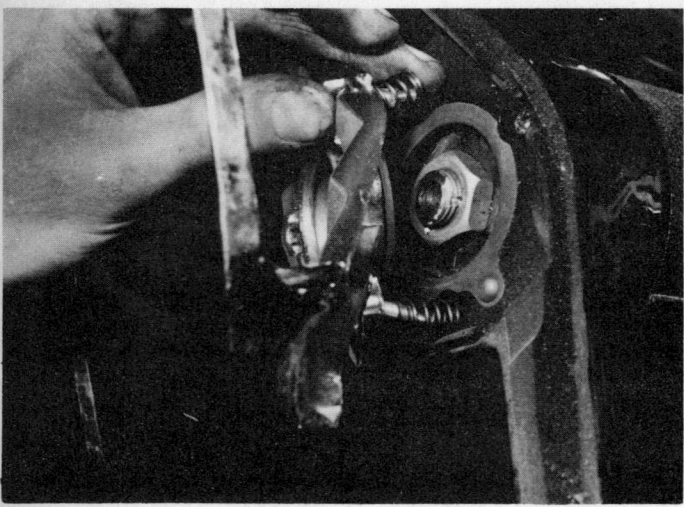

26.4 Removing the horn ring from the steering wheel – earlier models

27.3 Removing a rocker type switch

28 Headlights and bulbs – adjustment, removal and refitting

1 The headlamp assemblies must be removed to change the bulbs. First undo the screw holding the bezel at the top. On twin headlamp models there are two screws. Draw the bezel forward at the top a little and unhook it at the bottom (photos).

2 On all models the headlamp lens is secured by a retaining ring. Slacken the three screws on each lamp which hold the ring. Then turn the ring clockwise so that it can be detached together with the lamp. Do not confuse the three clamp screws with the two beam adjusting screws on each lamp.

3 To remove the bulb, first disconnect the terminal block from the bulb and then twist out the springs to release the bulb (photo).

4 Refitting is a reversal of the removal procedure. Do not use bulbs of different wattage values from those specified or the balance of the system will be upset.

5 Headlamp beam alignment is best carried out at a garage with the proper equipment. However, the owner is able to do reasonably accurate setting himself, provided he can arrange to line up the car on level ground facing a screen at a distance of $16\frac{1}{2}$ ft (5 m). The car should be unladen and the tyre pressures and suspension standing heights correct.

6 On the screen mark crosses in line with each headlamp centre. Taking single headlamp models first of all, mark a point (P) $4\frac{3}{4}$ in (12 cm) below each cross. With the headlamps switched on low beam the

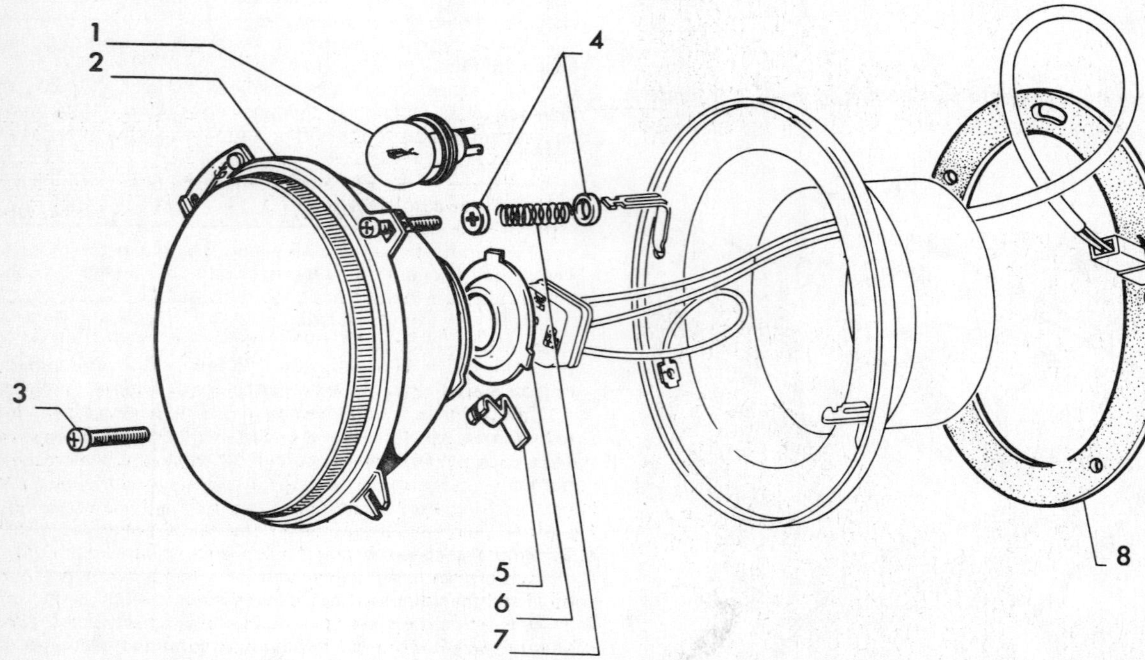

Fig. 10.9 Typical headlamp components (Sec 28)

| 1 Bulb | 3 Retaining screw | 5 Bulb holder clip | 7 Beam adjuster coil spring |
| 2 Lamp unit | 4 Beam adjuster screw | 6 Bulb holder | 8 Gasket |

Chapter 10 Electrical system

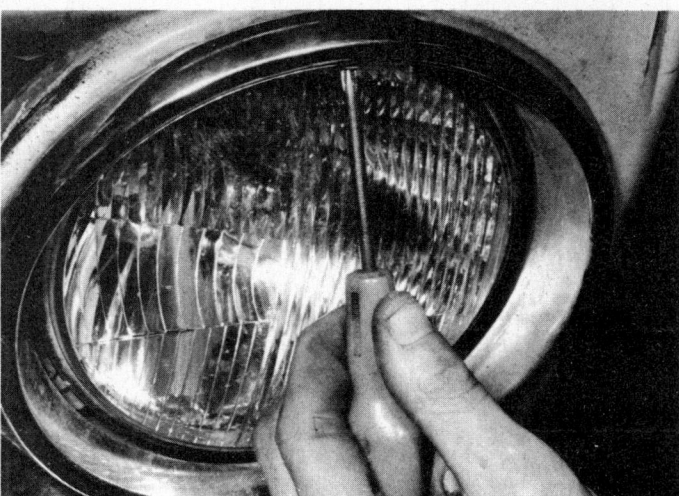

28.1A Removing the headlight bezel retaining screw

28.1B Lifting away the bezel

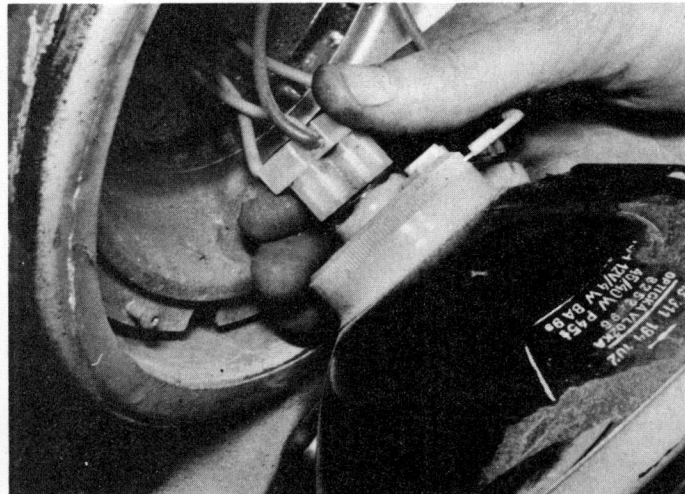

28.3 Disconnecting the terminal plug from the headlight unit

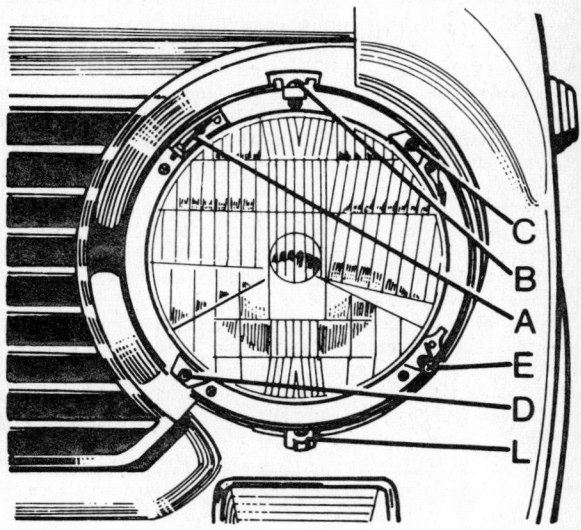

Fig. 10.10 Single headlamp retaining and adjusting screws (Sec 28)

A Spring retainer
B Bezel screw
C Beam adjuster screw (horizontal)
D Beam adjuster screw (vertical)
E Lock screw
L Lower retaining lug

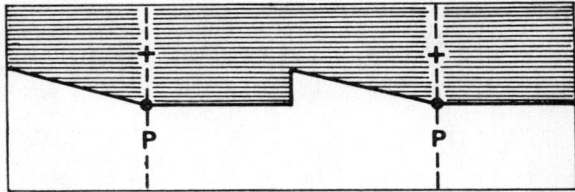

Fig. 10.11 Headlamp beam alignment diagram (RHD single) (Sec 28)

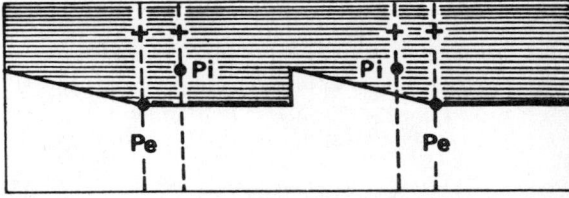

Fig. 10.12 Headlamp beam alignment diagram (RHD twin) (Sec 28)

light zone should correspond with this point as shown in Fig. 10.11.

7 Dual headlamps are arranged so that only the outer pair illuminates on dip. Having marked the four crosses on the wall, mark the outer lamp reference points 4 in (10 cm) below the crosses and the inner lamp reference points 2 in (5 cm) below the crosses. With the lamps on dip the outer lamp light zones should line up as indicated in Fig. 10.12. With all four lamps on, the centre points of the lamp pools of light must correspond with the reference points (Pe and Pi).

8 There are two lamp adjusting screws to each unit, one altering the vertical and the other the horizontal settings.

9 The beam setting charts show the light zones for cars where traffic drives on the left. For right-hand drive countries the angled lines from the reference points run to the right instead of the left.

10 Some models have quartz halogen headlamp bulbs. Take care not to touch the glass envelope of this type of bulb with the fingers as grease deposits will shorten the life of the bulb. If the bulb glass is accidentally touched with the fingers, clean it with methylated spirits before switching on the lamp.

Chapter 10 Electrical system

29 Front side and flasher light assemblies

1 Front parking light bulbs are accessible after removing the two lens mounting screws (photo). These bulbs are double filament, incorporating the flasher as well. The bayonet fitting is offset so that the bulb only goes in one way. This ensures that the 5 watt filament is used for the parking light circuit and the 21 watt filament for the flasher.

2 Cleanliness with electrical systems is essential. Clean the bulb terminals and holder terminals. The lens gasket should always be in good condition; this is essential to prevent foreign matter and water entering the light assembly and corroding the electrical connections.

3 The majority of electrical faults on a car are caused by the deterioration of electrical connections due to corrosion resulting from the entry of dirt and water into the electrical system components.

4 Since the bulbs rely on the electrical connection of the light assembly to the body, if corrosion has occurred it will be necessary to remove the basic shell of the light assembly from the body and clean the connection areas.

30 Rear light assemblies

1 The direction indicator bulb and the stop/tail light bulb on saloon cars are housed in a common cluster. The two separate lenses are interlocked and secured by a single central screw. When the screw is removed the lenses may be drawn out from the locating slots at each end. The bulbs can then be removed. The stop/tail lamp bulb is a double filament one with offset bayonet pins. It is the same as the side/flasher bulb used at the front.

2 *On estate cars,* the single rear light unit can be removed by undoing the top retaining screw and lifting the unit upwards to release it (photo).

3 To gain access to the bulbs, unscrew the plastic knob from the rear of the light unit and remove the lens. The bulbs can then be removed (photos).

4 *The number plate lamp* on 1200, 1300 and 1500 saloons is mounted in the rear bumper. The bulb holder snaps in and can be pulled out from underneath. *On 1600 saloons,* the number plate lamp is part of the rear lamp cluster. *On estate models,* the number plate

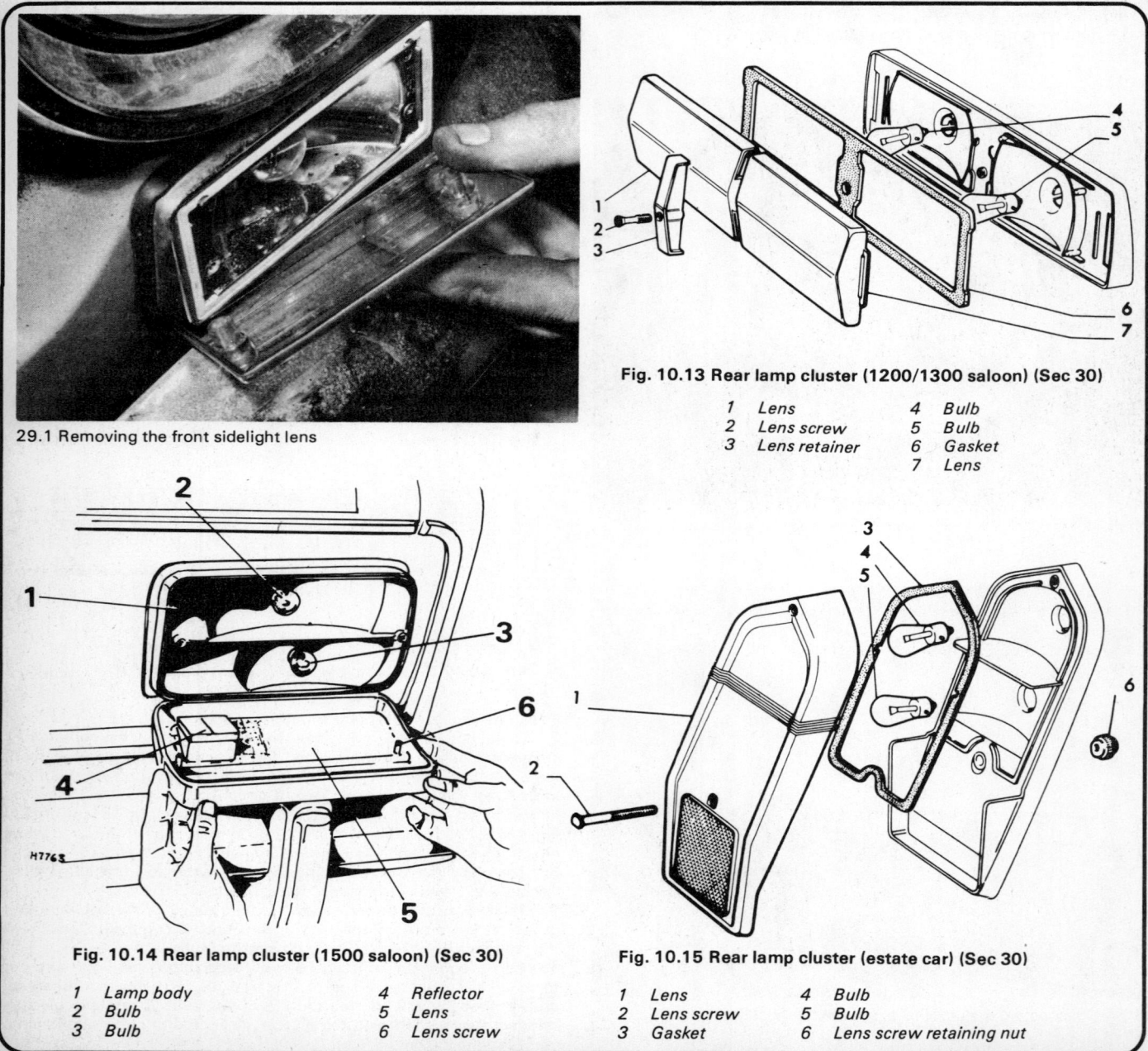

29.1 Removing the front sidelight lens

Fig. 10.13 Rear lamp cluster (1200/1300 saloon) (Sec 30)

1 Lens
2 Lens screw
3 Lens retainer
4 Bulb
5 Bulb
6 Gasket
7 Lens

Fig. 10.14 Rear lamp cluster (1500 saloon) (Sec 30)

1 Lamp body
2 Bulb
3 Bulb
4 Reflector
5 Lens
6 Lens screw

Fig. 10.15 Rear lamp cluster (estate car) (Sec 30)

1 Lens
2 Lens screw
3 Gasket
4 Bulb
5 Bulb
6 Lens screw retaining nut

Chapter 10 Electrical system

30.2 Removing the rear light retaining screw on estate car

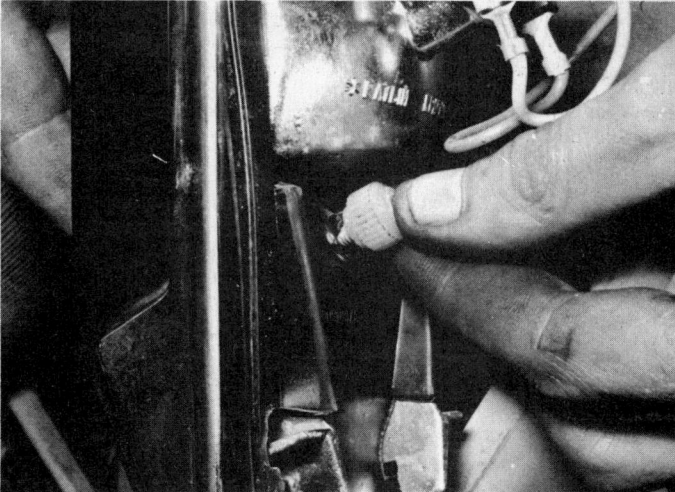
30.3A Removing the rear light lens retaining screw from the inside of the light unit (estate car)

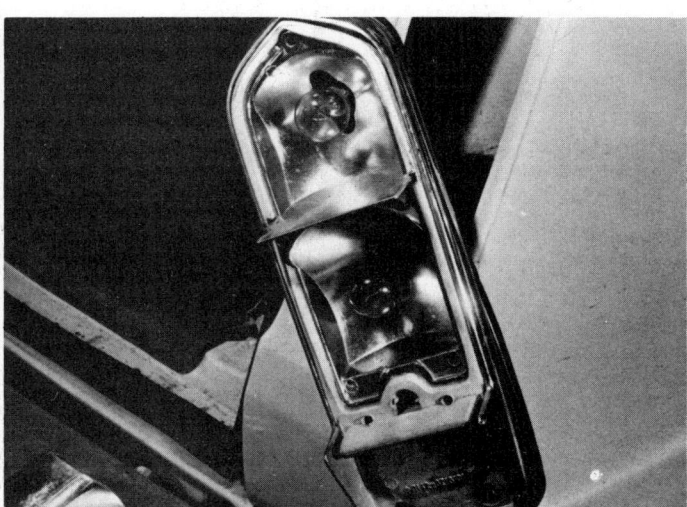
30.3B Rear light unit with lens removed (estate car)

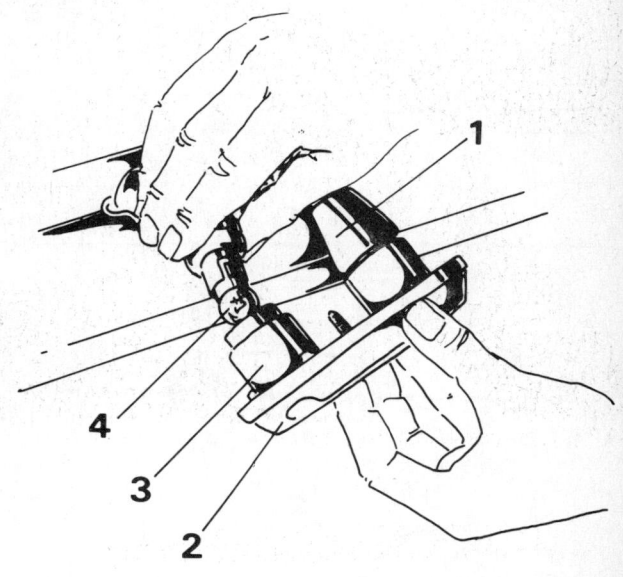

Fig. 10.16 Removing number plate lamp bulbs (saloon except 1600) (Sec 30)

1 Rubber dust excluder   3 Bulb case
2 Lamp body and lens   4 Bulb

light bulb can be renewed after removing the two lens retaining screws (photo).

5 *The reversing lamp bulbs* are accessible after removing the lens which is retained by screws.

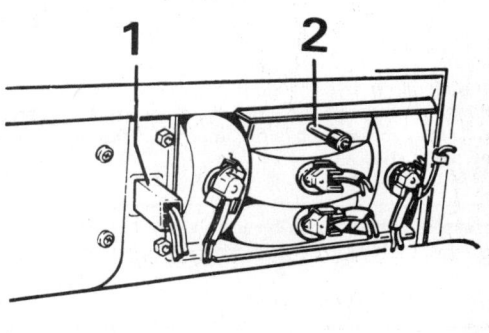
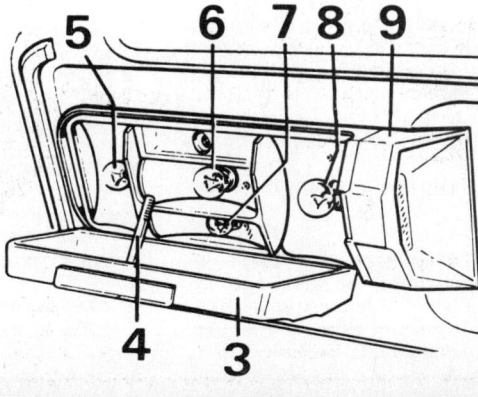

Fig. 10.17 Combined rear lamp and number plate lamp cluster (1600 saloon) (Sec 30)

1 Plug
2 Nut
3 Lens
4 Lens screw
5 Direction indicator bulb
6 Reverse lamp bulb
7 Tail lamp bulb
8 Stoplamp bulb
9 Number plate lamp

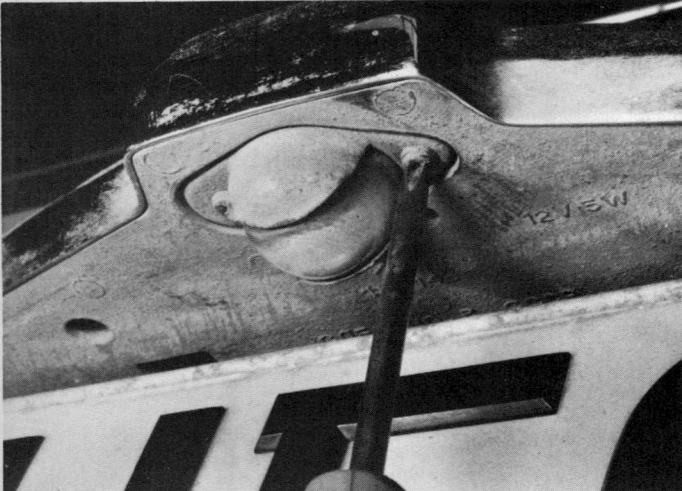

30.4 Removing number plate light retaining screws on estate car

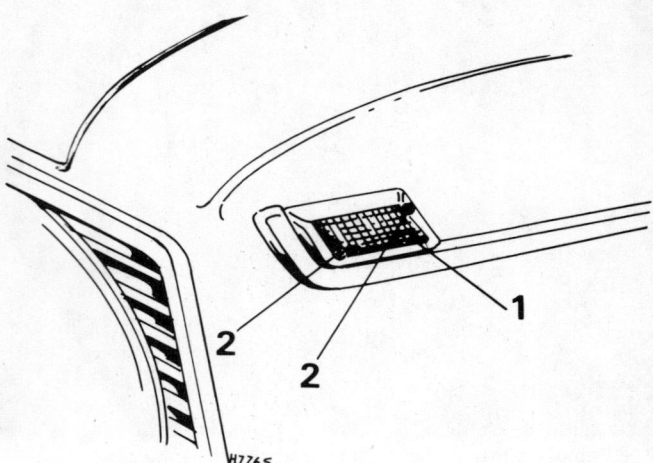

Fig. 10.19 Typical side flasher repeater lamp (Sec 31)

1 Lens   2 Lens retaining screws

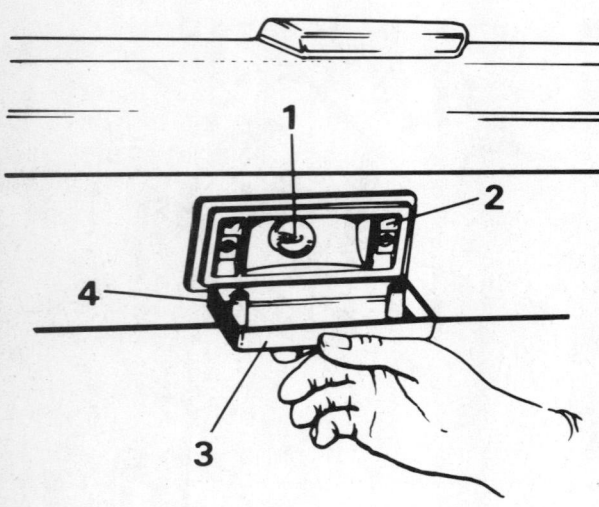

Fig. 10.18 Typical reversing lamp (not 1600) (Sec 30)

1 Bulb   3 Lens
2 Lamp body   4 Lens screw

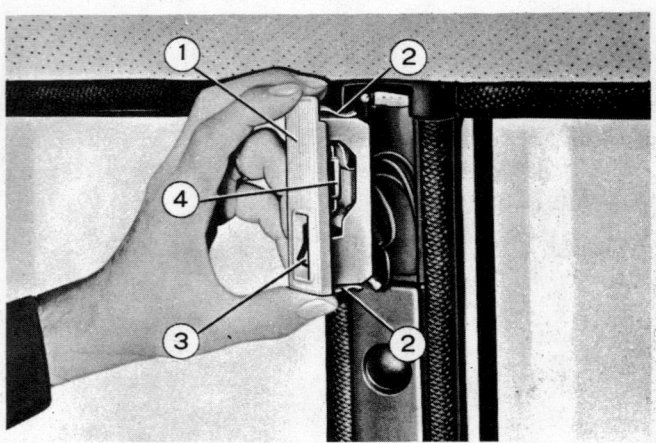

Fig. 10.20 Removing an interior lamp (Sec 32)

1 Lens   3 Switch
2 Retaining clip   4 Bulb

31 Flasher side repeater lights

1 It is a modern requirement to provide repeater flasher lights on each side of the vehicle.
2 The design of these lights is conventional comprising a lens, light shell, gasket and clamping member. The light shell projects through the body from the outside and the clamping member fits onto the inside to pull the shell onto the body. The lens pushes into the shell. The bulb is as usual a push/twist fit.
3 If the repeater light(s) fail, check the bulb, supply wire and its earth connection in that order.
4 On some models the lens is retained by one or two screws (see Fig. 10.19).

32 Interior lights (boot, glove compartment, engine compartment, passenger compartment)

1 *The boot light* operates when the parking lights are switched on and the lid is opened. The switch and bulb are held to the body by a screw. The bulb may be simply eased out of the clip on the bracket for renewal purposes.
2 *The glove compartment switch and bulb* are similar and work independently of the other lights.
3 *The engine compartment bulb* has a bayonet fitting.
4 *The passenger compartment lamps* are housed in the door pillars, one on each side of the car. The units are held in the recesses by spring leaf clips and can be removed by carefully easing them out with a flat blade. The festoon bulb can then be taken out.
5 *On 1600 models,* front door open warning lamps are incorporated in the door edges. Access to the bulb is obtained by extracting the two lens securing screws (Fig. 10.21).

33 Instrument cluster – removal, bulb renewal, and refitting

Rectangular type

1 The instruments are housed in a single unit which is easily removed. Disconnect the battery. Reach round behind the dashboard and press inwards the leaf spring at each side, so releasing the unit

Chapter 10 Electrical system

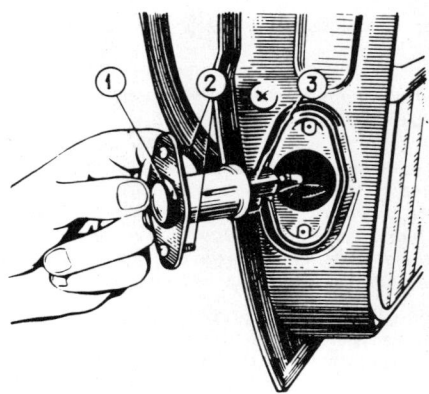

Fig. 10.21 Front door open warning lamp (Sec 31)

1 Lens
2 Screws
3 Bulb holder

which can then be pushed out a little way. When it is pulled rearwards far enough reach in and undo the knurled collar securing the speedometer cable. Pull out the two multi-socket connectors and the whole panel is then free (photo).

2 Individual bulbs for warning and illumination lamps are held into the rear by small bayonet holders which can be taken out after twisting them a ½ turn anticlockwise (photos).

3 The circuits to the warning lamps and instruments are on a printed circuit sheet which can only develop a fault if it is physically damaged or the fitting of an incorrect fuse in a circuit has caused it to burn out. Once all the lamps and the multi socket connectors have been removed the printed circuit can be taken off and renewed if necessary. Individual instruments may also be detached from the cluster and renewed as required. If any instrument or the printed circuit show signs of overheating or burning out it is essential to check the fuses, equipment and wiring of the circuit in question before renewing anything on the panel. Otherwise it is almost certain that the overheating or burn out will recur.

Round instrument type

4 On vehicles fitted with round instruments the complete panel assembly can be removed.

5 Begin by inserting a thin metal rod through the holes beneath the panel and push the retaining clips upwards (see Fig. 10.22).

6 Carefully pull the bottom of the instrument panel outwards as the clips are released and then withdraw the panel just enough to disconnect the speedometer cable and electrical wiring connectors. The instrument panel can now be removed from the vehicle.

7 The instrument panel warning and illumination bulbs can now be removed by pulling the bulb holders from the rear of the instruments.

8 To remove the individual instruments, release the clamping brackets and withdraw the instruments through the front of the panel.

9 The rocker switches are a spring fit and can also be withdrawn from the front of the panel.

10 Reassemble and refit the instrument panel assembly using the reverse of the removal procedure.

33.1 Instrument panel electrical plugs and speedometer drive cable

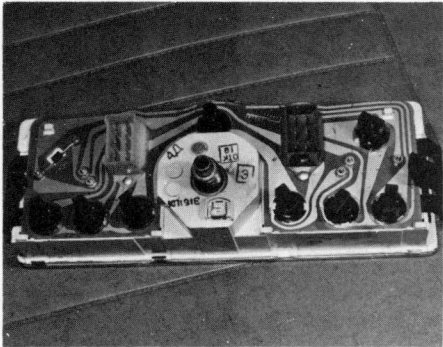
33.2A Rear of rectangular type instrument panel assembly with printed circuit

33.2B Removing a bulb holder from the rectangular type instrument panel

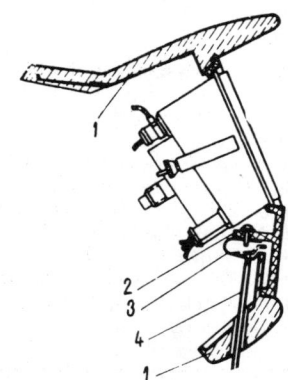

Fig. 10.22 Method of releasing instrument panel (round type) (Sec 33)

1 Fascia top panel
2 Instrument panel
3 Spring clip
4 Clip depressing tool (rod)

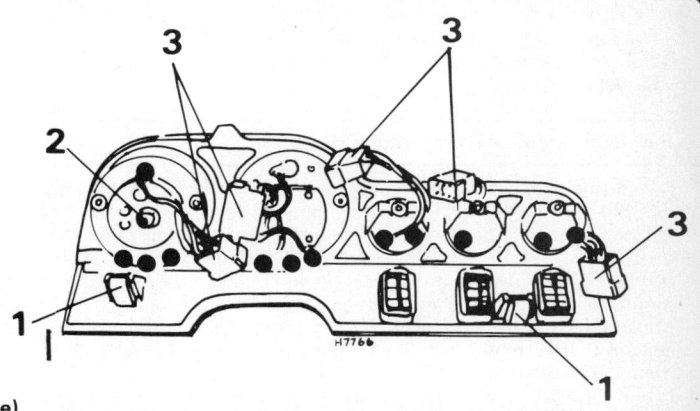

Fig. 10.23 Rear view of circular type instrument panel (Sec 33)

1 Retaining clip
2 Speedometer
3 Connector plugs

34 Speedometer cable

1   The speedometer cable may break or cause erratic operation of the speedometer. To investigate first pull out the instrument cluster as described in the previous Section. Then undo the cable securing collar and draw out the inner cable. If the cable is not broken wash it with petrol and with some fine emery smooth it down throughout its length to take off any high spots which may be causing it to 'twitch' in the outer cable. Re-grease it and replace it. If the cable is broken it will be necessary to undo the lower end where it connects to the gearbox in order to remove the other piece of cable.
2   Although it is not essential always to renew the cable outer when the drive cable breaks, it is conceivable that the outer cable could have been snagged by the broken ends of the inner cable. This may cause erratic running or even failure of a new cable fitted into it.
3   When fitting a new cable outer section take care that when passing it through the bulkhead and securing it that no severe bending occurs — otherwise it will be irretrievably kinked. Secure by its clips so that all the curves are not too acute.
4   A noisy speedometer or drive cable can usually be cured if the cable routing behind the instrument cluster is so arranged that the cable passes through the grommet in the U-shaped bracket attached to the steering column top bracket. In severe cases of noise, grind the corners of the inner cable collar to make the collar round in shape.
5   Never reset the trip recorder while the vehicle is moving.

35 Ignition switch (with steering lock) – removal and refitting

1   Disconnect the battery earth lead.
2   Remove the steering column cowls (4 screws).
3   Remove the rubber O-ring from the ignition switch.
4   Unscrew the cross-head screw from the side of the ignition switch.
5   Insert the ignition key, turn it to position 1, then turn it back to position 0 (ie horizontal) and withdraw it.
6   Using a thin length of dowel, depress the plate through the hole in the left-hand side of the lock mounting bracket.
7   Pull the ignition switch and lock from the bracket. If necessary, depress the plate from the top of the bracket when the switch is halfway out.
8   Note the position of the wires on the rear of the switch, then disconnect them.
9   Remove the ignition switch and lock.
10 Refitting is the reverse of the removal procedure.

36 Relays

1   On 1500 and 1600 models, a headlamp relay and a horn relay are mounted on the valance at the side of the engine compartment.
2   If the headlamps or horn fail to function, and the circuit fuses, wiring and bulbs (or horn) have all been examined and are in good condition, it must be assumed that the relay is at fault. A defective relay should be renewed as it is not repairable.
3   Refer to Section 20 for details of the wiper motor relay fitted to some models.

37 Radios and tape players – fitting (general)

A radio or tape player is an expensive item to buy and will only give its best performance if fitted properly. It is useless to expect concert hall performance from a unit that is suspended from the dash panel on string with its speaker resting on the back seat or parcel shelf! If you do not wish to do the installation yourself there are many in-car entertainment specialists who can do the fitting for you.

Make sure the unit purchased is of the same polarity as the car, and ensure that units with adjustable polarity are correctly set before commencing installation.

It is difficult to give specific information with regard to fitting, as final positioning of the radio/tape player, speakers and aerial is entirely a matter of personal preference. However, the following paragraphs give guidelines to follow, which are relevant to all installations.

Radios

The Lada range of cars all have a radio aperture provided, either in the dash panel or in the console (if fitted) between the dash panel and transmission tunnel. It is advisable to check the dimensions of the aperture before purchasing a radio to fit.

Should personal preference dictate the radio being fitted in another location either in or beneath the dash panel or in the console, the following points should be borne in mind before deciding exactly where to fit the unit:

(a) *The unit must be within easy reach of the driver wearing a seat belt*
(b) *The unit must not be mounted in close proximity to an electronic tachometer, the ignition switch and its wiring, or the flasher unit and associated wiring*
(c) *The unit must be mounted within reach of the aerial lead, and in such a place that the aerial lead will not have to be routed near the components detailed in the preceding paragraph (b)*
(d) *The unit should not be positioned in a place where it might cause injury to the car occupants in an accident; for instance, under the dash panel above the driver's or passenger's legs*
(e) *The unit must be fitted really securely*

Some radios will have mounting brackets provided together with instructions; others will need to be fitted using drilled and slotted metal strips, bent to form mounting brackets – these strips are available from most accessory shops. The unit must be properly earthed, by fitting a separate earthing lead between the casing of the radio and the vehicle frame.

Use the radio manufacturer's instructions when wiring the radio into the vehicle's electrical system. If no instructions are available refer to the relevant wiring diagram to find the location of the radio feed connection in the vehicle's wiring circuit. A 1–2 amp in-line fuse must be fitted in the radio's feed wire – a choke may also be necessary (see next Section).

The type of aerial used, and its fitted position, is a matter of personal preference. In general the taller the aerial, the better the reception. It is best to fit a fully retractable aerial – especially if a mechanical car-wash is used or if you live in an area where cars tend to be vandalised. In this respect electric aerials which are raised and lowered automatically when switching the radio on or off are convenient, but are more likely to give trouble than the manual type.

When choosing a site for the aerial the following points should be considered:

(a) *The aerial lead should be as short as possible – this means that the aerial should be mounted at the front of the car*
(b) *The aerial must be mounted as far away from distributor and HT leads as possible*
(c) *The part of the aerial which protrudes beneath the mounting point must not foul the roadwheels, or anything else*
(d) *If possible the aerial should be positioned so that the coaxial lead does not have to be routed through the engine compartment*
(e) *The plane of the panel on which the aerial is mounted should not be so steeply angled that the aerial cannot be mounted vertically (in relation to the 'end-on' aspect of the car). Most aerials have a small amount of adjustment available*

Having decided on a mounting position, a relatively large hole will have to made in the panel. The exact size of the hole will depend upon the specific aerial being fitted, generally, the hole required is of $\frac{3}{4}$ inch (19 mm) diameter. On metal bodied cars, a 'tank-cutter' of the relevant diameter is the best tool to use for making the hole. This tool needs a small diameter pilot hole drilled through the panel, through which the tool clamping bolt is inserted.

Fit the aerial according the manufacturer's instructions. If the aerial is very tall, or if it protrudes beneath the mounting panel for a considerable distance, it is a good idea to fit a stay between the aerial and the vehicle frame. This stay can be manufactured from the slotted and drilled metal strips previously mentioned. The stay should be securely screwed or bolted in place. For best reception it is advisable to fit an earth lead between the aerial and the vehicle frame.

It will probably be necessary to drill one or two holes through bodywork panels in order to feed the aerial lead into the interior of the car. Where this is the case ensure that the holes are fitted with rubber grommets to protect the cable, and to stop possible entry of water.

Positioning and fitting of the speaker depends mainly on its type. Generally, the speaker is designed to fit directly into the aperture already provided in the car. Where this is the case, fitting the speaker

Chapter 10 Electrical system

is just a matter of removing the protective grille from the aperture and screwing or bolting the speaker in place. Take great care not to damage the speaker diaphragm whilst doing this. It is a good idea to fit a gasket between the speaker frame and the mounting panel, in order to prevent vibration – some speakers will already have such a gasket fitted.

If a pod type speaker was supplied with the radio, this can be secured to the mounting panel with self-tapping screws.

When connecting a rear mounted speaker to the radio, the wires should be routed through the vehicle beneath the carpets or floor mats – preferably along the side of the floorpan, where they will not be trodden on by passengers. Make the relevant connections as directed by the radio manufacturer.

By now you will have several yards of additional wiring in the car. Use PVC tape to secure this wiring out of harm's way. Do not leave electrical leads dangling. Ensure that all new electrical connections are properly made (wires twisted together will not do) and completely secure.

The radio should now be working, but before you pack away your tools it will be necessary to trim the radio to the aerial. If specific instructions are not provided by the radio manufacturer proceed as follows. Find a station with a low signal strength on the medium-wave band, slowly turn the trim screw of the radio in, or out, until the loudest reception of the selected station is obtained – the set is then trimmed to the aerial.

Tape players

Fitting instructions for both cartridge and cassette stereo type players are the same and in general the same rules apply as when fitting a radio. Tape players are not usually prone to electrical interference like radio – although it can occur – so positioning is not so critical. If possible the player should be mounted on an even keel. Also it must be possible for a driver wearing a seat belt to reach the unit in order to change or turn over tapes.

For the best results from speakers designed to be recessed into a panel, mount them so that the back of the speaker protrudes into an enclosed chamber within the car (eg door interiors or the boot cavity).

To fit recessed type speakers in the front doors first check that there is sufficient room to mount the speakers in each door without it fouling the latch or window winding mechanism. Hold the speaker against the skin of the door, and draw a line around the periphery of the speaker. With the speaker removed draw a second cutting line, within the first, to allow enough room for the entry of the speaker back, but at the same time providing a broad seat for the speaker flange. When you are sure that the cutting line is correct, drill a series of holes around its periphery. Pass a hacksaw blade through one of the holes and then cut through the metal between the holes until the centre section of the panel falls out.

De-burr the edges of the hole and then paint the raw metal to prevent corrosion. Cut a corresponding hole in the door trim panel – ensuring that it will be completely covered by the speaker grille. Now drill a hole in the door edge and a corresponding hole in the door surround. These holes are to feed the speaker leads through – so fit grommets. Pass the speaker leads through the door trim, door skin and out through the holes in the side of the door and door surround. Refit the door trim panel and then secure the speaker to the door using self-tapping screws. **Note:** If the speaker is fitted with a shield to prevent water dripping on it, ensure that this shield is at the top.

Pod type speakers can be fastened anywhere offering a corresponding mounting point on each side of the car. Pod speakers sometimes offer a better reproduction quality if they face the rear window – which then acts as a reflector – so it is worthwhile to do a little experimenting before finally fixing the speaker.

38 Radios and tape players – suppression of interference

To eliminate buzzes and other unwanted noises, costs very little and is not as difficult as sometimes thought. With a modicum of common sense and patience and carrying out the instructions in the following paragraphs, interference can be virtually eliminated.

The first cause for concern is the alternator. The noise this makes over the radio is like an electric mixer and the noise speeds up when you rev up (if you wish to prove the point, you can remove the drivebelt and try it). The remedy for this is simple; connect a 1·0 – 3·0 mf capacitor between earth, probably the bolt that holds down the alternator base, and the large terminal on the alternator. This is most important for if you connect it to the small terminal, you will probably damage the alternator permanently (see Fig. 10.24).

A second common cause of electrical interference is the ignition system. Here a 1·0 mf capacitor must be connected between earth and the '+' terminal on the coil (see Fig. 10.25). This may stop the tick-tick-tick sound that comes over the speaker. Next comes the spark itself.

There are several ways of curing interference from the ignition HT system. One is to use carbon film HT leads but these have a tendency to snap inside and you do not know then, why you are firing on only half your cylinders. So the second, and more successful method is to use resistive spark plug caps (see Fig. 10.26) of about 10 000 to 15 000 ohm resistance. If, due to lack of room, these cannot be used, an alternative is to use in-line suppressors (Fig. 10.27) – if the interference is not too bad, you may get away with only one suppressor in the coil to distributor line. If the interference does continue (a clacking noise) then doctor all HT leads.

At this stage it is advisable to check that the radio is well earthed, also the aerial, and to see that the aerial plug is pushed well into the set and that the radio is properly trimmed (see preceding Section). In addition, check that the wire which supplies the power to the set is as short as possible and does not wander all over the car. It is a good idea to check that the fuse is of the correct rating. For most sets this will be about 1 to 2 amps.

At this point the more usual causes of interference have been suppressed. If the problem still exists, a look at the cause of interference may help to pinpoint the component generating the stray electrical discharges.

The radio picks up electromagnetic waves in the air; now some are made by radio stations and other broadcasters and some, not wanted, are made by the car. The car made signals are produced by stray electrical discharges floating around the car. Common producers of these signals are electric motors; ie the windscreen wipers, electric screen washers, electric window winders, heater fan or an electric aerial if fitted. Other sources of interference are electric fuel pumps, flashing turn signals, instruments. The remedy for these cases is shown in Fig. 10.28 for an electric motor whose interference is not too bad and Fig. 10.29 for instrument suppression. Turn signals are not normally suppressed. In recent years, radio manufacturers have included in the live line of the radio, in addition to the fuse, an in-line choke. If your installation lacks one of these, put one in as shown in Fig. 10.30.

All the foregoing components are available from radio shops or accessory shops. For a transistor radio, a 2A choke should be adequate. If you have an electric clock fitted this should be suppressed by connecting a 0·5 mf capacitor directly across it as shown for a motor in Fig. 10.28.

If after all this, you are still experiencing radio interference, first assess how bad it is, for the human ear can filter out unobtrusive unwanted noises quite easily. But if you are still adamant about eradicating the noise, then continue.

As a first step, a few 'experts' seem to favour a screen between the radio and the engine. This is OK as far as it goes, literally! – for the whole set is screened and if interference can get past that then a small piece of aluminium is not going to stop it.

A more sensible way of screening is to discover if interference is coming down the wires. First, take the live lead; interference can get between the set and the choke (hence the reason for keeping the wires short). One remedy here is to screen the wire and this is done by buying screened wire and fitting that. The loudspeaker lead could be screened also to prevent pick-up getting back to the radio – although this is unlikely.

Without doubt, the worst source of radio interferences comes from the ignition HT leads, even if they have been suppressed. The ideal way of suppressing these is to slide screening tubes over the leads themselves. As this is impractical, we can place an aluminium shield over the majority of the lead areas. In a vee- or twin-cam engine, this is relatively easy but for a straight engine the results are not particularly good.

Now for the really impossible cases, here are a few tips to try out. Where metal comes into contact with metal, an electrical disturbance is caused which is why good clean connections are essential. To remove interference due to overlapping or butting panels you must bridge the joint with a wide braided earth strap (like that from the frame to the engine/transmission). The most common moving parts

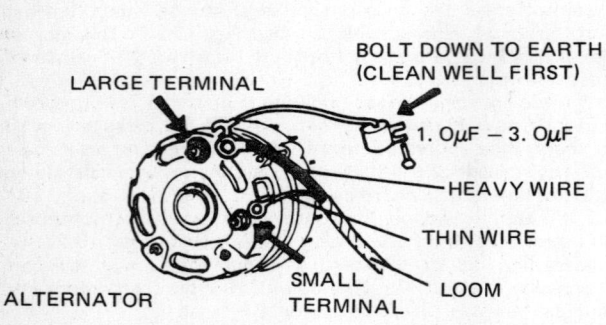

Fig. 10.24 The correct way to connect a capacitor to the alternator (Sec 38)

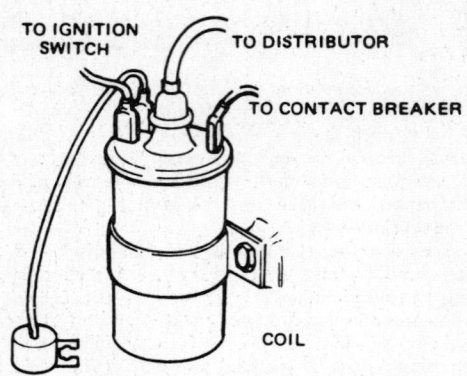

Fig. 10.25 The capacitor must be connected to the ignition switch (+) side of the coil (Sec 38)

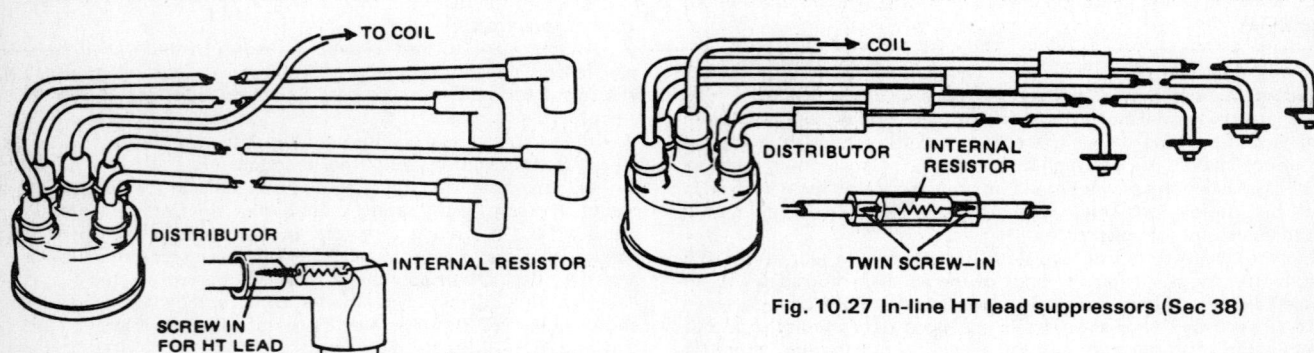

Fig. 10.26 Resistive spark plug caps (Sec 38)

Fig. 10.27 In-line HT lead suppressors (Sec 38)

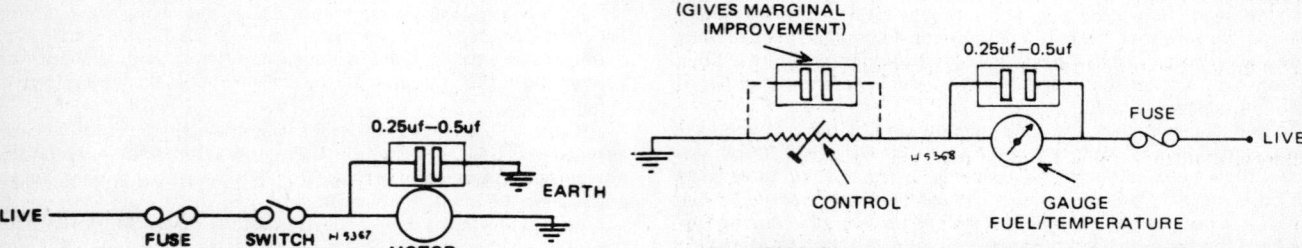

Fig. 10.28 Electric motor interference suppression (Sec 38)

Fig. 10.29 Gauge and control unit interference suppression (Sec 38)

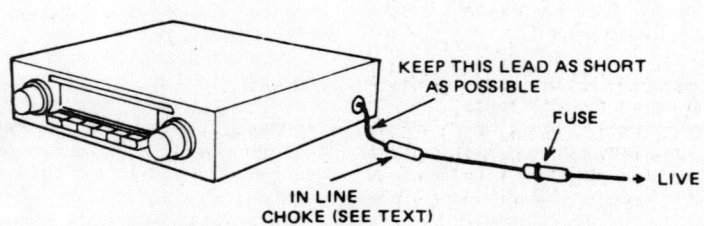

Fig. 10.30 Location of in-line choke (Sec 38)

Chapter 10 Electrical system

that could create noise and should be strapped are, in order of importance:

    (a) Silencer to frame
    (b) Exhaust pipe to engine block and frame
    (c) Air filter to frame
    (d) Front and rear bumpers to frame
    (e) Steering column to frame
    (f) Bonnet and boot lids to frame

These faults are most pronounced when (1) the engine is idling, (2) labouring under load. Although the moving parts are already connected with nuts, bolts, etc, these do tend to rust and corrode, thus creating a high resistance interference source.

If you have a ragged sounding pulse when mobile, this could be wheel or tyre static. This can be cured by buying some anti-static powder and sprinkling it liberally inside the tyres.

If the interference takes the shape of a high pitched screeching noise that changes its note when the car is in motion and only comes now and then, this could be related to the aerial, especially if it is of the telescopic or whip type. This source can be cured quite simply by pushing a small rubber ball on top of the aerial as this breaks the electric field before it can form; but it would be much better to buy yourself a new aerial of a reputable brand. If, on the other hand, you are getting a loud rushing sound every time you brake, then this is brake static. This effect is most prominent on hot dry days and is cured only by fitting a special kit, which is quite expensive.

In conclusion, it is pointed out that it is relatively easy and cheap to eliminate 95 per cent of all noises, but to eliminate the final 5 per cent is time and money consuming. It is up to the individual to decide if it is worth it. Please remember also, that you will not get concert hall performance from a cheap radio.

Finally at the beginning of this Section are mentioned tape players; these are not usually affected by interference but in a very bad case, the best remedies are the first three suggestions plus using a 3 – 5 amp choke in the live line and in incurable cases screen the live and speaker wires.

39 Fault diagnosis – electrical system

Symptom	Reason/s
Starter motor fails to turn engine	Battery discharged Battery defective internally Battery terminal leads loose or earth lead not securely attached to body Loose or broken connections in starter motor circuit Starter motor switch or solenoid faulty Starter brushes badly worn, sticking, or brush wires loose Commutator dirty, worn or burnt Starter motor armature faulty Field coils earthed
Starter motor turns engine very slowly	Battery in discharged condition Starter brushes badly worn, sticking, or brush wires loose Loose wires in starter motor circuit
Starter motor turns without turning engine	Starter motor pinion sticking on the screwed sleeve Pinion or flywheel gear teeth broken or worn
Starter motor noisy or excessively rough engagement	Pinion or flywheel gear teeth broken or worn Starter drive main spring broken Starter motor retaining bolts loose
Battery will not hold charge for more than a few days	Battery defective internally Electrolyte level too low or electrolyte too weak due to leakage Plate separators no longer fully effective Battery plates severely sulphated Alternator belt slipping Battery terminal connections loose or corroded Alternator not charging properly* Short in lighting circuit causing continual battery drain Regulator unit not working correctly
Ignition light fails to go out, battery runs flat in a few days	Drivebelt loose and slipping or broken Brushes worn, sticking, broken or dirty Brush springs weak or broken Alternator faulty*

*If all appears to be well but the alternator is still not charging, take the car to an automobile electrician to check the alternator and regulator.

Failure of individual electrical equipment to function correctly is dealt with alphabetically below. In cases of electrical failure it is always worth checking the obvious, such as blown fuses (particularly if associated equipment has also failed) and loose or broken wires.

Fuel gauge gives no reading	Fuel tank empty! Electric cable between tank sender unit and gauge earthed or loose Fuel gauge case not earthed Fuel gauge supply cable interrupted Fuel gauge unit broken
Fuel gauge registers full all the time	Electric cable between tank unit and gauge broken or disconnected

Chapter 10 Electrical system

Symptom	Reason/s
Horn operates all the time	Horn push either earthed or stuck down Horn cable to horn push earthed
Horn fails to operate	Blown fuse Cable or cable connection loose, broken or disconnected Horn has an internal fault
Horn emits intermittent or unsatisfactory noise	Cable connections loose Horn incorrectly adjusted
Lights do not come on	Blown fuse If engine not running, battery discharged Light bulb filament burnt out or bulbs broken Wire connections loose, disconnected or broken Lights switch shorting or otherwise faulty
Lights come on but fade out	If engine not running battery discharged
Lights give very poor illumination	Lamp glasses dirty Reflector tarnished or dirty Lamps badly out of adjustment Incorrect bulb with too low wattage fitted Existing bulbs old and badly discoloured Electrical wiring too thin not allowing full current to pass
Lights work erratically – flashing on and off, especially over bumps	Battery terminals or earth connections loose Lights not earthing properly Contacts in light switch faulty
Wiper motor fails to work	Blown fuse Wire connections loose, disconnected or broken Brushes badly worn Armature worn or faulty Field coils faulty
Wiper motor works very slowly and takes excessive current	Commutator dirty, greasy or burnt Drive to spindles bent or unlubricated Drive spindle binding or damaged Armature bearings dry or unaligned Armature badly worn or faulty
Wiper motor works slowly and takes little current	Brushes badly worn Commutator dirty, greasy or burnt Armature badly worn or faulty
Wiper motor works but wiper blades remain static	Linkage disengaged or faulty Drive spindle damaged or worn Wiper motor gearbox parts badly worn

See overleaf for wiring diagrams

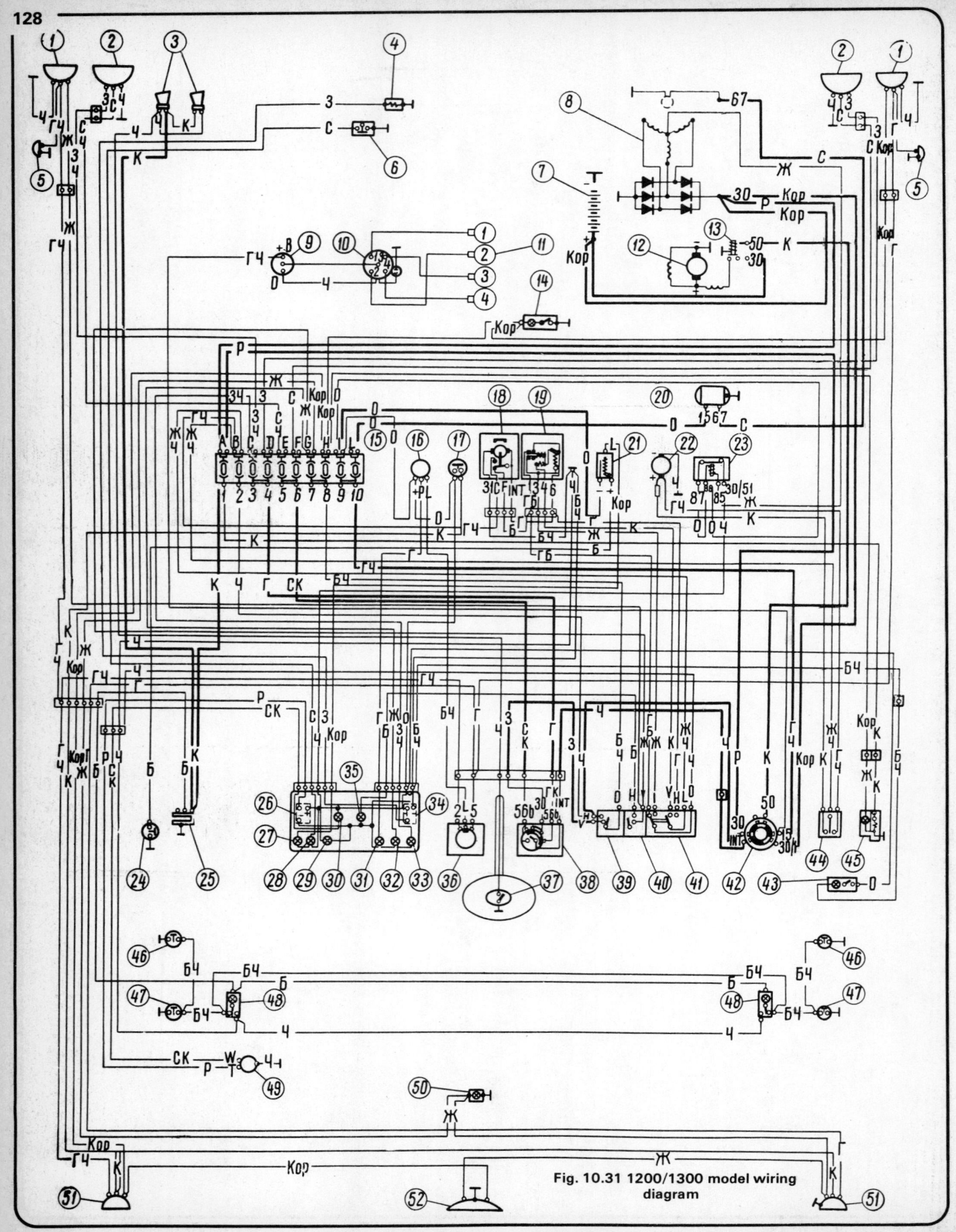

Fig. 10.31 1200/1300 model wiring diagram

Wiring diagram key – Fig. 10.31

1. Sidelight and direction indicator light bulbs
2. Headlamps (high and low beams)
3. Horns
4. Coolant temperature transmitter
5. Direction indicator side repeater lamps
6. Oil pressure warning light switch
7. Battery
8. Alternator
9. Ignition coil
10. Distributor
11. Spark plugs
12. Starter motor
13. Starter solenoid
14. Engine compartment lamp
15. Fuse box
16. Direction indicator flasher unit
17. Stoplight switch
18. Windscreen wiper motor
19. Windscreen wiper relay
20. Voltage reglator
21. Handbrake warning lamp flasher
22. Heater blower motor
23. Ignition warning lamp relay
24. Handbrake warning lamp switch
25. Service lamp socket
26. Fuel gauge
27. Handbrake warning lamp
28. Oil pressure warning lamp
29. Ignition warning lamp
30. Fuel level warning lamp
31. Direction indicator warning lamp
32. Sidelight warning lamp
33. High beam warning lamp
34. Coolant temperature gauge
35. Instrument lamp
36. Direction indicator switch
37. Horn push switch
38. Dipswitch
39. Headlight and sidelight switch
40. Instrument panel light switch
41. Windscreen wiper switch
42. Ignition switch
43. Glovebox lamp and switch
44. Heater blower switch
45. Cigarette lighter
46. Interior lamp switches – front doors
47. Interior lamp switches – rear doors
48. Interior lamps
49. Fuel gauge sender unit
50. Luggage boot lamp
51. Stop/tail and rear direction indicator lamps
52. Number plate lamp

Colour code

Б	– White
БЧ	– White with black tracer
Г	– Blue
ГБ	– Blue with white tracer
ГК	– Blue with red tracer
ГЧ	– Blue with black tracer
Ж	– Yellow
ЖЧ	– Yellow with black tracer
З	– Green
ЗЧ	– Green with black tracer
К	– Red
Кор	– Brown
О	– Orange
Р	– Pink
С	– Grey
СК	– Grey with red tracer
СЧ	– Grey with black tracer
Ч	– Black

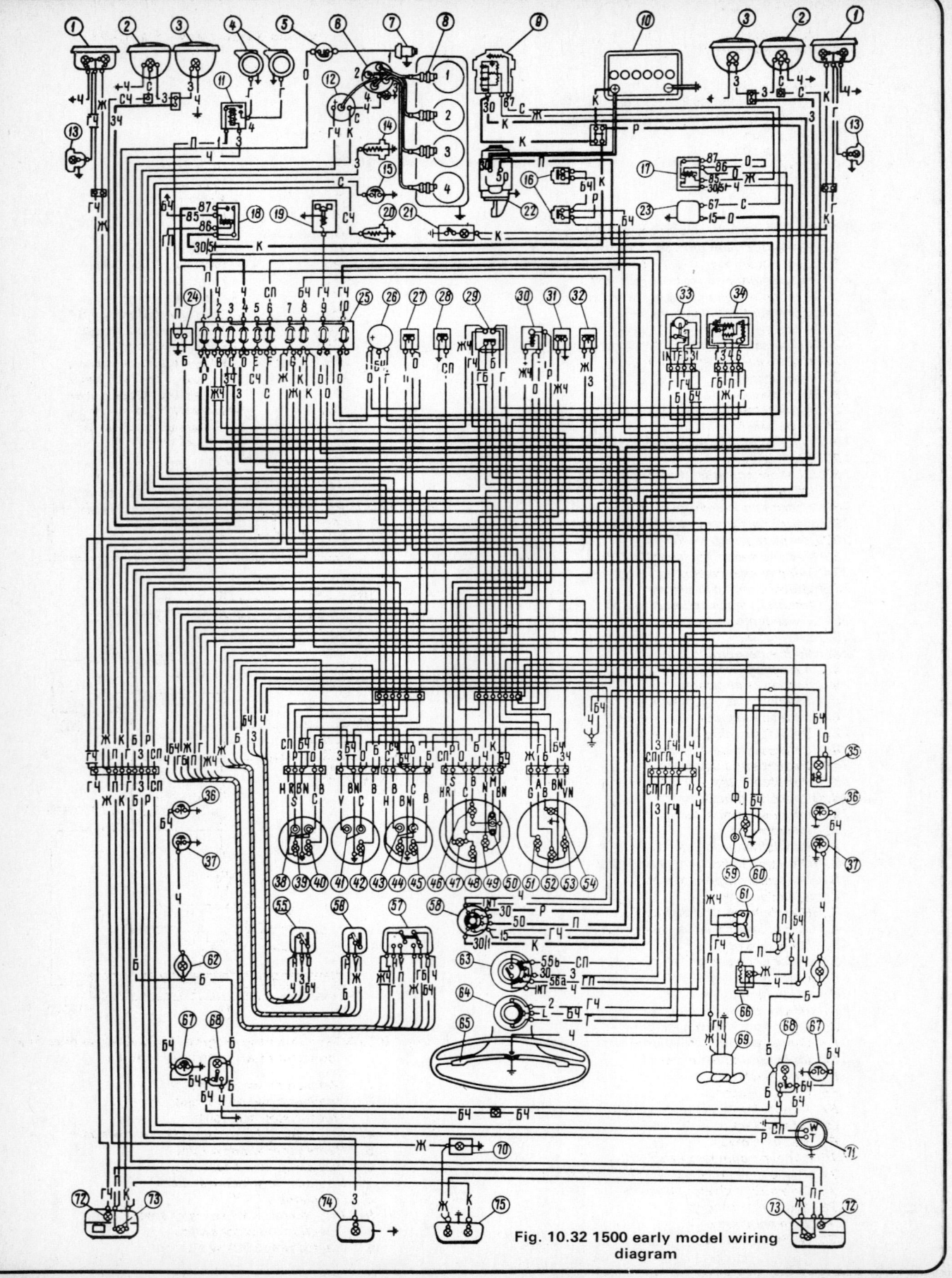

Fig. 10.32 1500 early model wiring diagram

Wiring diagram key – Fig. 10.32

1 Sidelights
2 Outer headlamps
3 Inner headlamps
4 Horns
5 Fan clutch thermostatic switch
6 Distributor
7 Fan clutch
8 Spark plugs
9 Alternator
10 Battery
11 Horn relay
12 Ignition coil
13 Direction indicator side repeater lamps
14 Coolant temperature transmitter
15 Oil pressure warning light switch
16 Brake fluid level warning light switch
17 Ignition warning lamp relay
18 Headlamp relay
19 Idle stop solenoid
20 Oil pressure sender unit
21 Engine compartment lamp
22 Starter motor
23 Voltage regulator
24 Service lamp socket
25 Fuse box
26 Direction indicator flasher unit
27 Stoplight switch
28 Choke warning light switch
29 Windscreen washer switch
30 Handbrake warning light flasher
31 Handbrake warning light switch
32 Reversing lamp switch
33 Windscreen wiper switch
34 Windscreen wiper relay
35 Glovebox lamp and switch
36 Interior lamp front door switch
37 Front door warning lamp switch
38 Fuel gauge
39 Fuel level warning lamp
40 Fuel gauge lamp
41 Temperature gauge
42 Temperature gauge lamp
43 Oil pressure gauge
44 Oil pressure warning lamp
45 Oil pressure gauge lamp
46 Tachometer lamp
47 Handbrake/brake fluid level warning lamp
48 Choke warning lamp
49 Ignition warning lamp
50 Tachometer
51 Sidelight warning lamp
52 Direction indicator warning lamp
53 High beam warning lamp
54 Speedometer lamp
55 Headlight and sidelight switch
56 Instrument panel light switch
57 Windscreen wiper switch
58 Ignition switch
59 Clock
60 Clock lamp
61 Heater blower switch
62 Front door warning lamp
63 Dipswitch
64 Direction indicator switch
65 Horn push switch
66 Cigarette lighter
67 Interior lamp rear door switch
68 Interior lamp
69 Heater blower
70 Luggage boot lamp and switch
71 Fuel gauge and level warning light sender
72 Direction indicator lamp
73 Stop and tail lamp
74 Reversing lamp
75 Number plate lamp

Colour code

Б	– White
Г	– Blue
Ж	– Yellow
З	– Green
К	– Brown
П	– Red
О	– Orange
Р	– Pink
С	– Grey
Ч	– Black
БЧ	– White with black tracer
ГБ	– Blue with white tracer
ГП	– Blue with red tracer
ГЧ	– Blue with black tracer
ЖЧ	– Yellow with black tracer
ЗЧ	– Green with black tracer
СП	– Grey with red tracer
СЧ	– Grey with black tracer

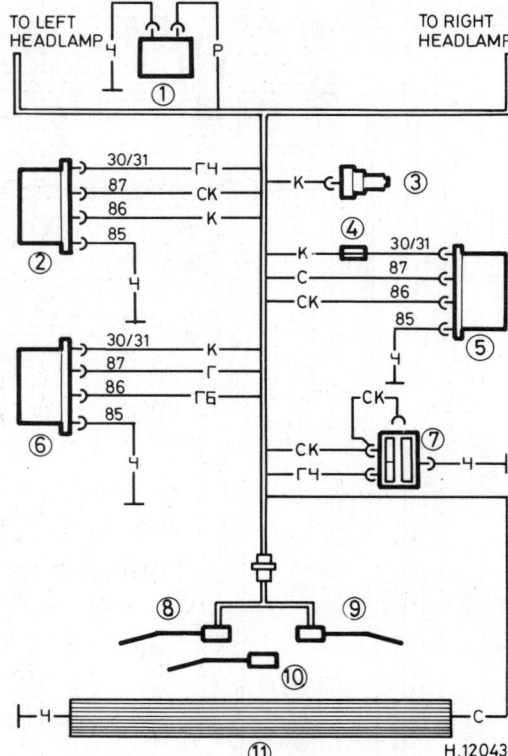

Fig. 10.32A Typical wiring diagram for additional electrical equipment fitted to later models

1 Windscreen washer motor
2 Headlamp low beam relay
3 Throttle stop solenoid
4 In-line fuse (16A) for heated rear window
5 Heated rear window relay
6 Headlamp high beam relay
7 Heated rear window switch
8 Headlamp switch
9 Windscreen washer/wiper switch
10 Direction indicator switch
11 Heated rear window element

132

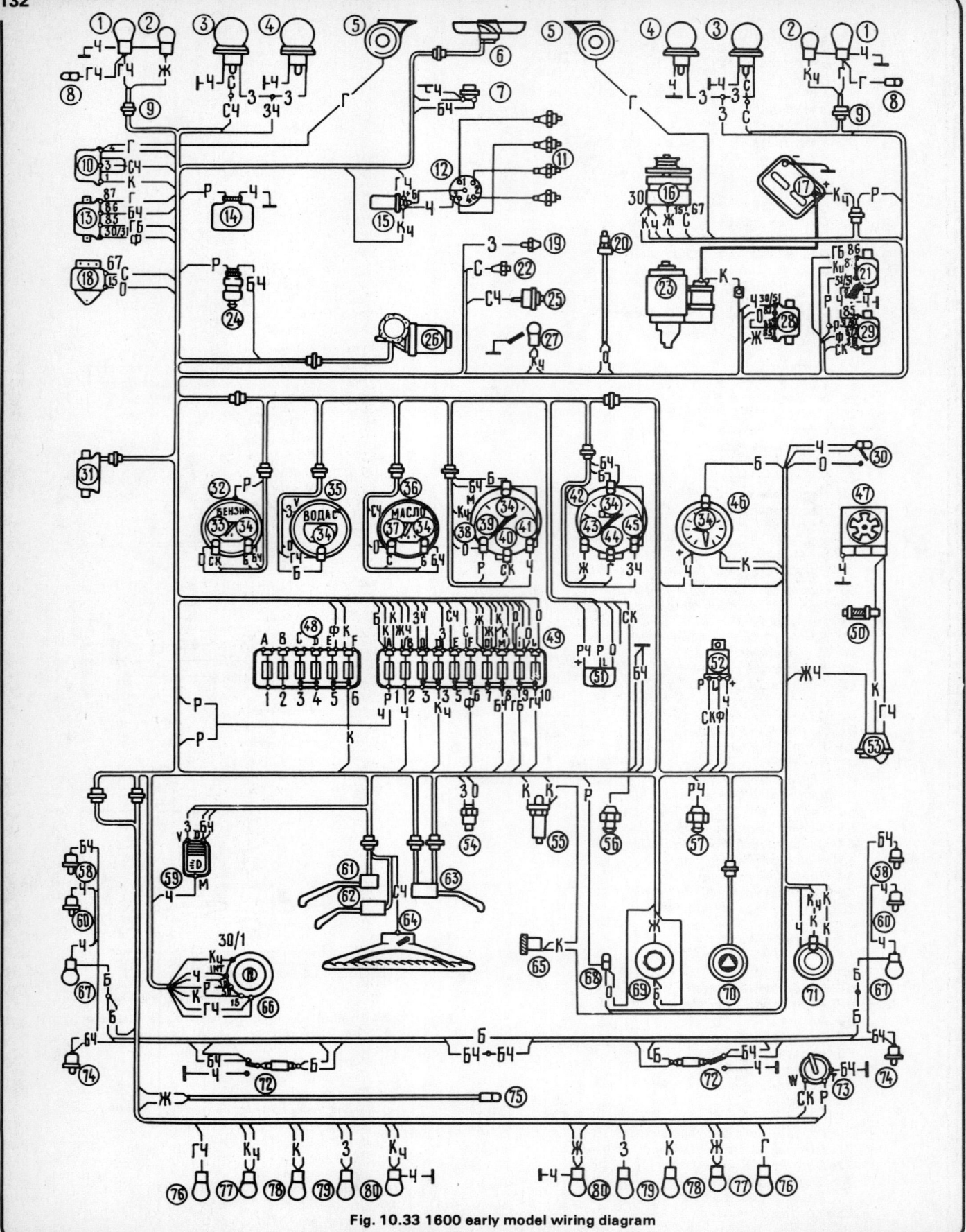

Fig. 10.33 1600 early model wiring diagram

Wiring diagram key – Fig. 10.33

1 Direction indicator lamps
2 Sidelights
3 Outer headlamps
4 Inner headlamps
5 Horns
6 Electric cooling fan
7 Thermostatic fan switch
8 Direction indicator side repeater lamps
9 Connectors
10 Horn relay
11 Spark plugs
12 Distributor
13 Cooling fan relay
14 Windscreen washer pump
15 Ignition coil
16 Alternator
17 Battery
18 Voltage regulator
19 Coolant temperature sender unit
20 Idle stop solenoid
21 High beam relay
22 Oil pressure warning lamp sender unit
23 Starter motor
24 Brake fluid level warning lamp sender
25 Oil pressure gauge sender unit
26 Windscreen wiper motor
27 Engine compartment lamp and switch
28 Ignition warning lamp relay
29 Low beam relay
30 Glovebox lamp and switch
31 Windscreen wiper relay
32 Fuel gauge
33 Fuel level warning lamp
34 Instrument lamps
35 Coolant temperature gauge
36 Oil pressure gauge
37 Oil pressure warning lamp
38 Tachometer
39 Handbrake warning lamp
40 Choke warning lamp
41 Ignition warning lamp
42 Speedometer
43 Sidelight warning lamp
44 Direction indicator warning lamp
45 High beam warning lamp
46 Clock
47 Heater blower
48 Fuse box
49 Fuse box
50 Heater blower series resistor
51 Handbrake warning lamp flasher
52 Direction indicator flasher
53 Heater blower switch
54 Reversing lamp switch
55 Stoplight switch
56 Choke warning lamp switch
57 Handbrake warning lamp switch
58 Interior lamp front door switch
59 Headlight and sidelight switch
60 Front door warning lamp switch
61 Dipswitch
62 Direction indicator switch
63 Windscreen wiper switch
64 Horn push switch
65 Service lamp socket
66 Ignition switch
67 Front door warning lamp
68 Handbrake/brake fluid level warning lamp
69 Instrument lighting switch
70 Hazard warning switch
71 Cigarette lighter
72 Interior lamp switches
73 Fuel gauge and level warning lamp sender
74 Interior lamp rear door switches
75 Luggage boot lamp
76 Direction indicator lamps
77 Tail lamps
78 Stoplamps
79 Reversing lamps
80 Number plate lamps

Colour code

Г   – Blue
В   – White
Ж   – Yellow
З   – Green
К   – Red
Кч   – Brown
О   – Amber
Р   – Pink
С   – Grey
Ч   – Black
Ф   – Violet
Бч   – White with black tracer
ГВ   – Blue with white tracer
Гч   – Blue with black tracer
Жч   – Yellow with black tracer
Зч   – Green with black tracer
Рч   – Pink with black tracer
Ск   – Grey with red tracer
Сч   – Grey with black tracer

Chapter 11 Suspension and steering

For modifications, and information applicable to later models, see Supplement at end of manual

Contents

Fault diagnosis – suspension and steering	25
Front hub bearings – adjustment and lubrication	4
Front hub – overhaul	5
Front shock absorber – removal, testing and refitting	2
Front suspension anti-roll bar – removal and refitting	12
Front suspension balljoints – inspection, removal and refitting	6
Front suspension coil spring – removal and refitting	3
Front suspension crossmember – removal and refitting	11
Front suspension lower control arm – overhaul	10
Front suspension lower control arm – removal and refitting	9
Front suspension upper control arm – overhaul	8
Front suspension upper control arm – removal and refitting	7
Front wheel alignment – adjustment	23
General description and maintenance	1
Idler arm – removal and refitting	21
Rear axle locating arms – removal, overhaul and refitting	15
Rear suspension coil spring – removal and refitting	14
Rear shock absorber – removal and refitting	13
Steering box – dismantling, examination and reassembly	19
Steering box – removal and refitting	18
Steering column and shaft – removal and refitting	17
Steering gear – adjustment	20
Steering rods and balljoints – general	22
Steering wheel – removal and refitting	16
Wheels and tyres	24

Specifications

Front suspension
Type .... Independent, upper and lower wishbone type suspension arms with coil springs, telescopic type shock absorbers and anti-roll bar

Coil springs free length:
    Yellow marking .... 14·2 in (360 mm)
    Green marking .... 14·2 in (360 mm)
Wheel bearing lubricant type/specification .... Multi-purpose lithium-based grease, to NLGI No 3 (Duckhams LB 10)

Rear suspension
Type .... Live axle located by upper and lower suspension arms, coil springs, telescopic shock absorbers and Panhard rod

Coil springs free length:
    Saloon .... 17·4 in (442 mm)
    Estate .... 17·9 in (455 mm)

Steering
Type .... Worm and sector
Ratio .... 16·4 : 1
No of turns lock to lock 3
Turning circle:
    Except 1600 .... 35·1 ft (10·7 m)
    1600 .... 36·74 ft (11·2 m)

Steering angles:

	Laden*	**Unladen**
Toe-in	3 mm ± 1 mm	4 mm ± 1 mm
Camber	0° 30′ ± 20′	0° 05′ ± 20′
Castor	4° ± 30′	3° 30′ ± 30′

Turning angle:
    Inner wheel .... 39° ± 1° 30′
    Outer wheel .... 30°
Steering box oil type/specification .... Hypoid gear oil, viscosity SAE 90 or 75W/90, to API GL5 (Duckhams Hypoid 75W/90S)
Steering box oil capacity .... 0·38 pints (0·215 litres)

*With 4 occupants and 110 lb (50 kg) luggage or equivalent

Wheels and tyres
Wheel size:
    1200, 1300 and 1500 models .... 4½J x 13
    1600 models .... 5J x 13
Tyre size:
    1200 and 1300 models .... 155 x 13 radial
    1500 and 1600 models .... 165 x 13 radial

Chapter 11 Suspension and steering

	Front	Rear
Tyre pressures (cold) lbf/in^2 (kgf/cm^2):		
1200 and 1300 saloon models	24 (1·7)	26 (1·8)
1200 and 1500 estate models	24 (1·7)	28 (1·96)
1500 and 1600 saloon models	24 (1·7)	27 (1·9)

Torque wrench settings — lbf ft — kgf m

Front suspension

	lbf ft	kgf m
Crossmember to side frame:		
Upper bolts	69	9·5
Lower bolts	58	8·0
Suspension upper arm pivot shaft	65	9·0
Suspension lower arm pivot shaft	72	10·0
Shock absorber upper mounting	11	1·5
Shock absorber lower mounting	44	6·0
Anti-roll bar nuts	13	1·8
Track-rod end ball-pin	25	3·5
Suspension swivel ball-pin	72	10·0
Steering arm to stub axle	44	6·0
Disc shield	25	3·5

Rear suspension

	lbf ft	kgf m
Shock absorber mountings	44	6·0
Suspension arm pivot bolts	58	8·0

Steering

	lbf ft	kgf m
Steering box mounting bolts	29	4·0
Steering balljoint taper pin	44	6·0
Steering wheel nut	36	5·0
Steering drop arm nut	174	24·0
Steering shaft coupling pinch-bolt	18	2·5
Idler arm bracket	29	4.0
Idler arm self-locking nut	72	10.0

1 General description and maintenance

1   In the independent front suspension system fitted, the front wheels, hubs and brakes are all mounted on a stub axle (steering knuckle) assembly. On the upper and lower ends of the stub axle there are balljoints which connect it to the upper and lower suspension control arm. The upper arm pivots on rubber bushes on a spindle which is attached to the bodyshell. The lower arm is similarly attached to the bodyshell and to a box crossmember which stiffens the body structure locally around the lower control arm attachment.
2   The coil spring, telescopic shock absorbers and anti-roll bar, all act on the lower control arm.
3   The steering system comprises an articulated steering column, a worm and roller steering box and track and link rods to the steering knuckle arms mounted on the stub axle member. There is a relay lever and idler arm assembly mounted on the opposite side of the vehicle to the steering box to ensure symmetry of the track-rod geometry.
4   The rear suspension comprises two coil springs and telescopic shock absorbers. Movement of the rear axle is controlled by two lower longitudinal radius rods, two upper longitudinal radius rods and a transverse (Panhard) rod.
5   At regular intervals, inspect the rubber gaiters on the steering and suspension balljoints for splits or damage and renew as necessary.
6   With the help of an assistant, check the steering balljoints, linkage and bushes for wear by gripping each member in turn while the adjacent component is moved and the steering wheel moved slightly in both directions. Any slackness or lost motion must be rectified immediately by renewal of the components concerned.
7   Check the steering box oil level at the intervals specified in Routine Maintenance. The best way to do this is to use a thin rod as a dipstick. Make a mark with a hacksaw 3/8 in (9.5 mm) from the bottom end of the dipstick and insert it into the filler plug hole of the steering box. Read off the oil level and top-up as necessary to bring the oil level up to the mark on the dipstick.
8   Heavy steering occurring on Lada models may be due to the steering column top bush requiring lubrication. To do this, release the steering column top bracket bolts and jack-up the front of the vehicle. Turn the steering wheel to left lock, then to right lock, and then set the front roadwheels in the straight ahead position. Remove the cowling from the upper end of the steering column and apply engine oil to the steering column top bush.
9   Periodically check that the pinch-bolt on the coupling between the column and the wormshaft is tightened to 18 lbf ft (2.5 kgf m).
10   Re-pack the front hub bearings and adjust as described in Section 4 at the intervals specified in Routine Maintenance.

2 Front shock absorber – removal, testing and refitting

1   The front shock absorbers may be checked without removing them from the car. If the ride in your car is rough, or if there is a tendency for the car to pitch excessively when on an undulating road, the shock absorbers are probably worn and should be renewed. A further test is to bounce the car at each corner: if the shock absorbers are in good condition, the car will come to rest as soon as the bouncing is stopped, going only once past the static position. To confirm that a shock absorber is no longer serviceable, remove the unit from the vehicle as described below and grip its lower mounting in a vice so that the shock absorber is in the vertical position. Fully extend and retract the shock absorber six to ten times. Any tendency for the action to be 'jumpy' or exceptionally stiff, or to have no resistance at all, will indicate the need for renewal.
2   Note that shock absorbers should only be renewed in pairs. The vehicle will not handle properly and may even be dangerous, if shock absorbers at different states of wear or of different rate are used on the front suspension.
3   There is no need to remove any major suspension components before extracting the shock absorbers from the front suspension.
4   Raise the front of the car onto car ramps or chassis stands. Chock the rear wheels and ensure that the vehicle is safely supported.
5   From inside the engine compartment, remove the nuts which secure the top of the shock absorbers to the bodyshell (photo).
6   Retrieve the washers and rubber bushes from the shock absorber spindle.
7   From underneath the car, remove the nut and long bolt which locates the bottom of the shock absorber in a bracket in the lower control arm.
8   The shock absorber may then be lowered through the aperture in the lower control arm.
9   Refitting is the exact reversal of removal, but remember to tighten the mounting nuts and bolts to the specified torques.

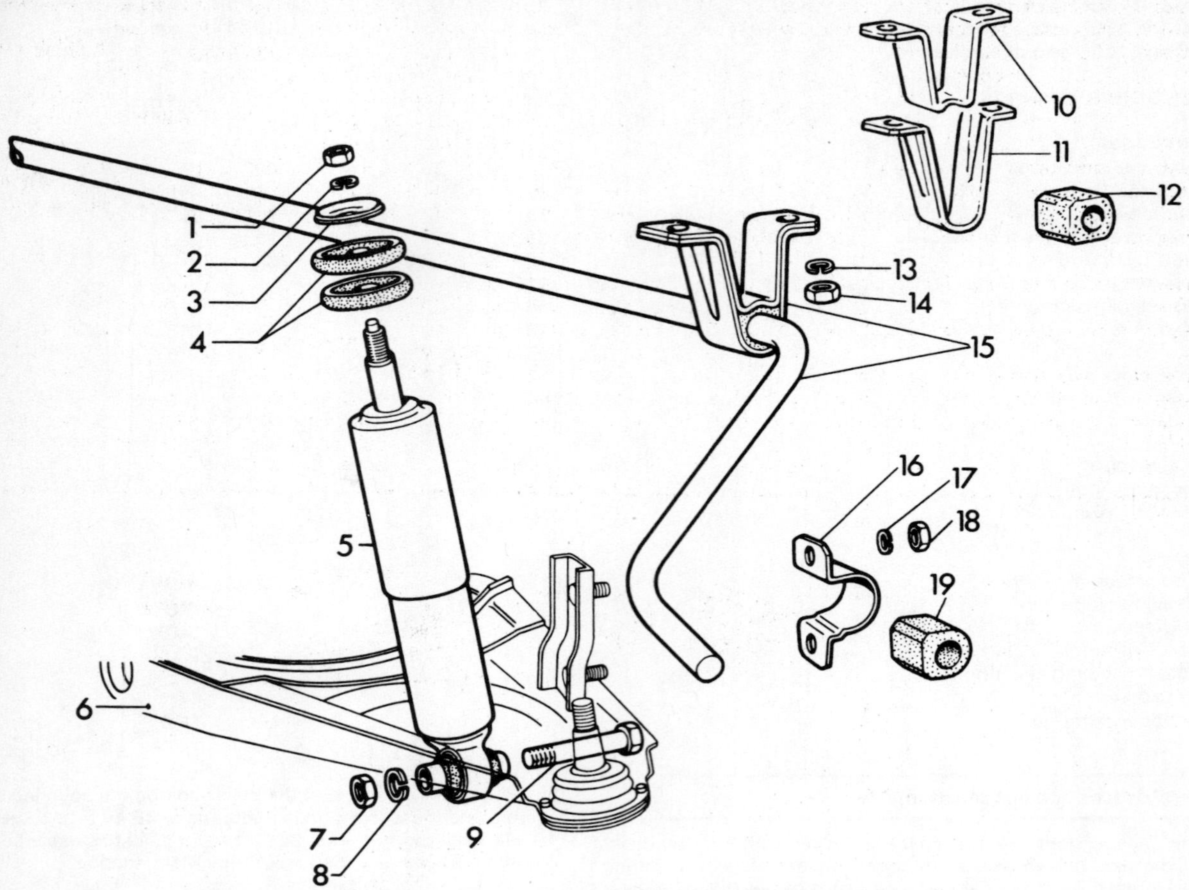

Chapter 11 Suspension and steering

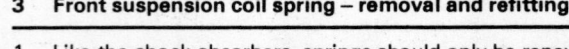

Fig. 11.1 Components of front shock absorber and anti-roll bar (Sec 2)

1 Nut	6 Suspension arm	11 Outer clamp	16 Clamp
2 Spring washer	7 Nut	12 Flexible mounting	17 Spring washer
3 Washer	8 Spring washer	13 Spring washer	18 Nut
4 Flexible mounting bushes	9 Bolt	14 Nut	19 Flexible mounting
5 Shock absorber	10 Inner clamp	15 Anti-roll bar and clamp	

2.5 Front shock absorber top retaining nut

3 Front suspension coil spring – removal and refitting

1 Like the shock absorbers, springs should only be renewed in pairs. It is worth noting however that coil suspension springs rarely give trouble and if the vehicle ground clearance is less than it should be, the shock absorbers will probably be suspect and not the spring. This may appear a little strange, but none the less it is true that if the shock absorber is worn or faulty, it will affect the steady height of the car on the suspension.
2 Section 2 of this Chapter details a few tests that can be carried out with the shock absorbers on the car to determine their condition.
3 Being satisfied that the springs do merit removal and closer inspection, begin by raising the front of the car onto chassis stands placed underneath the ends of the front suspension crossmember. Chock the rear wheels and ensure that the whole vehicle is safely supported.
4 Remove the shock absorbers as directed in Section 2.
5 The next task can only safely be completed with the aid of a spring compressor (Lada Tool No A74174) which is used to compress the spring in situ so that it can be extracted from the suspension linkage. A suitable alternative spring compressor can often be hired, or purchased from a motor accessory shop.
6 Assemble the tool into the spring and compress the spring to the point when there is no force on the lower control arm of the suspension.
7 Remove the two nuts which secure the anti-roll bar locating bush and cap to the lower control arm. Lift the anti-roll bar and associated components from the lower suspension arm.

Chapter 11 Suspension and steering

8 Unscrew the nut which secures the suspension lower balljoint to the stub axle. Using a balljoint separator, disconnect the balljoint from the stub axle, supporting the brake/stub axle assembly during the process to avoid strain on the brake hydraulic hose.
9 Finally the two nuts which secure the lower suspension arm pivot spindle to the crossmember may be removed to allow the spring compressing tool and lower arm to be lifted away from the vehicle together.
10 Retrieve the spacer shims that fit on the bolts between the lower arm pivot spindle and the crossmember. Store them so that they are replaced in the exact positions from which they are taken. The shims are the means by which the camber and castor angle of the front wheels are adjusted.
11 Recover the spring seatings from both upper and lower suspension arms.
12 The spring compressor can now be relaxed and removed from the spring.
13 Clean all the components for inspection.
14 Measure the free length of the spring and compare the result with the figure in the Specifications at the beginning of this Chapter. If either spring is too short, renew both springs.
15 It should be noted that two rates of springs are fitted to the suspension. They are identified by a green or yellow strip on the coils.
Note: *From 22 May 1977 all right-hand drive vehicles have a white or blue marked spring on the driver's side front only.*
16 Naturally both front springs must be of the same rate class, but also the rear springs must be matched as follows.
17 The rear springs are marked as are the front springs into two rate classes. (Note the springs are quite different, only the rates are similarly classified).
18 Yellow striped front springs may be fitted with either yellow or green striped rear springs. Green striped front springs may only be fitted with green striped rear springs, never with yellow striped rear springs.
19 Inspect the rubber spring seatings and if they are cracked or worn, renew them.
20 Having satisfied yourself as to the condition (used or new) of the springs, their rate class (yellow or green) and their compatibility with the rear springs, then the springs may be refitted to the car.
21 Refitting the front springs is quite straightforward, being a reversal of the removal procedure. However, ensure that correct spacer shims are fitted on the appropriate bolts between the lower arm pivot spindle and the bodyshell/crossmember. Make certain that all securing nuts have been tightened to their specified torques.

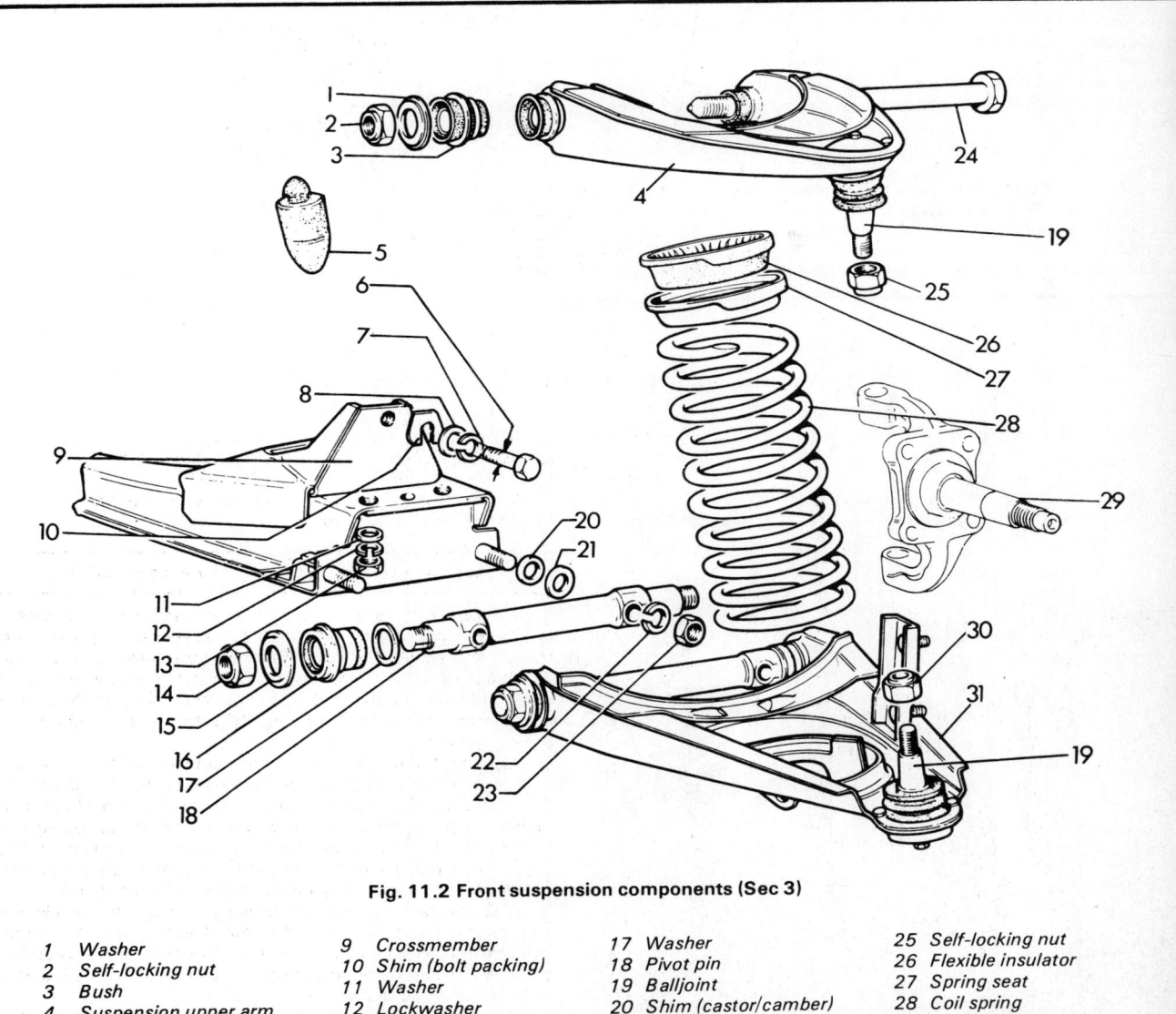

Fig. 11.2 Front suspension components (Sec 3)

1 Washer	9 Crossmember	17 Washer	25 Self-locking nut
2 Self-locking nut	10 Shim (bolt packing)	18 Pivot pin	26 Flexible insulator
3 Bush	11 Washer	19 Balljoint	27 Spring seat
4 Suspension upper arm	12 Lockwasher	20 Shim (castor/camber)	28 Coil spring
5 Buffer	13 Nut	21 Washer	29 Stub axle
6 Crossmember bolt	14 Self-locking nut	22 Lockwasher	30 Self-locking nut
7 Lockwasher	15 Washer	23 Nut	31 Suspension lower arm
8 Plain washer	16 Bush	24 Pivot pin	

4 Front hub bearings – adjustment and lubrication

1 The front hub bearings may be adjusted without removing the roadwheel or any other major item.
2 If the play on the front roller bearings is considered excessive, proceed as follows.
3 Place a jack under the front suspension and raise the roadwheel from the ground. Remove the wheel trim and then the cap on the end of the hub.
4 Wipe the exposed stub axle nut free from grease and loosen it completely. **Note**: *The right-hand stub axle nut has a left-hand thread.* Ensure that the previous crimping does not interfere with the nut's movement.
5 Tighten the nut as described in paragraph 18 of the next Section. Stake the nut and refit the dust cap.
6 At the specified intervals, the hub should be removed, the oil seal extractor and all lubricant cleaned away as described in paragraphs 1 to 9 of the next Section.
7 Repack the hub and roller races with multi-purpose grease. Refit the components and adjust the bearings as described in paragraph 18 of the next Section.

5 Front hub – overhaul

1 Place a jack beneath the lower control arm of the suspension, as near to the road wheel as practicable, and raise the wheel off the ground.
2 Undo the four bolts which secure the wheel to the hub and remove the wheel.
3 Undo the two bolts which secure the brake caliper block to the stub axle. Lift the caliper block away from the brake disc and carefully support it nearby, tying it to the upper central arm or spring will be acceptable.
4 In any event do not strain the flexible brake hose which connects the caliper to the braking system. Wedge the brake pads apart whilst the caliper assembly is separated from the hub assembly.
5 Unscrew the two spigot bolts which hold the brake disc to the hub and remove the disc.
6 Tap the dust cap from the end of the hub.
7 Unscrew and remove the hub nut, *noting that the right-hand stub axle nut has a left-hand thread.*
8 Support the hub assembly and pull it from the stub axle, taking care to catch the outer bearing and the thrust washer if it is displaced.
9 Prise out the oil seal from the inner end of the hub and remove the inner and outer roller bearings.
10 If the bearings are to be renewed, drive out the bearing outer tracks using a brass rod. If the bearings are only to be cleaned and re-packed with lubricant, then the outer tracks should not be removed.
11 Examine the bearing roller and ring surfaces for pitting, scoring and general deterioration. If wear or damage is found, renew both inner and outer bearings complete.
12 Examine the oil seal. If it is damaged or not in a virtually new condition renew the seal.
13 Once all the hub components have been inspected and renewed or accepted for continued service, assembly of the hub commences as follows.
14 Using a piece of tubing, drive the two bearing tracks into their seats. Make sure that they are fully seated and located the correct way round.
15 Press multi-purpose grease into the bearings and apply a liberal coating of grease to the interior of the hub. Fit the bearing, spacer and oil seal to the hub inboard end.
16 Hold the hub level and push it onto the stub axle, taking care not to damage the oil seal.
17 Fit the outboard bearing and the thrust washer and screw the nut into position.
18 Tighten the nut to a torque of 15 lbf ft (2.1 kgf m), then release the nut and tighten again to 5 lbf ft (0.7 kgf m). Finally slacken the nut $\frac{1}{12}$ th of a turn (30°) and stake the nut into the groove in the stub axle. To tighten the nut with the left-hand thread, if a suitable torque wrench is

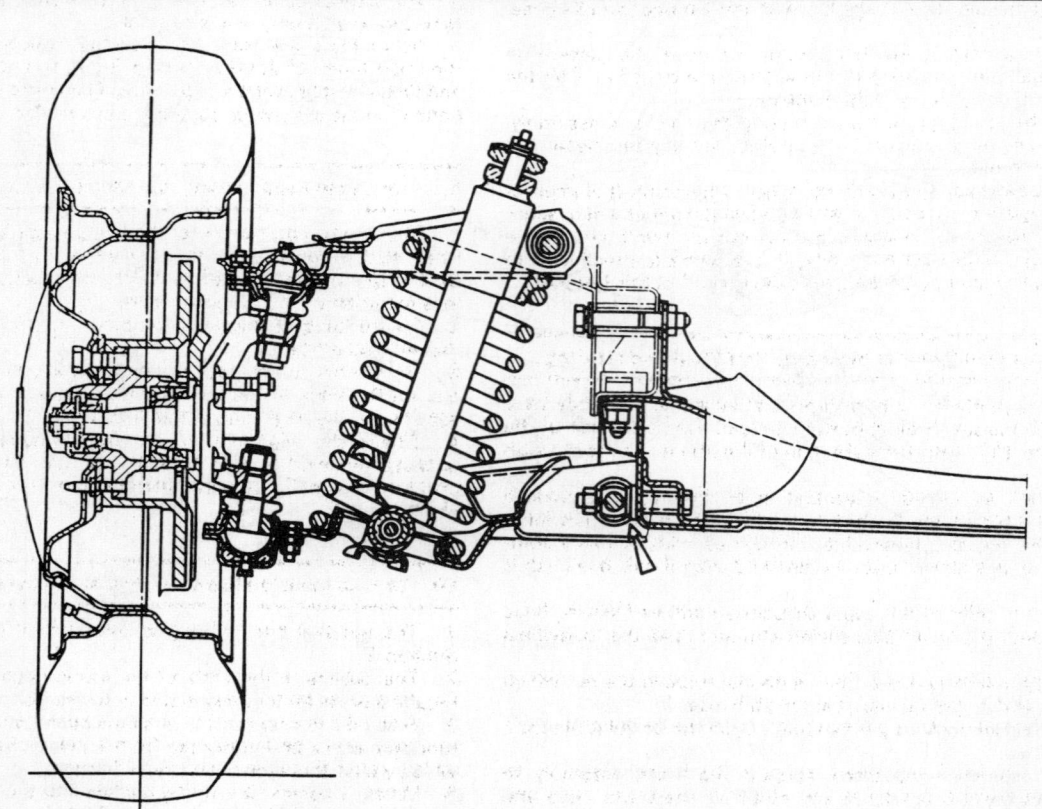

Fig. 11.3 Sectional view of one side of the front suspension. Arrow shows location of castor/camber adjusting shims (Sec 5)

not available, use a spanner to which is attached a spring balance. Provided the spring balance is attached to the spanner at a point 12 in (305 mm) from the centre of the stub axle, the setting will be fairly accurate.
19 Half fill the dust cap with grease and tap it squarely into position. Fit the disc, caliper and roadwheel.

6 Front suspension balljoints – inspection, removal and refitting

1 The condition of the balljoints on the top and bottom of the stub axle assembly can be checked by placing a jack under the lower suspension arm acting on the damper mounting, and raising the road wheel off the ground.
2 Grasp the wheel and try to move it up and down, side to side and finally in and out. If relative movement of the stub axle and lower suspension member is observed, the balljoint is worn and must be renewed.
3 The upper balljoint may be checked by grasping the upper suspension arm and stub axle and forcing them in opposition. If relative movement across the balljoint is observed then it too is worn and requires renewal.
4 A point to be mentioned before the removal task is commenced, is that when originally assembled, the balljoint housing was riveted to the appropriate suspension arm. Therefore, if it is the original joint that is to be replaced, the rivets must be drilled out, and the proper nuts, bolts and washers purchased with the new balljoint.
5 Proceed to remove the balljoints as follows. Raise the suspension and roadwheel as directed in paragraph 1 of this section. Remove the roadwheel.
6 Undo and remove the nut on the ball-pin of the joint. Using a suitable balljoint separator, disconnect the balljoint from the stub axle.
7 Support the wheel hub/brake assembly adequately to prevent the flexible brake hose and/or the steering linkage being strained.
8 Remove the three nuts and bolts, or centre punch and drill out the three rivets, which secure the balljoint housing onto the suspension arm.
9 Lift the old balljoint away together with the pin boot and its base. Discard the old balljoint.
10 Place the new balljoint and boot in position under the suspension arm with the ball-pin projecting in the appropriate direction. Bolt the joint and cover in position and tighten the nuts.
11 Insert the ball-pin into its mating hole in the stub axle assembly. Slip the pin washer and a new nut into position and tighten the pin nut to the specified torque.
12 Refit the roadwheel and check the wheel alignment. It is improbable that the replacement balljoint will significantly affect wheel alignment, but if it has been necessary to drill out old rivets and fit the replacement joint with nuts and bolts, it is a sensible precaution to check the wheel alignment as directed in Section 23 of this Chapter.

7 Front suspension upper control arm – removal and refitting

1 The upper suspension control arm is attached to the body by a long bolt which passes through bushed ends of the inner edge of the triangular suspension arm. The outer end of the arm holds the top stub axle balljoint.
2 The balljoint, as already described in Section 6, is renewable individually, and the rubber bushes which fit in the ends of the inner edge of the arm are also renewable, through as with so many components on cars they should both be renewed even if only one bush is suspect.
3 Commence to remove the upper suspension arm as follows. Place a jack underneath the lower suspension arm and raise the roadwheel off the ground.
4 Remove the roadwheel and then undo and remove the nut which retains the pin of the upper balljoint in the stub axle.
5 With a universal balljoint pin extractor, push the balljoint pin from the stub axle.
6 Once the pin has been freed, support the brake assembly to prevent it from moving outwards and straining the brake hose and steering linkage.
7 Next undo the self-locking nut on the long bolt which retains the arm to the body. Then extract the long bolt. The upper suspension arm is now free and can be taken to the work-bench for inspection and

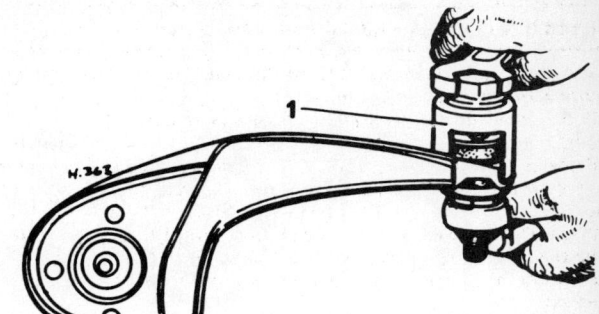

Fig. 11.4 Renewing the suspension upper arm bushes (Sec 8)

1 Special tool No A47046

renovation.
8 Refitting the suspension arm follows the reversal of the removal procedure. Remember to use new self-locking nuts on the balljoint pin, and long bolt associated with the arm. Those nuts should also be tightened to the torques specified at the beginning of this Chapter.

8 Front suspension upper control arm – overhaul

1 The balljoint can be removed and renewed as described in Section 6 of this Chapter.
2 The rubber bushes in the ends of the inside edge of the triangular arm need to be pushed out of the arm with a drift or mandrel acting from the middle of the arm. A tool can easily be improvised with a piece of tube, some washers and a long bolt.
3 As mentioned in Section 7 both bushes in the arm should be renewed, even if only one is suspect.
4 While the suspension arm is on the work-bench it is wise to take the opportunity of thoroughly cleaning it, inspecting for rust or cracks and finally rust proofing and painting. Obviously this arm is a vital component on the car and its condition must not be allowed to deteriorate.

9 Front suspension lower control arm – removal and refitting

1 The removal procedure for the lower suspension arm is exactly the same as the procedure for the removal of the front spring (see Section 3). The spring acts between seatings on the bodyshell and similar seatings mounted on the suspension arm.
2 The refitting of the lower suspension arm follows the reversal of the removal procedure.
3 Remember to use new self-locking nuts on the lower balljoint pin and on the ends of the pivot spindle. Tighten all nuts to the torques specified at the beginning of this Chapter.
4 Finally the alignment of the front wheels should be checked, because renewed bushes or balljoints may have altered the correct position sufficiently to merit different shims between the pivot spindle and bodyshell.

10 Front suspension lower control arm – overhaul

1 The removal and refitting of the balljoint has been described in Section 6.
2 The bushes in the ends of the inside edge of the triangular arm require a press for their extraction and insertion.
3 A spindle is employed to push the bush from the arm (Fig. 11.5). A tool may easily be improvised from a piece of steel tube, the bore of which would allow the bush to slip through.
4 When it comes to refitting bushes into the suspension arm, tools may be improvised from suitable steel tubing and washers (Fig. 11.6).
5 The opportunity should be taken to thoroughly clean the lower suspension arm when it is on the bench and to inspect it for rust and cracks. The arm can then be rust proofed and painted. If cracks or joint

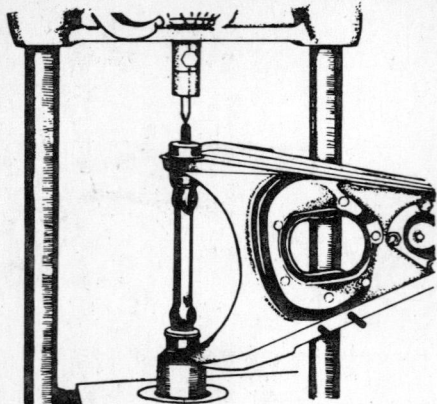

Fig. 11.5 Pressing out the suspension lower arm bushes (Sec 10)

separation are found the arm should be discarded and a new one fitted. The suspension arm is a vital component on the car and it must be kept in good condition.

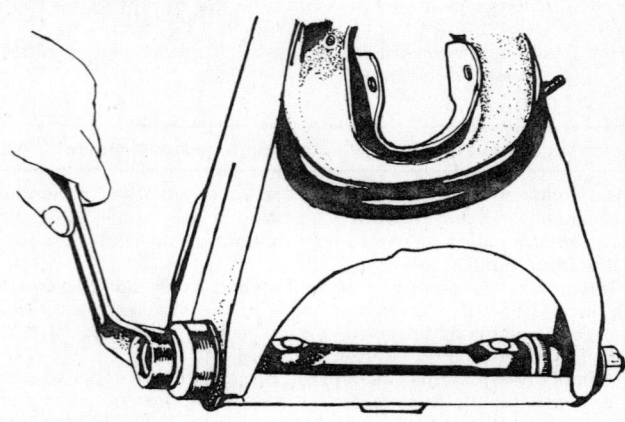

Fig. 11.6 Method of fitting the suspension arm bushes (Sec 10)

11 Front suspension crossmember – removal and refitting

1 This is a complex task and fortunately it is very rarely necessary since there is little to go wrong. The crossmember is a box beam which supports the lower suspension arms at its ends and the engine at mountings near the ends. The beam is secured to a strengthened area of bodyshell between the wheel arch and engine compartment.
2 Raise the front of the car onto chassis stands located just forward or rear of the crossmember location on the bodyshell. Remove the roadwheels.
3 Continue by removing both lower suspension arms as described in Sections 3 and 9 of this Chapter.
4 The engine should now be supported with a jack acting on the sump, or a sling around the front of engine connected to a hoist, so that the rubber engine suspension blocks may be detached from the crossmember.
5 A strut which spans the top of the engine compartment can be used to support the engine whilst the crossmember is detached.
6 The two nuts and single bolt which attach each end of the crossmember to the body can now be removed and the crossmember itself lifted clear. Retrieve the spacers and washers on the bolt and two nuts and store them so that they may be replaced in the exact position from which they were removed.
7 Refitting is the exact reversal of removal, but remember to use new spring locking washers on the crossmember attachments and new self-locking nuts on the lower suspension balljoint pins.
8 Tighten all nuts and bolts to the torques specified at the beginning of this Chapter.

12 Front suspension anti-roll bar – removal and refitting

1 This task is quite straightforward, the bar is located at four points. Its ends are attached to the lower suspension arms and two caps and split bushes hold the centre length of the bar to strengthened areas of the body beneath the forward end of the engine compartment.
2 Begin removal of the bar by raising the front of the car onto ramps or chassis stands placed beneath the ends of the suspension crossmember.
3 Continue by removing the nuts which secure the caps and bushes at the ends of the bar to brackets on the lower suspension arm.
4 Finally remove the nuts which secure the main bar support brackets to the body and lift away the bar together with all the bushes, caps and centre brackets.
5 The centre brackets come in two parts which are available individually as spares. the bar end bush caps are also available individually.
6 The bushes can be slid off the bar and renewed if necessary.
7 Refitting is the exact reversal of removal, but remember to use new spring washers behind the retaining nuts and tighten all nuts to the torques specified at the beginning of this Chapter.

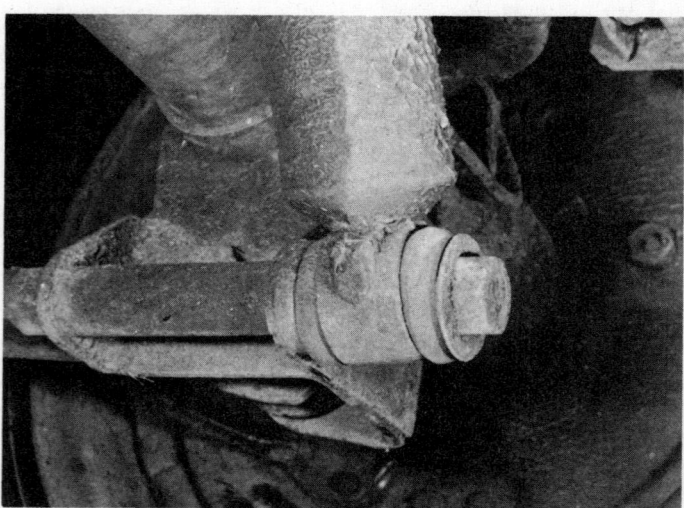

13.3 Rear shock absorber lower mounting bolt

13 Rear shock absorber – removal and refitting

1 The condition of the shock absorbers is indicated by the behaviour of the vehicle on the road. If there is an excessive pitching motion when motoring on an undulating road or an excessively bumpy ride, the shock absorbers merit renewal. Another indication of shock absorber malfunction is when the vehicle is pressed and bounced at each corner, and it does not immediately return to its usual rest position. If either the car is slow to return, or over returns and oscillates before recovering, the shock absorbers must be renewed.
2 Raise the rear of the car onto chassis stands located beneath the rear axle casing. Chock the front wheels and ensure that the vehicle is safely supported.
3 Remove the upper and lower shock absorber retaining nuts and withdraw the shock absorbers complete with rubber bushes (photo).
4 Refit the shock absorbers and rubber bushes using the reverse of the removal procedure.

14 Rear suspension coil spring – removal and refitting

1 Raise the rear of the car and support it on stands under the side frame members. Remove the rear wheels.
2 Undo the four nuts which hold the propeller shaft central bearing support flange to the body.
3 Jack-up the rear axle slightly to support its weight and remove the

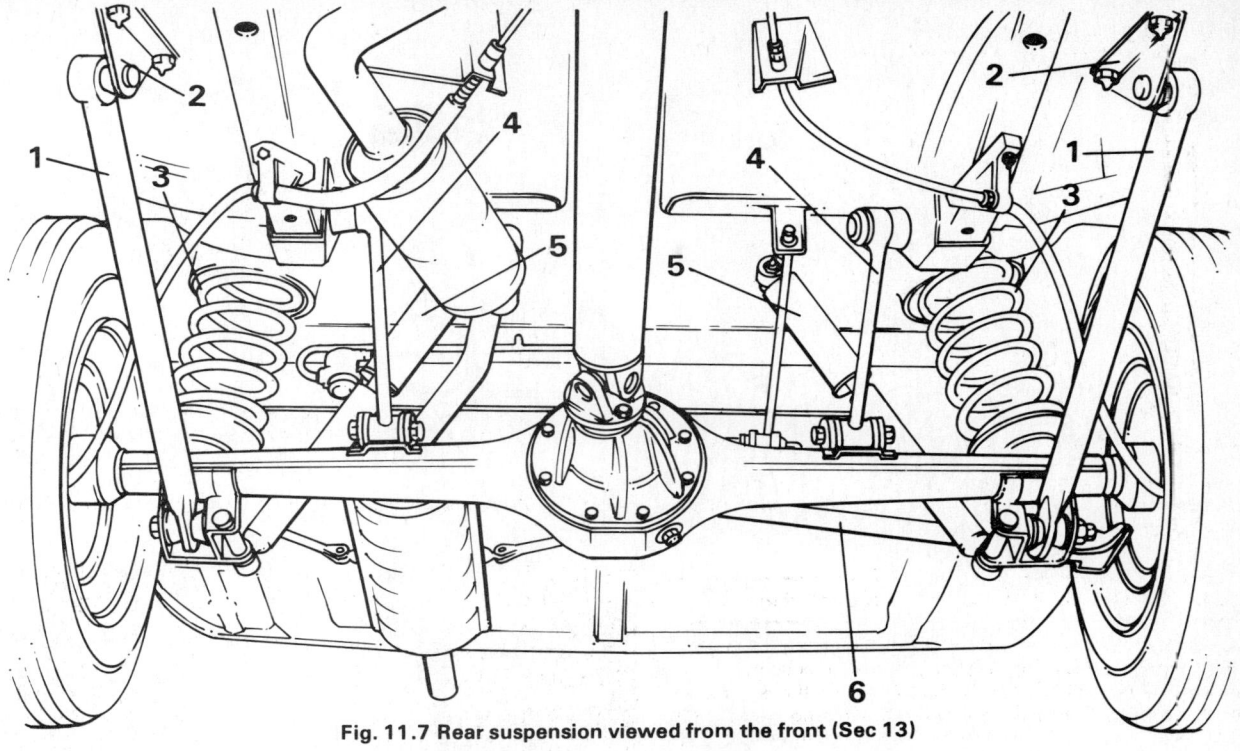

Fig. 11.7 Rear suspension viewed from the front (Sec 13)

1 Suspension lower arm
2 Bracket
3 Coil spring
4 Suspension upper arm
5 Shock absorber
6 Panhard rod

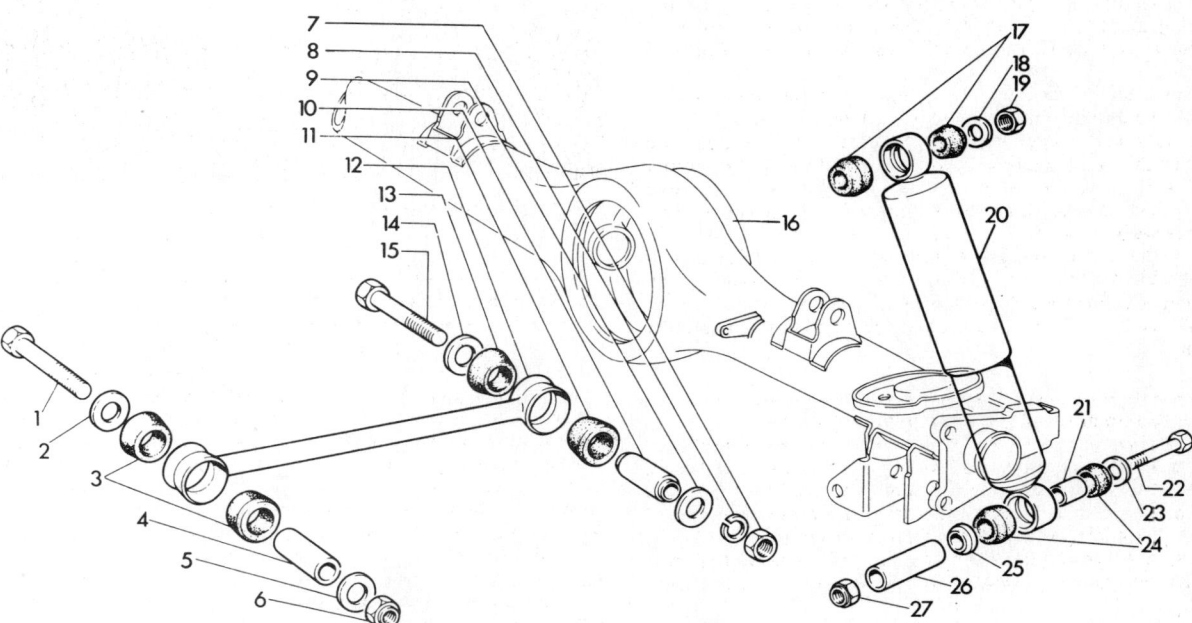

Fig. 11.8 Rear suspension upper arm and shock absorber (Sec 13)

1 Bolt
2 Washer
3 Flexible bush
4 Sleeve
5 Washer
6 Nut
7 Nut
8 Spring washer
9 Washer
10 Sleeve
11 Flexible bush
12 Suspension arm
13 Flexible bush
14 Washer
15 Bolt
16 Axle casing
17 Flexible bush
18 Washer
19 Nut
20 Shock absorber
21 Sleeve
22 Bolt
23 Washer
24 Flexible bushes
25 Spacer
26 Sleeve
27 Nut

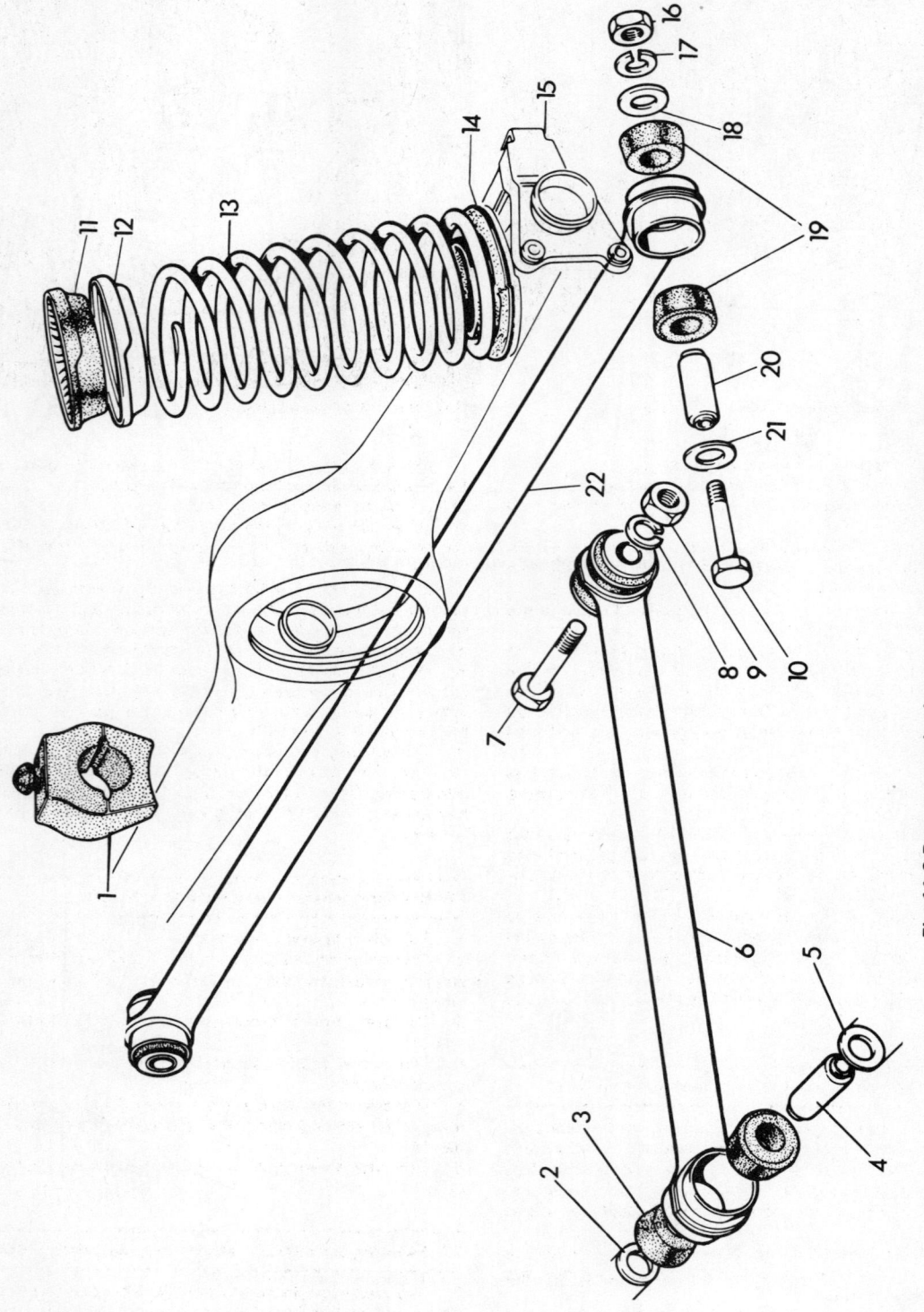

Fig. 11.9 Rear suspension lower arm and coil spring (Sec 14)

1 Buffer block
2 Washer
3 Flexible bush
4 Sleeve
5 Washer
6 Suspension arm
7 Bolt
8 Spring washer
9 Nut
10 Bolt
11 Rubber insulator
12 Spring seat
13 Coil spring
14 Rubber insulator
15 Axle casing
16 Nut
17 Spring washer
18 Washer
19 Flexible bushes
20 Sleeve
21 Washer
22 Panhard rod

Chapter 11 Suspension and steering

14.5 Connection of rear brake regulator link rod to axle casing

16.4 Steering wheel retaining nut

lower ends of the shock absorbers from the axle brackets.
4   Disconnect the upper radius rods from the axle brackets.
5   Disconnect the brake regulator link from the lug on the axle (photo).
6   Support the axle with a jack and then disconnect the two main radius arms at the attachments to the body and also the transverse Panhard rod from the body bracket.
7   Carefully lower the axle on the jack just sufficiently to enable the springs to be withdrawn.
8   Once the springs are clear of the car, they can be brushed clean and inspected carefully for rust and cracks. The free length of the springs should also be checked; details are included in the Specifications at the beginning of this Chapter. The springs should be renewed if their condition is less than perfect, as they are stressed and vital parts on the car.
9   Attention should be paid to the colour markings on the spring. There will be either a yellow or green stripe on the spring coils; these indicate the class which the rate of the spring is in.
10 Both rear springs should be the same class (colour); the rules set down in Section 3 should be observed when the springs are being renewed and the matching of the front and rear springs need consideration.
11 Refitting is a reversal of the removal procedure. Take care to reseat the springs correctly and do not tighten the nuts on the radius arm and Panhard rod fixing points until the whole assembly is refitted and the vehicle lowered to the ground. The vehicle should then be loaded and the suspension arm nuts should then be tightened to the specified torques.

15 Rear axle locating arms – removal, overhaul and refitting

1   There are several groups of locating arms acting on the rear axle. To begin with there is the main pair of rods attached to anchoraes bolted onto the body and to points welded to the axle casing. Then there is the transverse stabilising rod, known as a Panhard rod, and finally there is a pair of rods above the axle which accept the torque from the roadwheels.
2   All rods have a similar design, being basically a thick walled tube with bush housings welded to each end. Rubber bushes press into each housing around a tube spacer.
3   Never attempt to remove any more than one rod at a time, so if the task is to renew all the rubber bushes in the ends of the locating arms, remove one rod, renew the bushes, refit it and then proceed to the next arm.
4   The procedure for the removal of the rods is as follows: Raise the rear of the car onto chassis stands located beneath the body frame which will not result in interference with later tasks. Do not put the chassis stands underneath the axle as it is necessary for the suspension linkage to be unloaded.

5   Place a pair of jacks beneath the rear axle casing and raise to take the weight of the axle assembly. The suspension rods should now be unloaded and ready for removal.
6   Undo the nuts on the long bolts which pass through each end of the locating rod, extract the long bolts and pull the rod out of the anchorage.
7   The rod may now be brushed clean and inspected. It should be perfectly straight and the ends round. Check the welded ends carefully for cracks and corrosion. As with springs, the rods must be in perfect condition. Renew suspect or bent rods immediately.
8   With the rod on a bench, press the centre tube spacer inside the rubber bushes out with a suitable drift and prise the old bushes out with a screwdriver. New bushes may be pressed into position followed by the central spacer tube.
9   The refitting procedure is simply the reversal of that for removal. Always use new self-locking nuts or spring lockwashers as appropriate. Tighten the nuts ands bolts to the torques specified at the beginning of this Chapter, when the car has been lowered to the ground.

16 Steering wheel – removal and refitting

1   For safety reasons, first disconnect the battery earth terminal.
2   On earlier models prise out the plastic cover from the centre of the steering wheel, and remove the screws retaining the horn ring to the wheel.
3   On later models remove the screws securing the horn pad to the wheel.
4   The wheel retaining nut can now be removed using a suitable socket (photo).
5   Make sure the front wheels are in the straight ahead position and remove the steering wheel from the splines by striking the rim with both hands.
6   Refit the steering wheel using the reverse of the removal procedure.

17 Steering column and shaft – removal and refitting

1   The steering column is bolted by a bracket to the dash panel at the upper end and clamped to the steering box wormshaft by a splined collar at the lower end.
2   Once the lower coupling and upper bracket bolts have been loosened, and the wires disconnected from the ignition switch and headlamp flasher switch, the whole column complete with steering wheel can be removed from inside the car.
3   Although it is possible to line up the splined sections of the shaft later it is always good practice to mark each connection before separating it so that no mistake is made.

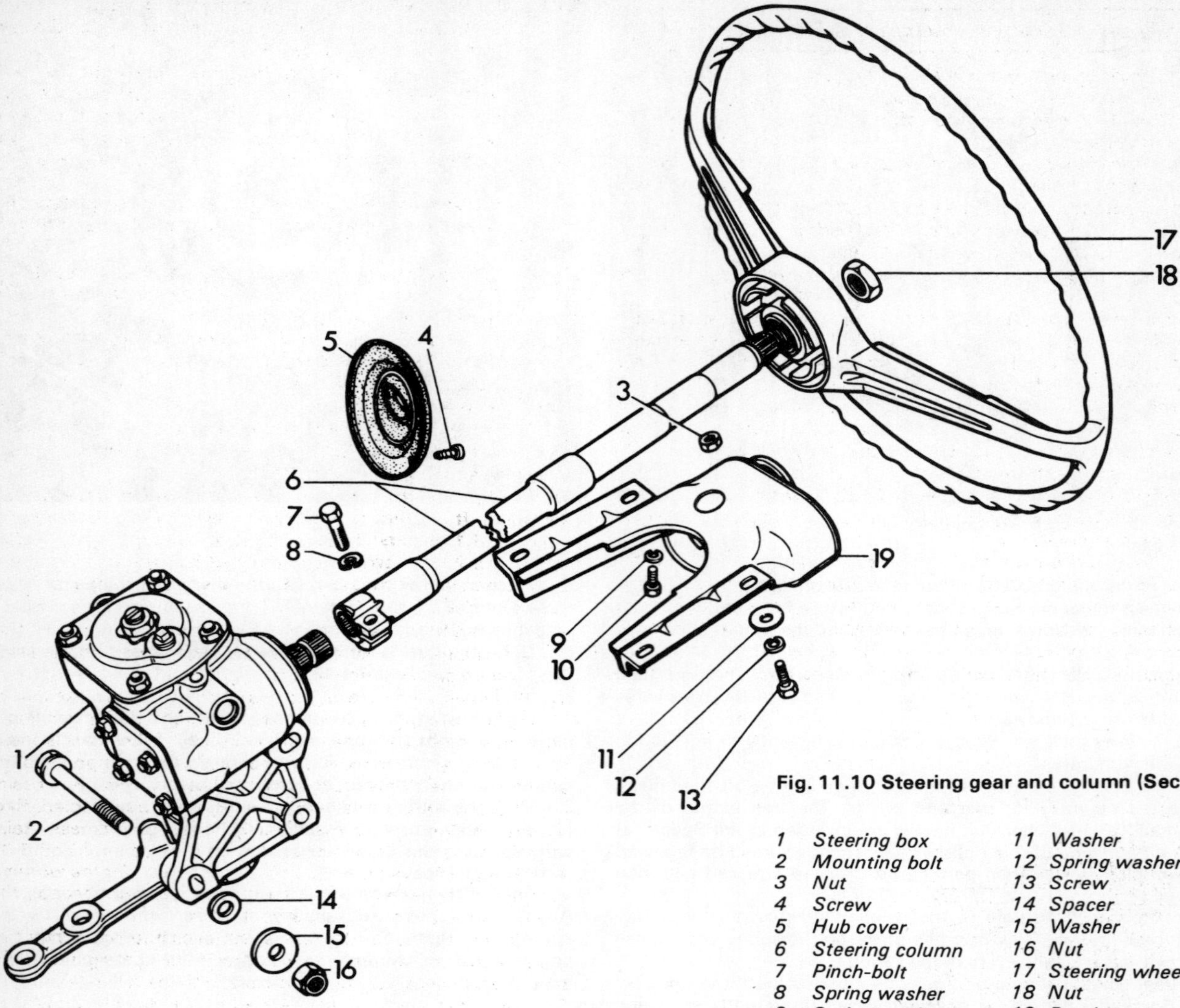

Chapter 11 Suspension and steering

Fig. 11.10 Steering gear and column (Sec 17)

1 Steering box
2 Mounting bolt
3 Nut
4 Screw
5 Hub cover
6 Steering column
7 Pinch-bolt
8 Spring washer
9 Spring washer
10 Screw
11 Washer
12 Spring washer
13 Screw
14 Spacer
15 Washer
16 Nut
17 Steering wheel
18 Nut
19 Bracket

4 When refitting the column, assemble all the mounting bolts loosely to start with so that the steering can be turned from side to side with the front wheels off the ground. Then tighten the mounting bracket bolts. This procedure will ensure that the steering column is correctly aligned.

18 Steering box — removal and refitting

1 Begin by raising the front of the car onto ramps or chassis stands located beneath the ends of the suspension crossmember.
2 Undo the nuts which retain the steering linkage balljoints into the drop arm on the steering box. The balljoints may be extracted from the drop arm either with a universal extractor or with a sprocket puller (photo).
3 From inside the car, remove the nut and bolt which secure the lower yoke on the steering column to the steering box wormshaft.
4 Finally undo the nuts and bolts which retain the steering box to the body and lift the box away from the car, taking it to a clean bench for dismantling and repair.
5 Refitting the steering box follows the reversal of the removal procedure. Use new spring lockwashers and self-locking nuts as appropriate and tighten all nuts and bolts to the torques specified at the beginning of this Chapter.

18.2 Steering box drop arm and balljoints

Chapter 11 Suspension and steering 145

19 Steering box – dismantling, examination and reassembly

1 Clean the exterior of the box and wipe dry with a non-fluffy rag.
2 Remove the oil drain/filler plug and allow the oil to drain out.
3 Mount the box firmly in a vice and unscrew the nut which holds the drop arm to the sector shaft.
4 Using a universal puller draw the drop arm from the spline on the sector shaft. Retrieve the tag washer. There is a master spline on these two parts so that they may only be fitted together in one way.
5 Undo and remove the four bolts which secure the top cover onto the box casing. Lift away the sector shaft cover and roller sector from the steering box. Retrieve the gasket fitted between the top cover and casing.
6 Remove the needle roller bearings in which the sector shaft ran. If absolutely necessary, they may be pushed out of the bore in the steering box casing using a soft drift. The oil seal in the base of the bore should be prised out with a screwdriver.
7 The top cover can be separated from the sector shaft once the locknut on the gear meshing adjusting screw in that cover has been undone and the adjusting screw turned sufficiently to work right through and disengage the cover.
8 The screw is retained in the sector shaft by a threaded insert. It will not normally be necessary to remove this insert to extract the gear mesh adjusting screw.
9 If play is found between the roller gear on the sector and the spindle which retains it to the shaft, it will be necessary to drift the spindle out to free the roller gear. The needle roller bearings in the roller, and the surface of the spindle, can then be inspected for wear.
10 The steering box wormshaft and worm gear can be extracted from the box once the four bolts which retain the end cover in place have been removed. Take care to collect the shims which fit between the cover and the box casing, they are the means by which the endplay on the wormshaft bearing is adjusted.
11 Once the cover is off the box, the wormshaft can be pushed through and out of the box, together with a spacer and bearing.
12 Prise the wormshaft oil seal out of its seating in the box casing with a screwdriver, and then tap the outer track of the wormshaft bearing out of its seating in the box. Retrieve the spacer and worm gear positioning shims.
13 All the steering box components are now completely dismantled and should be washed in paraffin and dried in preparation for inspection.
14 The splined ends of the worm and sector shafts should be checked for wear. Both shafts have surfaces which serve as the inner track for their respective bearing assemblies. The sector shaft shank surface should be closely inspected for scoring, pitting and other surface deterioration. The diameter of the shank should be between 1.1295 and 1.1286 inches in the bearing region. Renew the shaft and bearings if found to be worn.
15 Inspect the ball grooves on each side of the worm gear on the wormshaft for signs of pitting and wear. The gear surfaces should be inspected similarly.
16 Finally the bush in the top cover which accepts the top end of the sector shaft should be examined. Its internal diameter should be between 1.1298 and 1.1306 inches: if worn, prise it out of the cover and renew it. Remember to check the diameter of the top end of the sector shaft which runs in the bush; it should be between 1.1259 and 1.1286 inches.
17 Steering box reassembly is the reversal of dismantling. The following paragraphs indicate how the two major shafts on the steering box are set correctly in their bearings.
18 Begin with the steering box wormshaft. Always use a new oil seal in the box seating and retain the bearing balls in position on their respective races with medium grease whilst the shaft is inserted into the box casing. Once the shaft and associated components have been assembled, the shims fitting between the end cover and casing which adjust the running of the shaft bearings can be selected. Reduce the number of shims to increase running torque or increase their number to reduce the torque, which should be between 1.7 and 5.6 lbf in.
19 Having achieved the correct running torque for the wormshaft, and completed the assembly of the input shaft components, the sector shaft may be refitted into the steering box.
20 Lightly tap the needle race assemblies into position in the bore in the steering box, and fit a new oil seal in the seating in the base of the bore. Insert the sector shaft complete with the roller gear fitted.
21 Lower the top cover and its gasket over the box and turn the gear mesh adjusting screw to draw the cover onto the sector shaft and steering box casing.
22 Lift the cover and sector shaft temporarily to disengage the roller from the worm. Turn the worm so that when the roller is lowered into mesh, it lies exactly midway along the worm gear.
23 Screw the mesh adjusting screw so that the top cover can seat fully on the box casing. Bolt the cover into position and then without turning either shaft, push the drop arm onto the splined end of the sector shaft so that it lies parallel with the mounting base of the box. Fit a new spring lockwasher and screw the retaining nut into position.
24 Turn the mesh adjusting screw to eliminate any backlash of the roller and worm gear, and then turn the wormshaft, firstly to check that the sector shaft components and drop arm have been correctly assembled into the steering box, and secondly to check that the roller moves evenly along the worm gear.
25 The range of drop arm movement required is shown in Fig. 11.12. If it is found that as the wormshaft is turned and the roller moves along the worm gear, that the roller becomes progressively tight towards one end and slack towards the other, the worm gear wormshaft is not centred properly. This can be corrected by changing the thickness of the shims behind the upper wormshaft bearing outer track. Remember that a corresponding change must be made to the shims behind the end cover which govern the running torque of the wormshaft.
26 Finally, having satisfied yourself that the steering box components are properly adjusted and set, fit the locking nut onto the mesh adjusting screw, tighten and retain in position by bending a tab on the washer onto one of the nut faces. Tighten the drop arm retaining nut to the specified torque, and check the tightness of the top and cover bolts. Refill the steering box with the specified grade of oil. The box is now ready to be refitted to the car.

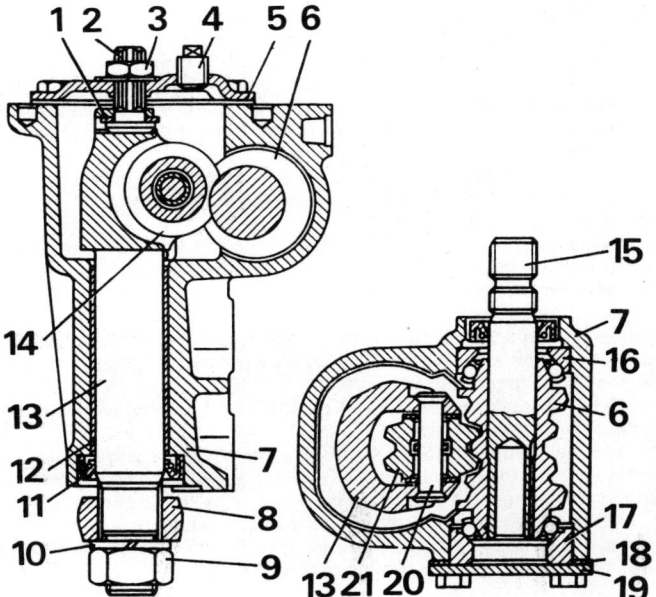

Fig. 11.11 Sectional views of the steering box (Sec 19)

1 Sector shaft adjusting screw shim
2 Sector shaft adjusting screw
3 Locknut
4 Oil filler plug
5 Top cover
6 Worm
7 Steering box
8 Drop arm
9 Nut
10 Spring washer
11 Oil seal
12 Bronze bush
13 Sector shaft
14 Roller
15 Wormshaft
16 Upper bearing
17 Lower bearing
18 Shim
19 End cover
20 Roller spindle
21 Roller

146 Chapter 11 Suspension and steering

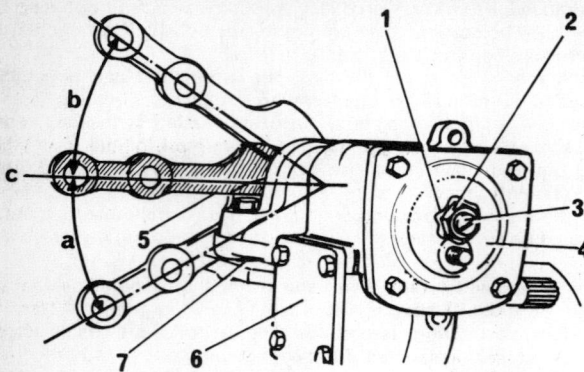

Fig. 11.12 Steering gear adjustment diagram (Sec 19)

1 Lockwasher
2 Locknut
3 Adjusting screw
4 Top cover
5 Drop arm
6 End cover
7 Mounting plate for testing
a and b 30° 40' ± 1° 40'
c Central position of gear

20 Steering gear – adjustment

1 There is no real advantage in trying to tackle this task with the box in place on the car. The steering rods need to be detached from the drop arm on the steering box, and the steering column should also be detached from the box when the running of the gears is to be adjusted. Therefore, it will be easier and the job will be completed more accurately if the box is removed as described in Section 18.
2 The running of the steering gears can now be adjusted as described in paragraphs 18, 24 and 25 of Section 19.
3 If there is still some slackness, sloppiness or stiffness in the steering box gear system, then the box should be dismantled and inspected as detailed in Section 19.

21 Idler arm – removal and refitting

1 The idler arm pivot incorporates a damping device which accepts shocks and road vibration coming from the wheels and thereby serves to protect the steering system. From the driver's point of view, the steering will feel light and smooth with very little road vibration being fed back to the steering wheel.
2 The idler arm and pivot are mounted in the opposite position to the steering box on the bodyshell (photo). The idler arm and drop arm on the steering box are linked by a fixed length steering rod.
3 Adjustable length track-rods connect the idler arm and drop arm to the steering arms on the stub axle assemblies.

21.1 Steering idler pivot and arm

4 Removal of the idler arm and pivot is quite straightforward; begin by raising the front of the car onto a pair of car ramps or chassis stands. Chock the rear wheels and check that the car is safely supported; you will be working underneath it!
5 Continue by detaching the track-rod and link rod from the idler arm. Use a universal ball-pin extractor to press the balljoint from the idler arm once the retaining nuts have been undone.
6 The idler arm and pivot may be unbolted from the bodyshell.
7 Clamp the idler arm/pivot bracket in a vice, remove the split pin and nut, followed by the two washers noting the direction of fitting. The idler arm and pivot shaft can now be removed from the bracket.
8 Prise out the seals then, using a socket or drift of suitable diameter, drive the bushes out of the bracket. Examine the bushes for excessive clearance between the shaft and bushes, or whether they are out of round. If wear is evident, replace the bushes.
9 Examine the pivot shaft and idler arm for damage or distortion and replace if necessary. Separate the shaft and arm by undoing the self-locking nut and removing the arm followed by the washer and seal noting the direction of fitting.
10 Refitting is a reversal of the removal procedure but note the following points:
 a) If the idler arm and pivot shaft have been separated, the self-locking nut should be tightened to the specified torque wrench setting
 b) Coat the bushes with a suitable grease before reassembly
 c) Tighten the top nut until the idler arm, when horizontal, can turn when a force of 1 to 2 kgf (2.2 to 4.4 lbf) is applied, then fit a new split pin
 d) Use new self-locking nuts on the balljoint pins and bolts which secure the pivot housing to the bodyshell
 e) Tighten all nuts and bolts to the torque wrench settings specified at the beginning of this Chapter

22 Steering rods and balljoints – general

1 The steering drop arm is connected to the idler arm by a fixed length relay rod. The two track-rods are connected to the drop and idler arms on the one end and on the other end to the steering arms.
2 To allow for front wheel toe-in adjustment, each end of the track-rods has an internal thread into which the balljoint member is screwed. A clamp around each end tightens the split track-rod onto the balljoint member and locks the two together.
3 Regular inspections should be made of the balljoints on each end of the track-rods and relay rod. If the joints are found to be sloppy and worn they must be renewed.
4 Removal of the balljoint assemblies from the track rod is straightforward. With the rod assembly on a bench, measure the distance between the centres of the balljoints at each end of the rod. Slacken the end clamps and unscrew the joint from the rod, counting the number of turns taken. Screw a new joint into the rod to the same position as the old one had occupied. Check the balljoint centre distance and tighten the rod clamps.
5 Refit the track-rods to the vehicle and check the front wheel toe-in as described in Section 23.
6 The balljoints on the ends of the centre relay rod cannot be removed since they are integral with the rod. If one or both joints are worn it will be necessary to purchase a new relay rod assembly.
7 As usual, new self-locking nuts should be used when refitting rod assemblies back onto the car, and all nuts should be tightened to the torques specified at the beginning of this Chapter.

23 Front wheel alignment – adjustment

1 Provided that any repair work on the steering and suspension system involved only the renewal of joints and/or bushes and not disturbance of the lengths of the track rods, then you should be able to drive carefully to your nearest Lada dealer, where the wheels may be aligned using the specialised equipment which is essential for this task.
2 Of all the settings to be considered, castor angle, camber angle and toe-in, it will only be the latter which is likely to be seriously affected during repair work on the car. As long as the same shims are used on reassembly as were found on the front suspension lower control arm pivot spindle, then the camber and castor angles should be good enough for the drive to the local Lada dealer.

Chapter 11 Suspension and steering

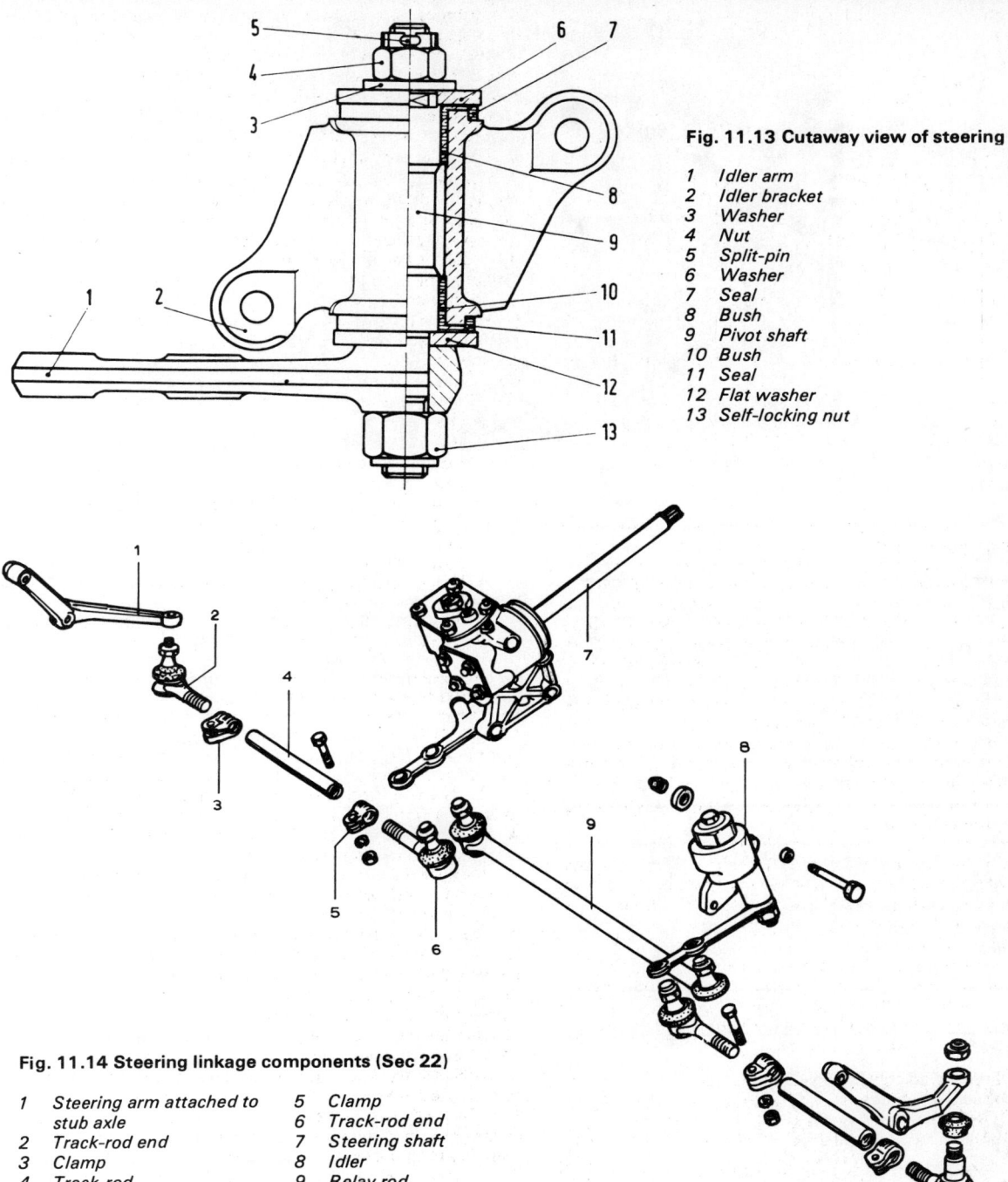

Fig. 11.13 Cutaway view of steering idler (Sec 21)

1. Idler arm
2. Idler bracket
3. Washer
4. Nut
5. Split-pin
6. Washer
7. Seal
8. Bush
9. Pivot shaft
10. Bush
11. Seal
12. Flat washer
13. Self-locking nut

Fig. 11.14 Steering linkage components (Sec 22)

1. Steering arm attached to stub axle
2. Track-rod end
3. Clamp
4. Track-rod
5. Clamp
6. Track-rod end
7. Steering shaft
8. Idler
9. Relay rod

3 To check the toe-in, the car should be positioned on a flat and level surface and loaded with four persons or their equivalent weight. The tyres should be at their correct pressures.
4 The front wheel toe-in is the difference in the distance between the edges of the wheel rims when measured at hub height at the front and then at the rear of the roadwheels.
5 Check the toe-in against the figure given in the Specifications at the beginning of this Chapter using a suitable gauge or rod adjustable for length.
6 Adjust the toe-in by loosening the clamps on each end of the track-rods, then rotate each rod equally to move the balljoints out or in as appropriate. The track-rod assembly behaves as a turnbuckle because the threads in which the end balljoints seat are of opposite hands (photo).
7 Once the correct toe-in has been achieved, tighten the trackrod end clamps and drive carefully to your local Lada garage where an accurate job of the wheel alignment can be made.
8 The alignment of the front wheels may be unwittingly altered when the wheels accidently hit a kerb or similar. Always keep a watch on tyre wear because it will indicate whether the wheel is correctly aligned or not. Excessive wear on the inner or outer edges of both front tyres indicates that the toe-in setting is incorrect. Scrubbing of the tyre, accompanied by feathering on the edges of the tread and uneven wear across the tyre, indicates error in camber angle. A tendency for the car to pull either to the left or right can be due to a brake fault, an error in castor angle, or one tyre tread being significantly deeper than the other.
9 It must be appreciated that in all probability it will not be just one setting in error, but a combination of errors; and this is why it is advisable to entrust the wheel alignment task to your local Lada garage.

Chapter 11 Suspension and steering

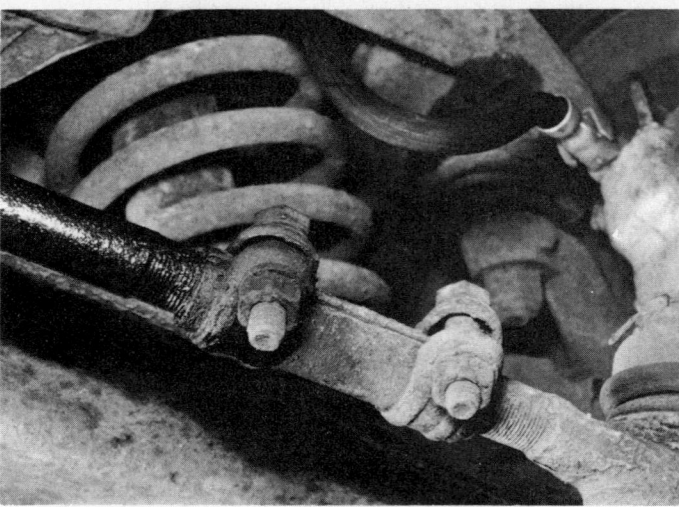
23.6 Securing clamps on adjustable trackrod

24 Wheels and tyres

1 Radial tyres are standard fitting and should never be mixed with crossply. It is also bad practice to mix tread patterns from side to side so avoid this. If for any emergency reason a mixture of tyres is necessary the only possible combination is to have two radials on the rear and two crossplys on the front.

2 A fully loaded estate can handle in a very peculiar manner if the tyres and pressures are incorrect.

3 Wheel checking and balancing should always be carried out when new tyres are fitted. Furthermore, wheels should be checked for balance after driving over bad roads or when inadvertent kerb thumping has occurred. Distorted wheels often go unnoticed by the driver and can contribute to excessive tyre wear and poor handling in certain situations.

25 Fault diagnosis – suspension and steering

Before assuming that the steering mechanism is at fault when mishandling is experienced make sure that the trouble is not caused by:

(a) Binding brakes
(b) Incorrect mix of radial and crossply tyres
(c) Incorrect tyre pressures
(d) Misalignment of the bodyframe and rear axle

Symptom	Reason/s
Steering wheel can be moved considerably before any sign of movement of the wheels is apparent	Wear in the steering linkage, gear and column coupling
Vehicle difficult to steer in a consistent straight line – wandering	As above Wheel alignment incorrect (indicated by excessive or uneven tyre wear) Front wheel hub bearings loose or worn Worn balljoints, track-rods or suspension arms
Steering stiff and heavy	Incorrect wheel alignment (indicated by excessive or uneven tyre wear) Excessive wear or seizure in one or more of the joints in the steering linkage or suspension arm balljoints Excessive wear in the steering gear unit Dry steering column top bush
Wheel wobble and vibration	Roadwheels out of balance Roadwheels buckled Wheel alignment incorrect Wear in the steering linkage, suspension arm balljoints or suspension arm pivot bushes Broken front spring
Excessive pitching and rolling on corners and during braking	Defective dampers and/or broken spring

Chapter 12 Bodywork and underframe

For modifications, and information applicable to later models, see Supplement at end of manual

Contents

Bonnet and boot lids – adjustment and removal	15
Doors – checking and adjustment	13
Doors – removal and dismantling	14
Fascia panel – removal and refitting	12
Front and rear bumpers – removal and refitting	9
General description	1
Heating and ventilation system – description	17
Heating and ventilation system – removal, overhaul and refitting	18
Maintenance – bodywork and underframe	2
Maintenance – interior	3
Maintenance – locks and hinges	7
Maintenance – vinyl roof covering	4
Major body damage – repair	6
Minor body damage – repair	5
Radiator grille – removal and refitting	8
Rear window and fixed side windows (estate) – removal and refitting	11
Tailgate – removal and refitting	16
Windscreen – removal and refitting	10

1 General description

The combined body and chassis underframe is an all welded unitary structure of sheet steel. The rear axle is attached to the body by arms bolted directly to it on rubber bushes, and a detachable crossmember across the bottom of the engine compartment provides support for the engine and front suspension. A second detachable item is a central crossmember bridging the transmission tunnel which is the rear support of the engine/gearbox unit.

The estate model has four side doors and a full width single rear door hinged at the top.

2 Maintenance – bodywork and underframe

The general condition of a vehicle's bodywork is the one thing that significantly affects its value. Maintenance is easy but needs to be regular. Neglect, particularly after minor damage, can lead quickly to further deterioration and costly repair bills. It is important also to keep watch on those parts of the vehicle not immediately visible, for instance the underside, inside all the wheel arches and the lower part of the engine compartment.

The basic maintenance routine for the bodywork is washing – preferably with a lot of water, from a hose. This will remove all the loose solids which may have stuck to the vehicle. It is important to flush these off in such a way as to prevent grit from scratching the finish. The wheel arches and underframe need washing in the same way to remove any accumulated mud which will retain moisture and tend to encourage rust. Paradoxically enough, the best time to clean the underframe and wheel arches is in wet weather when the mud is thoroughly wet and soft. In very wet weather the underframe is usually cleaned of large accumulations automatically and this is a good time for inspection.

Periodically, except on vehicles with a wax-based underbody protective coating, it is a good idea to have the whole of the underframe of the vehicle steam cleaned, engine compartment included, so that a thorough inspection can be carried out to see what minor repairs and renovations are necessary. Steam cleaning is available at many garages and is necessary for removal of the accumulation of oily grime which sometimes is allowed to become thick in certain areas. If steam cleaning facilities are not available, there are one or two excellent grease solvents available such as Holts Engine Cleaner or Holts Foambrite which can be brush applied. The dirt can then be simply hosed off. Note that these methods should not be used on vehicles with wax-based underbody protective coating or the coating will be removed. Such vehicles should be inspected annually, preferably just prior to winter, when the underbody should be washed down and any damage to the wax coating repaired using Holts Undershield. Ideally, a completely fresh coat should be applied. It would also be worth considering the use of such wax-based protection for injection into door panels, sills, box sections, etc, as an additional safeguard against rust damage where such protection is not provided by the vehicle manufacturer.

After washing paintwork, wipe off with a chamois leather to give an unspotted clear finish. A coat of clear protective wax polish, like the many excellent Turtle Wax polishes, will give added protection against chemical pollutants in the air. If the paintwork sheen has dulled or oxidised, use a cleaner/polisher combination such as Turtle Extra to restore the brilliance of the shine. This requires a little effort, but such dulling is usually caused because regular washing has been neglected. Care needs to be taken with metallic paintwork, as special non-abrasive cleaner/polisher is required to avoid damage to the finish. Always check that the door and ventilator opening drain holes and pipes are completely clear so that water can be drained out. Bright work should be treated in the same way as paint work. Windscreens and windows can be kept clear of the smeary film which often appears, by the use of a proprietary glass cleaner like Holts Mixra. Never use any form of wax or other body or chromium polish on glass.

3 Maintenance – interior

Mats and carpets should be brushed or vacuum cleaned regularly to keep them free of grit. If they are badly stained remove them from the vehicle for scrubbing or sponging and make quite sure they are dry before refitting. Seats and interior trim panels can be kept clean by wiping with a damp cloth and Turtle Wax Carisma. If they do become stained (which can be more apparent on light coloured upholstery) use a little liquid detergent and a soft nail brush to scour the grime out of the grain of the material. Do not forget to keep the headlining clean in the same way as the upholstery. When using liquid cleaners inside the vehicle do not over-wet the surfaces being cleaned. Excessive damp could get into the seams and padded interior causing stains, offensive odours or even rot. If the inside of the vehicle gets wet accidentally it is worthwhile taking some trouble to dry it out properly, particularly where carpets are involved. *Do not leave oil or electric heaters inside the vehicle for this purpose.*

4 Minor body damage – repair

The colour bodywork repair photographic sequences between pages 32 and 33 illustrate the operations detailed in the following sub-sections.
Note: *For more detailed information about bodywork repair, the Haynes Publishing Group publish a book by Lindsay Porter called The Car Bodywork Repair Manual. This incorporates information on such aspects as rust treatment, painting and glass fibre repairs, as well as details on more ambitious repairs involving welding and panel beating.*

Repair of minor scratches in bodywork

If the scratch is very superficial, and does not penetrate to the metal of the bodywork, repair is very simple. Lightly rub the area of the scratch with a paintwork renovator like Turtle Wax New Color Back, or a very fine cutting paste like Holts Body + Plus Rubbing Compound, to remove loose paint from the scratch and to clear the surrounding bodywork of wax polish. Rinse the area with clean water.

Apply touch-up paint, such as Holts Dupli-Color Color Touch or a paint film like Holts Autofilm, to the scratch using a fine paint brush; continue to apply fine layers of paint until the surface of the paint in the scratch is level with the surrounding paintwork. Allow the new paint at least two weeks to harden; then blend it into the surrounding paintwork by rubbing the scratch area with a paintwork renovator or a very fine cutting paste, such as Holts Body + Plus Rubbing Compound or Turtle Wax New Color Back. Finally, apply wax polish from one of the Turtle Wax range of wax polishes.

Where the scratch has penetrated right through to the metal of the bodywork, causing the metal to rust, a different repair technique is required. Remove any loose rust from the bottom of the scratch with a penknife, then apply rust inhibiting paint, such as Turtle Wax Rust Master, to prevent the formation of rust in the future. Using a rubber or nylon applicator fill the scratch with bodystopper paste like Holts Body + Plus Knifing Putty. If required, this paste can be mixed with cellulose thinners, such as Holts Body + Plus Cellulose Thinners, to provide a very thin paste which is ideal for filling narrow scratches. Before the stopper-paste in the scratch hardens, wrap a piece of smooth cotton rag around the top of a finger. Dip the finger in cellulose thinners, such as Holts Body + Plus Cellulose Thinners, and then quickly sweep it across the surface of the stopper-paste in the scratch; this will ensure that the surface of the stopper-paste is slightly hollowed. The scratch can now be painted over as described earlier in this Section.

Repair of dents in bodywork

When deep denting of the vehicle's bodywork has taken place, the first task is to pull the dent out, until the affected bodywork almost attains its original shape. There is little point in trying to restore the original shape completely, as the metal in the damaged area will have stretched on impact and cannot be reshaped fully to its original contour. It is better to bring the level of the dent up to a point which is about 1/8 in (3 mm) below the level of the surrounding bodywork. In cases where the dent is very shallow anyway, it is not worth trying to pull it out at all. If the underside of the dent is accessible, it can be hammered out gently from behind, using a mallet with a wooden or plastic head. Whilst doing this, hold a suitable block of wood firmly against the outside of the panel to absorb the impact from the hammer blows and thus prevent a large area of the bodywork from being 'belled-out'.

Should the dent be in a section of the bodywork which has a double skin or some other factor making it inaccessible from behind, a different technique is called for. Drill several small holes through the metal inside the area – particularly in the deeper section. Then screw long self-tapping screws into the holes just sufficiently for them to gain a good purchase in the metal. Now the dent can be pulled out by pulling on the protruding heads of the screws with a pair of pliers.

The next stage of the repair is the removal of the paint from the damaged area, and from an inch or so of the surrounding 'sound' bodywork. This is accomplished most easily by using a wire brush or abrasive pad on a power drill, although it can be done just as effectively by hand using sheets of abrasive paper. To complete the preparation for filling, score the surface of the bare metal with a screwdriver or the tang of a file, or alternatively, drill small holes in the affected area. This will provide a really good 'key' for the filler paste.

To complete the repair see the Section on filling and re-spraying.

Repair of rust holes or gashes in bodywork

Remove all paint from the affected area and from an inch or so of the surrounding 'sound' bodywork, using an abrasive pad or a wire brush on a power drill. If these are not available a few sheets of abrasive paper will do the job just as effectively. With the paint removed you will be able to gauge the severity of the corrosion and therefore decide whether to renew the whole panel (if this is possible) or to repair the affected area. New body panels are not as expensive as most people think and it is often quicker and more satisfactory to fit a new panel than to attempt to repair large areas of corrosion.

Remove all fittings from the affected area except those which will act as a guide to the original shape of the damaged bodywork (eg headlamp shells etc). Then, using tin snips or a hacksaw blade, remove all loose metal and any other metal badly affected by corrosion. Hammer the edges of the hole inwards in order to create a slight depression for the filler paste.

Wire brush the affected area to remove the powdery rust from the surface of the remaining metal. Paint the affected area with rust inhibiting paint like Turtle Wax Rust Master; if the back of the rusted area is accessible treat this also.

Before filling can take place it will be necessary to block the hole in some way. This can be achieved by the use of aluminium or plastic mesh, or aluminium tape.

Aluminium or plastic mesh or glass fibre matting, such as the Holts Body + Plus Glass Fibre Matting, is probably the best material to use for a large hole. Cut a piece to the approximate size and shape of the hole to be filled, then position it in the hole so that its edges are below the level of the surrounding bodywork. It can be retained in position by several blobs of filler paste around its periphery.

Aluminium tape should be used for small or very narrow holes. Pull a piece off the roll and trim it to the approximate size and shape required, then pull off the backing paper (if used) and stick the tape over the hole; it can be overlapped if the thickness of one piece is insufficient. Burnish down the edges of the tape with the handle of a screwdriver or similar, to ensure that the tape is securely attached to the metal underneath.

Bodywork repairs – filling and re-spraying

Before using this Section, see the Sections on dent, deep scratch, rust holes and gash repairs.

Many types of bodyfiller are available, but generally speaking those proprietary kits which contain a tin of filler paste and a tube of resin hardener are best for this type of repair, like Holts Body + Plus or Holts No Mix which can be used directly from the tube. A wide, flexible plastic or nylon applicator will be found invaluable for imparting a smooth and well contoured finish to the surface of the filler.

Mix up a little filler on a clean piece of card or board – measure the hardener carefully (follow the maker's instructions on the pack) otherwise the filler will set too rapidly or too slowly. Alternatively, Holts No Mix can be used straight from the tube without mixing, but daylight is required to cure it. Using the applicator apply the filler paste to the prepared area; draw the applicator across the surface of the filler to achieve the correct contour and to level the filler surface. As soon as a contour that approximates to the correct one is achieved, stop working the paste – if you carry on too long the paste will become sticky and begin to 'pick up' on the applicator. Continue to add thin layers of filler paste at twenty-minute intervals until the level of the filler is just proud of the surrounding bodywork.

Once the filler has hardened, excess can be removed using a metal plane or file. From then on, progressively finer grades of abrasive paper should be used, starting with a 40 grade production paper and finishing with 400 grade wet-and-dry paper. Always wrap the abrasive paper around a flat rubber, cork, or wooden block – otherwise the surface of the filler will not be completely flat. During the smoothing of the filler surface the wet-and-dry paper should be periodically rinsed in water. This will ensure that a very smooth finish is imparted to the filler at the final stage.

At this stage the 'dent' should be surrounded by a ring of bare metal, which in turn should be encircled by the finely 'feathered' edge of the good paintwork. Rinse the repair area with clean water, until all of the dust produced by the rubbing-down operation has gone.

Spray the whole repair area with a light coat of primer, either Holts Body + Plus Grey or Red Oxide Primer – this will show up any imperfections in the surface of the filler. Repair these imperfections with fresh filler paste or bodystopper, and once more smooth the surface with abrasive paper. If bodystopper is used, it can be mixed with cellulose thinners to form a really thin paste which is ideal for filling small holes. Repeat this spray and repair procedure until you are satisfied that the surface of the filler, and the feathered edge of the paintwork are perfect. Clean the repair area with clean water and allow to dry fully.

The repair area is now ready for final spraying. Paint spraying must be carried out in a warm, dry, windless and dust free atmosphere. This condition can be created artificially if you have access to a large indoor working area, but if you are forced to work in the open, you will have to pick your day very carefully. If you are working indoors, dousing the floor in the work area with water will help to settle the dust which would otherwise be in the atmosphere. If the repair area is confined to one body panel, mask off the surrounding panels; this will help to minimise the effects of a slight mis-match in paint colours. Bodywork fittings (eg chrome strips, door handles etc) will also need to be masked off. Use genuine masking tape and several thicknesses of newspaper for the masking operations.

Before commencing to spray, agitate the aerosol can thoroughly,

Chapter 12 Bodywork and underframe

then spray a test area (an old tin, or similar) until the technique is mastered. Cover the repair area with a thick coat of primer; the thickness should be built up using several thin layers of paint rather than one thick one. Using 400 grade wet-and-dry paper, rub down the surface of the primer until it is really smooth. While doing this, the work area should be thoroughly doused with water, and the wet-and-dry paper periodically rinsed in water. Allow to dry before spraying on more paint.

Spray on the top coat using Holts Dupli-Color Autospray, again building up the thickness by using several thin layers of paint. Start spraying in the centre of the repair area and then work outwards, with a side-to-side motion, until the whole repair area and about 2 inches of the surrounding original paintwork is covered. Remove all masking material 10 to 15 minutes after spraying on the final coat of paint.

Allow the new paint at least two weeks to harden, then, using a paintwork renovator or a very fine cutting paste such as Turtle Wax New Color Back or Holts Body+Plus Rubbing Compound, blend the edges of the paint into the existing paintwork. Finally, apply wax polish.

5 Major body damage – repair

Beause the body is built on the monocoque principle and is integral with the underframe, major damage must be repaired by competent mechanics with the necessary welding and hydraulic straightening equipment.

If the damage has been serious it is vital that the body is checked for correct alignment as otherwise the handling of the car will suffer and many other faults such as excessive tyre wear and wear in the transmission and steering may occur.

There is a special body jig which most large body repair shops have and to ensure that all is correct it is important that this jig be used for all major repair work.

6 Maintenance – locks and hinges

Once every 6 months or 6000 miles (10 000 km) the door, bonnet and boot or tailgate hinges should be lubricated with a few drops of engine oil. Door striker plates can be given a thin smear of grease to reduce wear and ensure free movement.

7 Radiator grille – removal and refitting

1 Unscrew the self-tapping screws securing the top of the grille to the body.
2 Pull the top of the grille forwards and then lift it upwards to clear the lower retaining tabs.
3 Refit the grille using the reverse of the removal procedure.

8 Front and rear bumpers – removal and refitting

1 The front bumper is held in the centre by two nuts which fit on the bumper studs through mounting brackets. At the outer ends of the bumper a bolt from inside the wing screws into the spacer attached to the bumper. Removal of the two bolts and the two nuts enables the bumper to be taken off.
2 The rear bumper is mounted in the same manner except that the fixing bolts for the outer ends are accessible inside the luggage compartment on saloon versions.
3 Overriders are fitted on both bumpers and are secured by two nuts and bolts to the bolts being anchored in the overriders.

9 Windscreen – removal and refitting

1 To remove the windscreen first remove the windscreen wiper arms. The glass is taken out from inside at the upper corners (Fig. 12.2). This releases the inner lip of the weatherstrip from the aperture flange and the whole screen can be removed with the strip still attached.
2 The weatherstrip may be re-used provided it is not split or

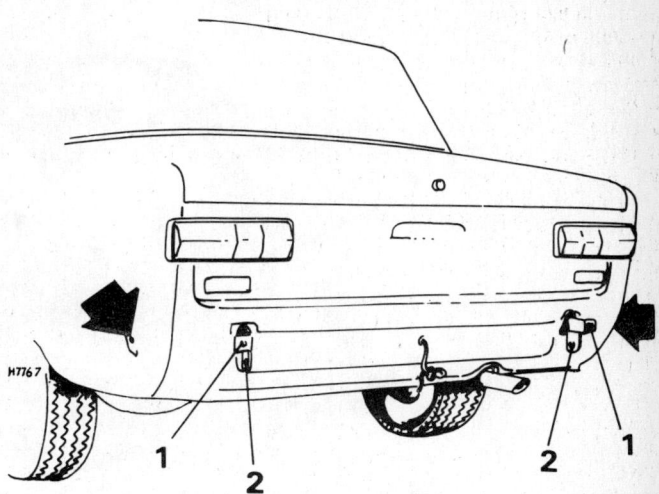

Fig. 12.1 Typical bumper mounting brackets (Sec 9)

1 Brackets *2 Towing eyes*
   End fixing screws (arrowed)

otherwise damaged and the rubber still retains reasonable flexibility. It is essential to clean away all traces of sealing compound from both the weatherstrip and the aperture flange. The moulding must be removed from the weatherstrip before fitting the weatherstrip to the glass.
3 Te replace the screen first fit the weatherstrip to the glass.
4 Next fit a piece of strong thin cord (fishing line is ideal) into the body mating groove of the weatherstrip, leaving two ends crossed over at the bottom centre.
5 Place the glass in position centrally and whilst an assistant holds it in place, draw on the cords from inside so that the lip of the weatherstrip is lifted over the aperture flange. There are two problems which usually arise. One is to avoid cutting the cord on the edge of the flange, and the other is to avoid slicing off the lip of the weatherstrip with the cord. For this reason the angle at which the cord is withdrawn must be carefully maintained and no hurried jerks or excessive force must be used. If necessary use a blunt blade of some sort to assist in the tough spots. Above all do not try and rush it. A smear of soapy water round the seal lip will also assist.
6 Provided the weatherstrip is in good condition and the aperture flange is not distorted in any way there should be a watertight seal. If experience shows that water is seeping in use a proprietary sealer of the thin clear type which can be injected with a thin flat nozzle. Inject this between the glass and weatherstrip on the outside all the way round. If this is inadequate inject more between the flange and the weatherstrip on the outside all the way round.
7 Finally, refit the bright trim to the weatherstrip. A special tool with a loop and wheel to open the groove whilst the moulding is pressed in helps to carry out this operation speedily – however, with patience and care an old screwdriver, with the corners of the blade rounded off, will enable the moulding to be fitted in.

10 Rear window and fixed side windows (estate) – removal and refitting

1 The operations are very similar to those described in the preceding Section. In the case of the rear window, disconnect the leads from the heater element terminals, and push the glass out by applying pressure at the *lower* corners.

11 Fascia panel – removal and refitting

1 At each end of the fascia panel remove the screw holding the plastic trim to the windscreen pillar and then slide the trim ends up a little to disengage them from the pillars.
2 Remove the cowls round the direction indicator switch by undoing the four screws.

Chapter 12 Bodywork and underframe

Fig. 12.2 Pushing out the windscreen (Sec 10)

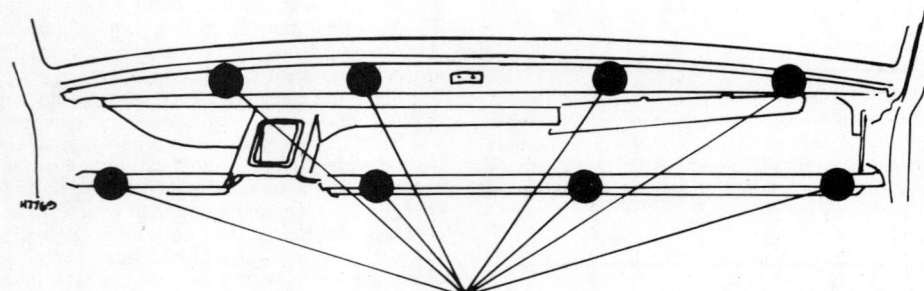

Fig. 12.3 Captive nut locations (fascia panel removed) (Sec 12)

3 Take out the instrument cluster as described in Chapter 10.
4 Push out the cluster of three switches (lights, wiper, panel) as described in Chapter 10 and remove all the wire connections (make sure you note their positions).
5 Disconnect the wire to the glove compartment light and take away the windscreen washer pipes.
6 Remove the ashtray.
7 Remove the four screws holding the lower edge of the fascia panel to the dash panel.
8 Through the glove compartment and instrument cluster apertures undo the four nuts securing the upper edge of the fascia panel to the dash panel. Take out also the two screws securing the glove compartment partition panel so that the heater ventilator cables can be disconnected from the levers which are attached to the trim panel.
9 Carefully lift the fascia panel away.
10 When refitting the fascia panel make sure the heater control cables are properly fitted to the levers.

12 Doors – checking and adjustment

1 Body rattles are often due to the doors and this can usually be traced to hinges, the door latch or the window mechanism inside the door.
2 Assuming that there has been no damage to the door or body it should be positioned centrally in its frame with an equal gap all round. If this is not the case the hinges may have moved on their attachments

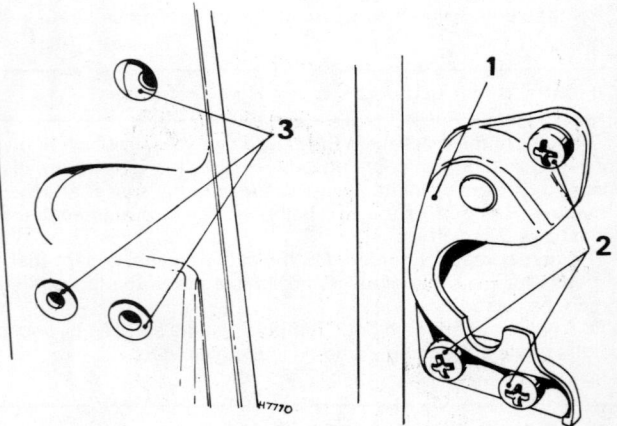

Fig. 12.4 Door latch striker plate (Sec 13)

1 Striker plate
2 Retaining screws
3 Movable anchor plate for striker screws

Chapter 12 Bodywork and underframe

to the door pillar. The hinges should be marked around first with a pencil so that subsequent movements may be obvious. The two countersunk cross-head securing screws are very tight and an impact screwdriver is essential both for loosening and tightening them. Move the door on the hinges until it is central in its aperture.

3   When the door is shut it should be flush with the body panel and not be movable. Also no excessive force should be needed to close the door – slamming should not be necessary – and hard pressure should not be needed on the catch release button to open it. The latch plate screwed to the door post can be moved in, out or up and down to improve these conditions. Mark the plate outline in pencil for a reference point and then slacken the three screws so that it can be moved. Move it in the direction required a small amount only and retighten the screws. Keep making small adjustments and testing until the desired position is reached.

4   Rattles which may be coming from inside the door will require some dismantling and this is dealt with in the next Section.

13 Doors – removal and refitting

1   If the doors are to be taken off the car, first mark the hinge positions. Disconnect the door check link and undo the four hinge screws with an impact screwdriver. Lift the door off.
2   Work on latches, windows and window winder mechanisms can all be carried out with the door in position on the car.
3   The window winder handles are held to the shafts with spring wire clips. To remove these press back the escutcheon behind the handle and then use a length of wire with a hook at one end to pull them off. The handle can then be taken off (photos).
4   The inside remote door latches are fixed to the door inner panel. Remove the escutcheons by levering them off with a flat blade. Unscrew the armrest from the panel (photos).
5   To remove the inner trim panel use a flat blade between the trim and door and lever out the press clips along the sides and bottom edge. The panel can then be drawn down and taken away.
6   The window mechanism is operated by a cable and there are two points of adjustment to ensure the window rises evenly and squarely (Figs. 12.8 and 12.9).
7   To remove the window the sill mouldings must be removed and the window lowered. The glass support channel is then detached from the cable clip and the lower guide channel taken out.
8   The quarter light can be removed after the main window lower guide channel is removed and the glass lowered. The screws in the quarter light door frame are taken out and the quarter light can then be taken out complete (Figs. 12.10 and 12.11).

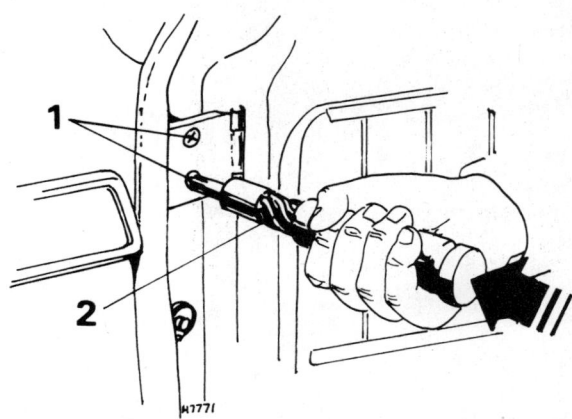

Fig. 12.5 Removing door hinge fixing screws (Sec 14)

1   Screws   2   Impact screwdriver

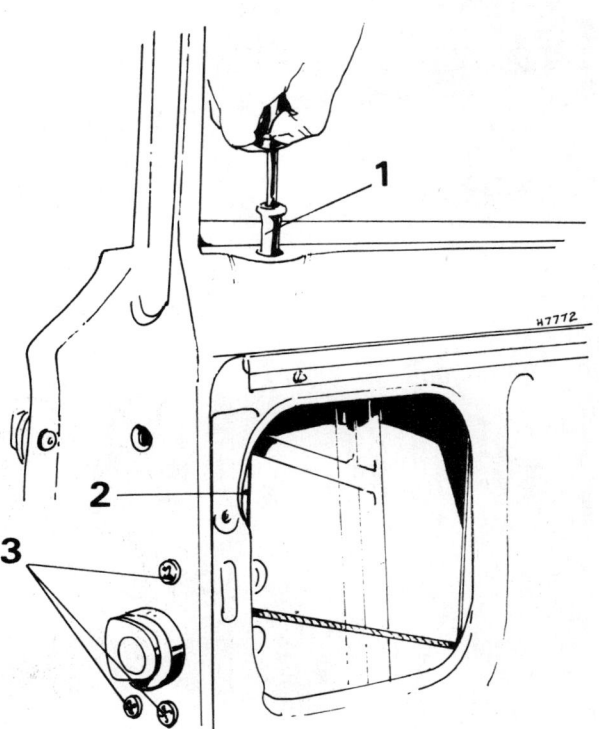

Fig. 12.6 Door latch components (Sec 14)

1   Door lock plunger   3   Lock securing screws
2   Plunger rod

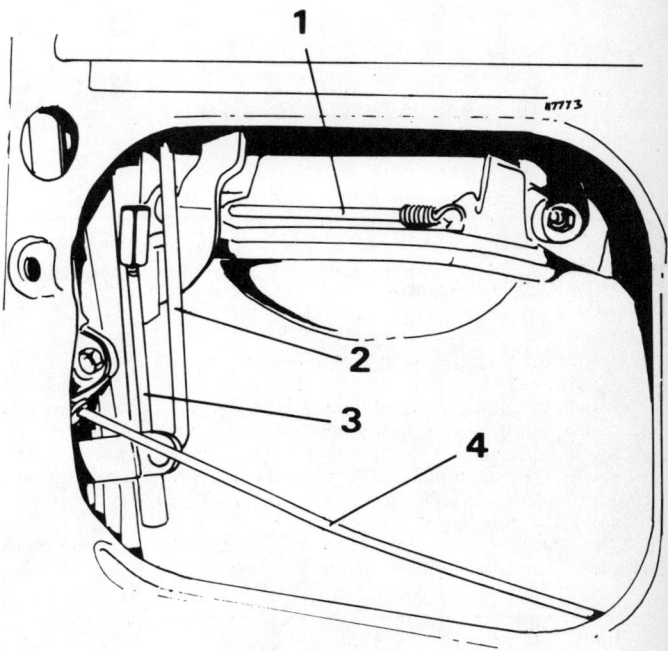

Fig. 12.7 Door interior linkage (Sec 14)

1   Door handle link rod   4   Interior remote control
2   Door lock linkage   handle linkage
3   Lock rod

14.3a View of window regulator handle retaining clip

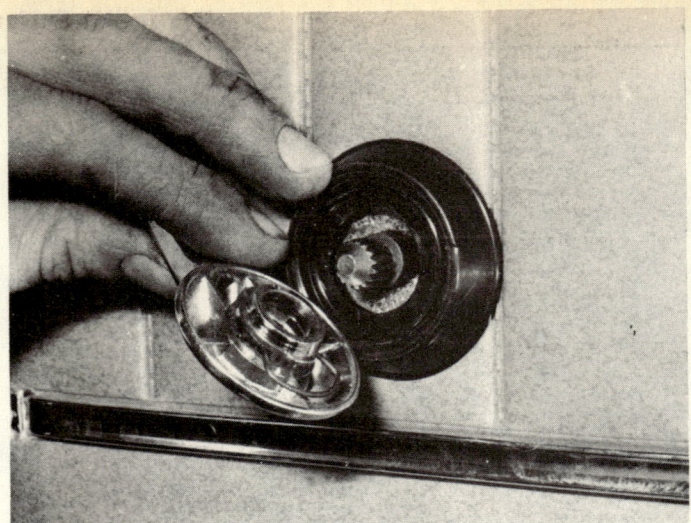
14.3b Removing the window regulator handle

14.4a Interior remote control bar lock handle

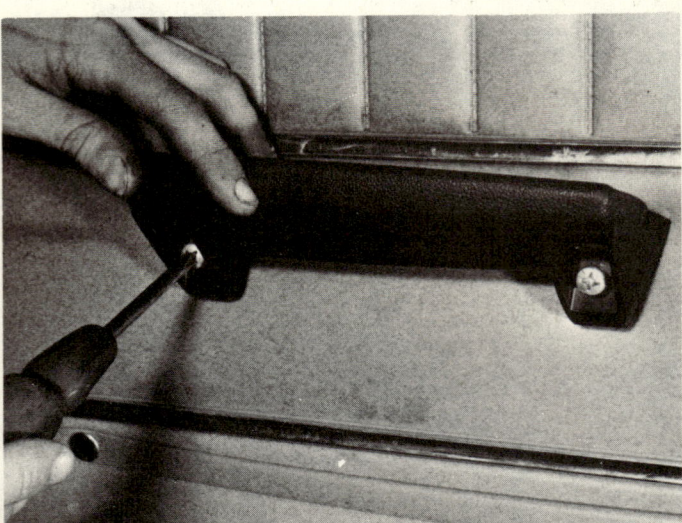
14.4b Extracting the door armrest screws

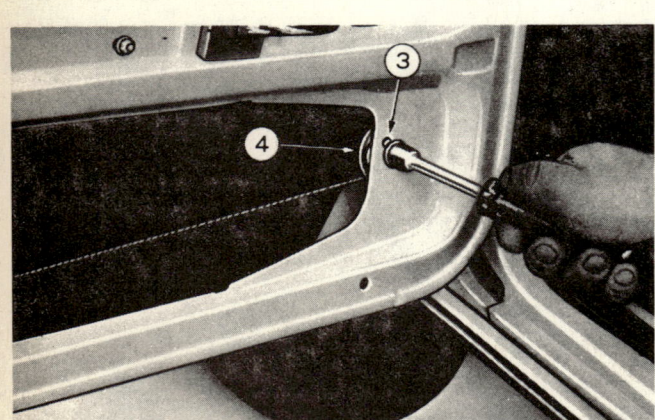
Fig. 12.8 Adjusting the position of the window regulator pulley (Sec 14)

3 Pulley   4 Cable

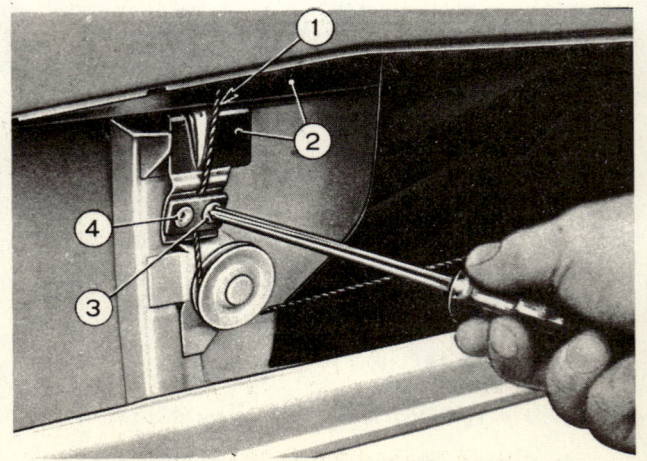
Fig. 12.9 Adjusting the window regulator cable clamp (Sec 14)

1 Cable   2 Glass channel and stop   3 and 4 Cable clamp screws

Chapter 12 Bodywork and underframe

Fig. 12.10 Removing the quarter light retaining screws (Sec 14)

Fig. 12.11 Lifting out the quarter light assembly (Sec 14)

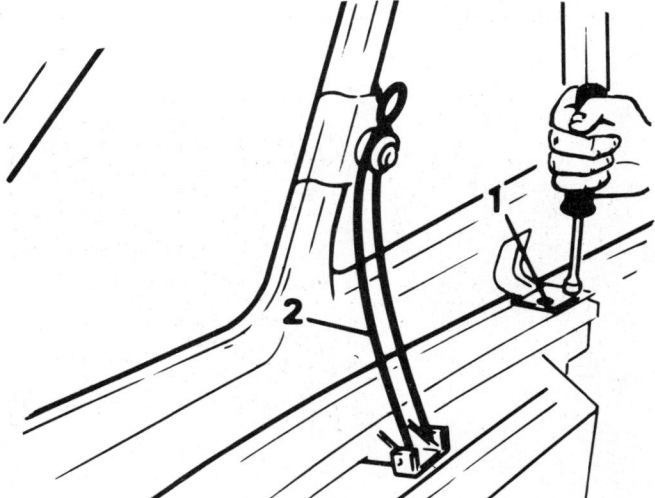

Fig. 12.12 Removing the bonnet (Sec 15)

1   Hinge bolts   2   Bonnet stay

14 Bonnet and boot lids – adjustment and removal

1 The bonnet lid hinges at the front and each hinge is held to the body cross panel by two screws. The hinge holes are slotted so that the position of the lid can be altered as necessary. To remove the lid, squeeze the legs of the wire prop together so as to release it from the bracket. Undo the hinge screws completely to lift the bonnet clear (Fig. 12.12).

2 The boot lid is similar and the nuts securing the hinges to the lid may be slackened in order to reposition the lid in the aperture. The latch plate and catch positions are also adjustable if required in order to enable the lid to be closed without undue force.

15 Tailgate – removal and refitting

1 To remove the tailgate from the estate car version, open it to its fullest extent and disconnect the wires from the window heater element terminals.

2 Mark round the hingeplates on the inside of the tailgate and then with the help of an assistant, unbolt the hinges and lift the tailgate away.

3 It is unlikely that the hinges themselves will require removing but should this be necessary, the rear end of the head lining will have to be detached and the tension on the hinge arms released before they can be withdrawn. Releasing the torsion bar tension can be dangerous even if the special tool (A77777) is being used.

4 Refit the tailgate by reversing the removal operations. Adjust the position of the tailgate within the body aperture if necessary by moving the tailgate within the limits of the hingeplate bolt holes.

16 Heating and ventilation system – description

1 The heater/ventilation unit is mounted centrally below the dash panel and incorporates a heater radiator supplied with hot water from the engine cooling system. A valve can shut off the supply of hot water and is controlled by a lever on the dash panel. Below the heater is a housing which contains an electrically powered blower fan. Air enters the car through two adjustable diffusers in the panel trim and through a shutter in the bottom of the fan housing. If the shutter is closed all air enters through the diffusers. Outside air enters through the grille in front of the windscreen into a plenum chamber and can then pass through a flap controlled aperture via the heater. In warm weather, the heater valve being off, cool air will enter the car, if necessary with assistance from the blower. When the air inlet flap is closed no air passes through the heater housing (Fig. 12.13).

17 Heating and ventilation system – removal, overhaul and refitting

1 Any repairs to the fan or heater radiator will require removal of the units from the car. First drain the engine cooling system making sure that the heater valve lever is in the 'hot' position so that all water will drain from the heater as well.

2 Loosen the hose clips on the pipes to the heater matrix.

3 Inside the engine compartment remove the rubber seals around the heater pipes by undoing the screws and washers securing them.

4 Slacken the clip holding the cable to the water valve and disconnect the cable.

5 Disconnect the blue wire for the blower fan at the fan switch and the red resistor wire.

6 The lower half of the unit, the fan housing, is held to the upper part by four spring clips. Remove these and the fan housing can be taken off. The fan earth wire is connected to one of the heater mounting nuts.

7 The heater radiator is held to the body by four nuts. After removing these, lower the matrix sufficiently to slacken the clamp screws which hold the operating cable to the air intake flap. The whole unit may then be removed.

8 The fan motor may be dismantled for examination and cleaning and renewal of carbon brushes. The water valve also may be taken off the heater matrix for checking and cleaning (Fig. 12.14).

9 When refitting the heater unit it is important to make sure that all

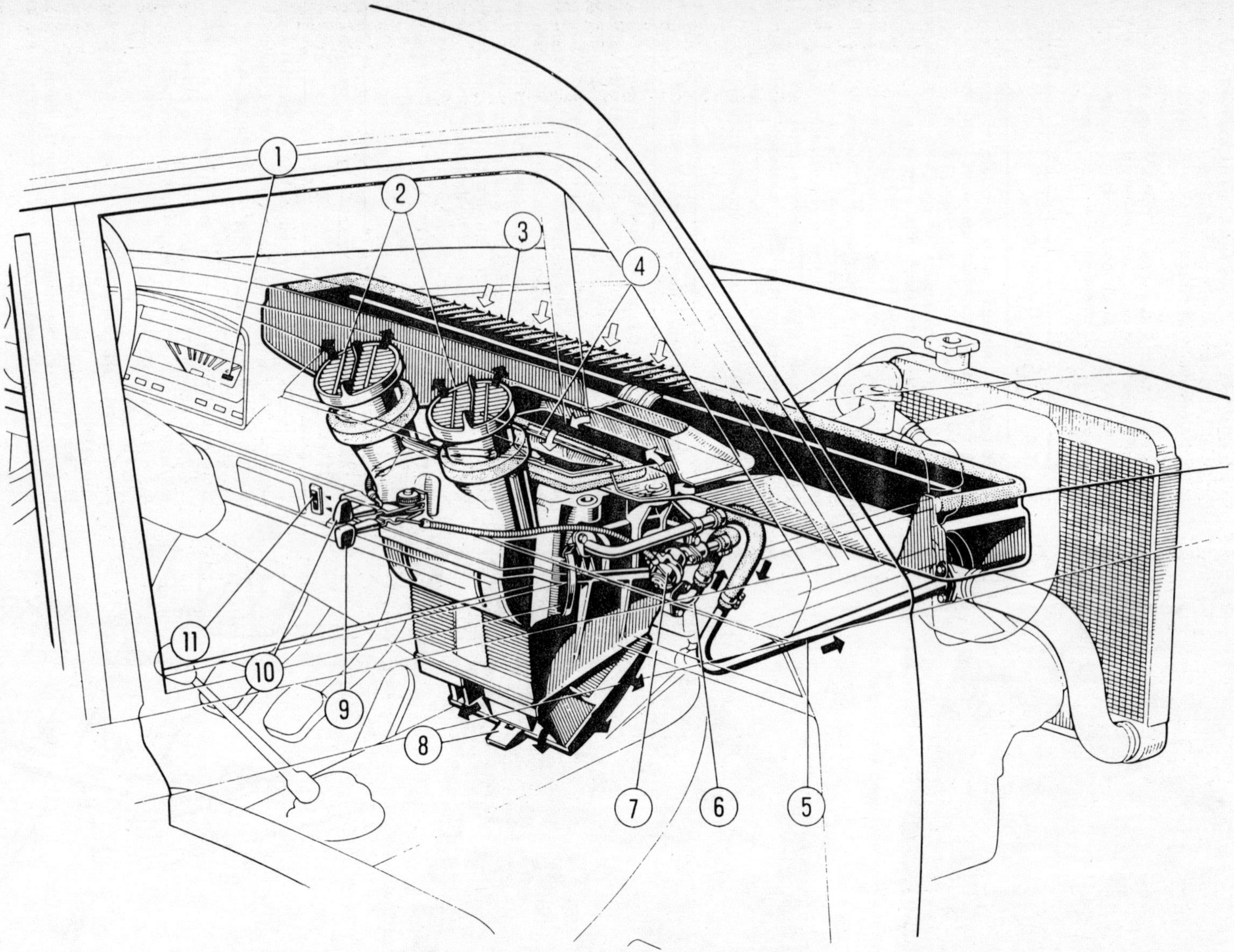

Fig. 12.13 Layout of heater/ventilation system (Sec 17)

1 Coolant temperature warning lamp
2 Adjustable air diffusers
3 Air intake grille
4 Heater flap
5 Coolant return pipe
6 Coolant delivery pipe
7 Coolant flow control valve
8 Lower shutter (air inlet)
9 Coolant valve lever
10 Heater flap lever
11 Blower fan switch

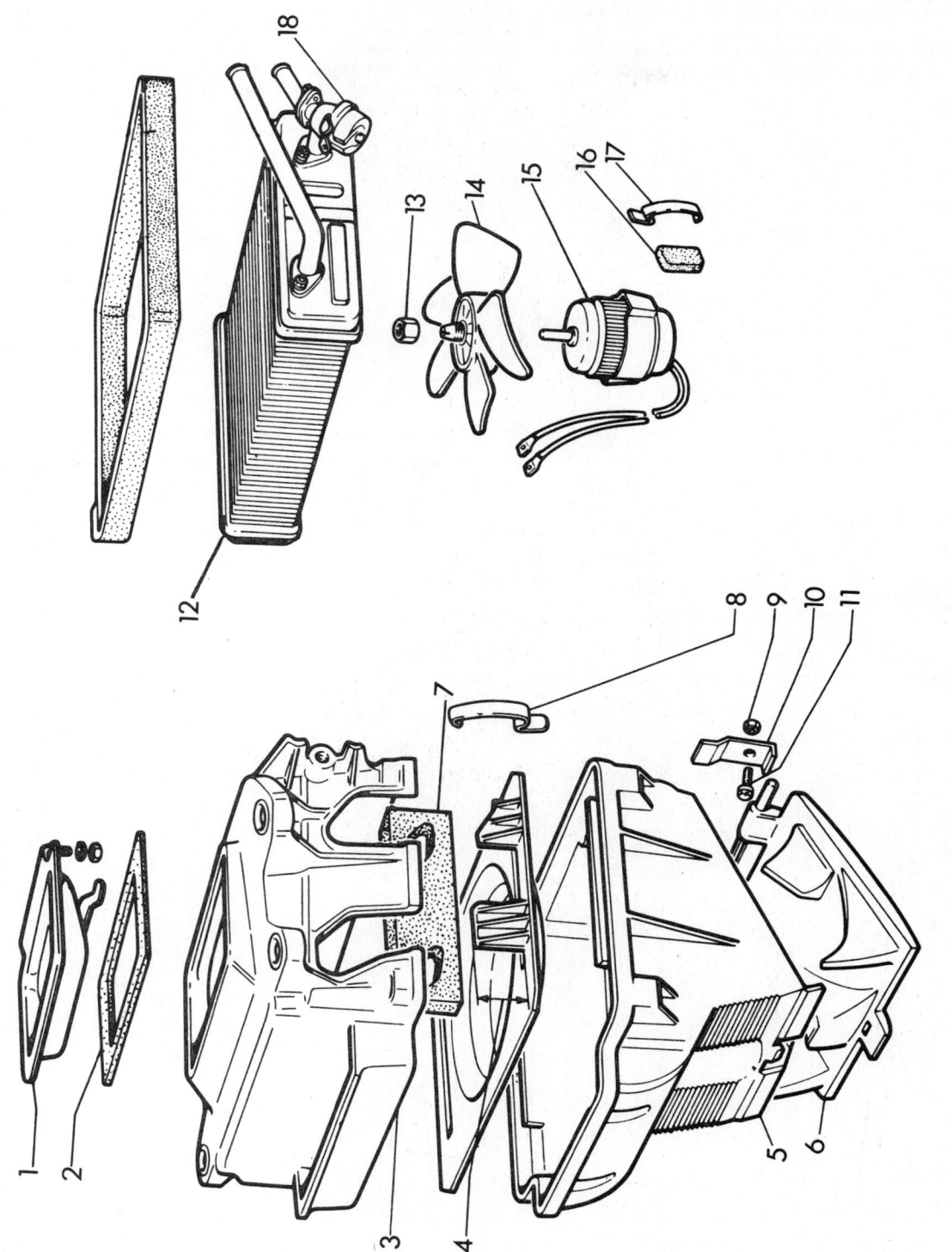

Fig. 12.14 Components of the heater (Sec 18)

1 Air inlet flap
2 Flap seal
3 Heater matrix housing
4 Fan mounting
5 Fan casing
6 Lower shutter (air outlet)
7 Heater hose grommet
8 Casing clips
9 Nut
10 Shutter pivot clip
11 Screw
12 Heater matrix
13 Fan securing nut
14 Fan blades
15 Fan motor
16 Insulating pad
17 Clip
18 Coolant flow control valve

water seals are perfectly secure and that the fan motor runs freely. Test it on the bench in advance. Make sure that the fan electrical connections are made and that the gasket between the fan housing and body is correctly positioned.

10 When refilling the cooling system make sure that the water valve is open and after running the engine for a short time check the coolant level.

11 Poor heating performance in cold weather is usually due to a faulty thermostat in the engine cooling system. Check this first before anything else. Details are given in Chapter 2.

12 Failure of the blower fan may be due to a blown fuse. If this is the case the windscreen wipers and other items on fuse No 1 will not work either (see Chapter 10). Check which item is blowing the fuse before assuming the fan motor is at fault.

Chapter 13 Supplement:
Revisions and information on later models

Contents

Introduction	1
Specifications	2
Routine maintenance	3
Engine (1300cc Riva)	4
General description	
Timing belt – renewal	
Timing cover oil seals – renewal	
Cylinder head – removal and refitting	
Camshaft front oil seal – renewal	
Auxiliary driveshaft – removal and refitting	
Cooling system	5
Cooling system – draining and filling	
Electric cooling fan relay – description	
Fuel and exhaust systems	6
Air cleaner (Riva models) – description and adjustment	
Air cleaner (except Riva models) – adjustment	
Ozone carburettor – description	
Ozone carburettor – idle speed and mixture adjustment	
Electronically controlled idle circuit – checking	
Carburettor – overhaul and adjustments	
Accelerator linkage – removal and refitting	
Choke (cold start) control cable – renewal	
Unleaded fuel	
Ignition system	7
Distributor – description	
Ignition timing – checking and adjustment	
Ignition switch (with steering lock) – description, removal and refitting	
Clutch	8
Clutch – inspection and renovation	
Release lever – removal and refitting	
Gearbox (five-speed)	9
Description	
Removal and refitting	
Overhaul	
Mainshaft – dismantling	
Inspection and renewal/components	
Mainshaft – reassembly	
Reassembly	
Propeller shaft	10
Universal joints – overhaul	
Rear axle	11
Axleshaft bearings – checking	
Pinion oil seal – renewal	
Braking system	12
Brake fluid warning switch – checking	
Rear brake drums – removal	
Rear brake shoes (except Riva 1500 and 1600 models) – renewal	
Rear brakes (Riva and later models) – description	
Rear brake shoes (Riva and later models) – renewal	
Rear brake wheel cylinder (Riva and later models) – removal, refitting and overhaul	
Rear brake pressure compensator valve – adjustment	
Hydraulic system – additional bleeding methods	
Handbrake warning switch – removal and refitting	
Brake pedal free movement – adjustment	
Brake servo unit – adjustment	
Brake servo unit – adjustment	
Brake servo unit and non-return valve – checking	
Electrical system	13
Alternator (Riva models) – description	
Fuses and relays (early Riva models)	
Fuses and relays (later Riva models)	
Ignition warning lamp	
Headlamp wiper motor – description, removal and refitting	
Washer pump (electric) – description, renewal and refitting	
Washer pump (foot-operated) – description	
Headlight bulb (Riva models) – removal and refitting	
Front parking bulb (Riva models) – removal and refitting	
Front direction indicator bulb (Riva models) – removal and refitting	
Rear lamp cluster bulbs (Riva models) – removal and refitting	
Number plate lamp (Riva models) – removal and refitting	
Hydraulic headlamp adjuster (Riva models) – description	
Battery (Riva models)	
Instrument panel (Riva models) – removal and refitting	
Headlamp (Riva models) – removal and refitting	
Headlamp dim-dip system	
Suspension and steering	14
Steering column and shaft (Riva type) – description	
Steering column (Riva type) – removal and refitting	
Steering column and shaft (Riva models) – overhaul	
Steering box (Riva models) – removal and refitting	
Steering box (Riva models) – overhaul	
Steering idler and arm – checking	
Wheels and tyres – general care and maintenance	
Bodywork and underframe	15
Front and rear bumpers (Riva models) – removal and refitting	
Facia panel (Riva models) – removal and refitting	
Doors (Riva models) – dismantling and reassembly	
Centre console – removal and refitting	
Seats – removal and refitting	
Seat belts	
Radiator grille (1300 and 1600 Riva) – removal and refitting	
Grab rails – removal and refitting	
Remote control exterior mirror (except Riva models) – removal and refitting	
Remote control exterior mirror (Riva models) – removal and refitting	
Bonnet strut (later models)	
Tailgate struts (later models)	
Heating and ventilation system (Riva models) – description	
Heating and ventilation unit (Riva models) – removal and refitting	
Heating and ventilation unit (Riva models) – dismantling and reassembly	
Plastic components	

1 Introduction

This Supplement includes revised and additional information which has become available since the manual was first published. In particular it includes information on the Riva models introduced in 1983 and 1984.

In order to use the Supplement to its best advantage it is recommended that it is always referred to before the preceding Chapters. In this way, any revised information can be incorporated into the work procedure.

Project vehicles

The cars used in the preparation of this Supplement, and appearing in many of the photographic sequences, were a Lada Riva 1300 saloon and a Lada 1600 SLX saloon.

Chapter 13 Supplement

2 Specifications

The Specifications below are revisions of, or supplementary to, those at the beginning of the preceding Chapters

Engine

Pistons and connecting rods (Riva models)

Piston-to-piston clearance (assembly limits)	0.06 to 0.08 mm
Piston ring end gap (assembly limits)	0.25 to 0.40 mm

Oil pump (Riva models)

Oil pump gear face clearance in pump body:
Assembly limits	0.066 to 0.161 mm
Wear limit	0.2 mm

Torque wrench settings

	lbf ft	kgf m
Cylinder head bolts:		
1st stage	29	4.0
2nd stage	83	11.5
Timing belt tensioner	22	3.0
Auxiliary and camshaft (timing belt)	58	8.0
Auxiliary and camshaft (timing chain)	35	4.9
Crankshaft dog nut	88	12.2

Cooling system

Capacity

All models from approximately 1976 on	17.3 pints (9.85 litres)

Radiator

Cap opening pressure (early 1300 cc models)	11.4 lbf/in² (0.8 kgf/cm²)

Thermostat (all models from approximately 1976 on)

Opening temperature	77 to 86°C (170.6 to 186.8°F)
Minimum valve travel	0.236 in (6.0 mm)

Fuel and exhaust systems

Carburettor (Ozone) – 2105-1107010 (1300 cc engine)

	Primary	Secondary
Venturi diameter	22	25
Main fuel jet	1.07	1.62
Main air jet	1.70	1.70
Emulsion tube	F15	F15
Idle fuel jet	0.50	0.60
Idle air jet	1.70	0.70
Accelerator pump jet	0.40	–
Accelerator pump volume ejected (10 strokes)	6.75 to 7.25 cc	
Econostat fuel jet	–	1.50
Econostat air jet	–	1.20
Econostat emission jet	–	1.50
Float setting (float-to-top cover gasket)	6.25 to 6.75 mm	
Maximum float stroke	7.75 to 8.25 mm	
Choke pull-down clearance	5.00 to 5.50 mm	
Throttle valve clearance (fast idle opening)	0.7 to 0.8 mm	
Fast idle speed	1500 to 2000 rpm	
Idle speed	820 to 900 rpm	
CO level at idle	1.5 to 2.5%	

Carburettor (Ozone) – 2107-1107010-40 (1500 cc engine)

	Primary	Secondary
Venturi diameter	22	25
Main fuel jet	1.12	1.50
Main air jet	1.50	1.50
Emulsion tube	F15	F15
Idle fuel jet	0.50	0.60
Idle air jet	1.70	0.70
Accelerator pump jet	0.40	–
Accelerator pump volume ejected (10 strokes)	6.75 to 7.25 cc	
Econostat fuel jet	–	1.50
Econostat air jet	–	1.20
Econostat emulsion jet	–	1.50
Float setting (float-to-top cover gasket)	6.25 to 6.75 mm	
Maximum float stroke	7.75 to 8.25 mm	
Choke pull-down clearance	5.25 to 5.75 mm	

Throttle valve clearance (fast idle opening) ... 0.9 to 1.0 mm
Fast idle speed ... 1500 to 2000 rpm
Idle speed ... 820 to 900 rpm
CO level (at idle) ... 1.5 to 2.5%

Ignition system
Coil (later models)
Primary winding resistance at 20°C ... 3.07 to 3.50 ohm
Secondary winding resistance at 20°C ... 5400 to 9200 ohm

Distributor (later models)
Initial advance setting (vacuum hose disconnected) ... 5 to 7° BTDC
Centrifugal advance ... 10 to 12° at 3000 rpm
Vacuum advance (max) ... 10 to 14°
Breaker point gap ... 0.014 to 0.018 in (0.35 to 0.45 mm)
Dwell angle ... 55 ± 3°
Condenser insulation resistance at 100 ± 2°C at 100 volts DC ... 4.0 M ohm

Clutch (Riva models)
Driven plate
Diameter:
    1200, 1300 and 1500 saloon models ... 7.87 in (200.0 mm)
    1300 and 1500 estate and van models ... 8.1 in (206.0 mm)
Minimum friction lining surface-to-rivet dimension ... 0.008 in (0.2 mm)
Maximum friction lining run-out ... 0.02 in (0.5 mm)

Clutch adjustment
Pedal free travel ... 0.98 to 1.38 in (25.0 to 35.0 mm)
Master cylinder pushrod free play ... 0.004 to 0.020 in (0.1 to 0.5 mm)

Torque wrench settings
	lbf ft	kgf m
Clutch pedal pivot	14.5	2.0
Master cylinder	10.9	1.5

Gearbox
Ratios (Riva models)
1200 4-speed:
    1st ... 3.75 : 1
    2nd ... 2.30 : 1
    3rd ... 1.49 : 1
    4th ... 1 : 1
    Reverse ... 3.87 : 1
1300, 1500 saloon and 1600 4-speed:
    1st ... 3.67 : 1
    2nd ... 2.10 : 1
    3rd ... 1.36 : 1
    4th ... 1 : 1
    Reverse ... 3.53 : 1
1500 estate and 1600 SLX 5-speed:
    1st ... 3.67 : 1
    2nd ... 2.10 : 1
    3rd ... 1.36 : 1
    4th ... 1.00 : 1
    5th ... 0.82 : 1
    Reverse ... 3.53 : 1

Torque wrench settings
	lbf ft	kgf m
Reverse lamp switch	32	4.4
Layshaft 5th gear extension bolt	45	6.2
5th/reverse fork lockbolt	28	3.9
Extension housing nuts	25	3.5
Gear lever turret bolts	25	3.5
Output drive coupling nut	64	8.8

Propeller shaft
Universal joint
Spider axial clearance ... 0.0004 to 0.0016 in (0.01 to 0.04 mm)
Circlip thicknesses:
    Plain ... 0.059 in (1.50 mm)
    Dark brown ... 0.060 in (1.53 mm)
    Dark blue ... 0.061 in (1.56 mm)
    Black ... 0.063 in (1.59 mm)
    Yellow ... 0.064 in (1.62 mm)

Rear axle
Final drive ratio (Riva models)
1200 and 1300 saloon and estate models	4.1 : 1, 4.3 : 1 or 4.4 : 1
1500 and 1600 saloon and estate models	3.9 : 1, 4.1 : 1
1300 van models	4.1 : 1
1500 van models	3.9 : 1
5-speed gearbox models	3.9 : 1

Axial shaft/bearing
Maximum endplay .... 0.028 in (0.7 mm)

Braking system
Rear drum brakes (Van models)
Drum diameter .... 9.76 in (248.0 mm)

Torque wrench setting
	lbf ft	kgf m
Rear wheel cylinder inner stop screw (Riva models)	2.9 to 5.1	0.4 to 0.7

Electrical system
Alternator
Type	AC T222 with built-in voltage regulator
Output	45 or 47 amp at 5000 rpm
Regulated voltage	13.5 to 14.6 volts

Fuses (Riva models)
Fuse No	Rating	Circuit protected
1	8A	Reversing lights, heater, heated back window warning lamp and relay
2	8A	Windscreen and headlamp wiper and washer motors and relays, also glovebox on pre 1983 models
3		Not used
4		Not used
5	16A	Heated rear window and relay
6	8A	Cigarette lighter, inspection lamp socket
7	16A	Horns
8	8A	Hazard warning lights and relay
9	8A	Alternator field winding
10	8A	Direction indicators and warning lamp, instrument panel warning lamps, carburettor pneumatic valve, handbrake warning lamp and relay
11	8A	Stop-lights, interior light
12	8A	RH headlight main beam and washer relay
13	8A	LH headlight main beam, main beam warning lamp
14	8A	LH front sidelight, RH rear tail light, number plate lights, engine compartment light, sidelight warning lamp
15	8A	RH front sidelight, LH rear tail light, instrument panel cigarette lighter, and glovebox illumination on 1983 on models
16	8A	RH headlight dipped beam and wiper relay
17	8A	LH headlight dipped beam, rear foglight and warning lamp

Bulbs (Riva models)
	Wattage
Headlight	55/60
Sidelight	4
Direction indicator	21
Rear foglight	21
Rear stop/tail	21/5
Reversing light	21
Number plate light	5
Interior light	5
Side repeater light	4

Suspension and steering
Rear suspension
Coil springs free length:
    Models 2103, 2105, 2106, 2107 and derivatives .... 17.087 in (434 mm)

Steering
Turning circle (all later saloons, estates and vans)	36 ft 9 in (11.2 m)	
Steering angles:		
King pin inclination	6° 04′	
Toe-in (Models 2101, 21011, 2102, 21021, 2103):	Laden	Unladen
With knuckle arm part number 2101-3001030/1	4 mm ± 1 mm	9 mm ± 1 mm
With knuckle arm part number 2101-3001030/1-01	3 mm ± 1 mm	4 mm ± 1 mm

Wheels and tyres
Wheel size:
    All Riva models .... 5J x 13

Chapter 13 Supplement

Tyre size:
 All Riva models .... 165SR13 or 175/70SR13 radial ply

Torque wrench settings   lbf ft   kgf m
Front suspension
Track rod end ball-pin (Riva models) 37   5.1
Steering
Relay rod ball-pin 37   5.1
Intermediate shaft clamp bolt 27   3.7

3 Routine maintenance

On Riva models routine maintenance is as given at the front of the manual, but with the following changes.

At 12 000 mile (20 000 km) intervals or every twelve months

Check function of idling speed economizer and electro-pneumatic valve

Under bonnet view of a Riva 1300

1 Headlamp rear cover
2 Headlamp wiper motor
3 Horn
4 Top hose
5 Radiator filler cap
6 Distributor cap
7 Engine oil dipstick
8 Radiator
9 Battery
10 Inlet manifold
11 Carburettor
12 Brake fluid reservoir
13 Clutch fluid reservoir
14 Wiper motor
15 Brake vacuum servo unit
16 Clutch master cylinder
17 Engine oil filler cap
18 Bonnet lock
19 Engine compartment light
20 Crankcase ventilation oil separator
21 Steering idler arm
22 Fusebox
23 Electronic control box for idling economizer
24 Expansion tank filler cap
25 Ignition coil
26 Windscreen washer pump
27 Fuel pump
28 Headlamp washer pump

Front underbody view of a Riva 1300

1 Exhaust downpipe
2 Steering box and Pitman arm
3 Alternator
4 Alternator/water pump drivebelt
5 Engine oil drain plug
6 Relay rod
7 Anti-roll bar
8 Oil filter
9 Front suspension lower control arm
10 Track rod
11 Clutch slave cylinder
12 Gearbox drain plug
13 Gearbox support crossmember
14 Propeller shaft
15 Exhaust middle section

Rear underbody view of a Riva 1300

1. Coil spring
2. Differential unit
3. Lower arm
4. Handbrake cable
5. Exhaust middle section
6. Propeller shaft
7. Handbrake equalizer
8. Upper arm
9. Brake hydraulic flexible hose
10. Brake hydraulic rigid pipe
11. Shock absorber
12. Brake pressure regulator valve rod
13. Panhard rod
14. Rear axle drain plug
15. Rear axle casing
16. Rear axle filler plug
17. Exhaust mounting
18. Exhaust rear section

At 18 000 mile (30 000 km) intervals or every eighteen months

Check timing belt for deterioration, cracks, and rubber separation (1300 cc engine)

4 Engine (1300 cc Riva)

General description

1 Although this engine is very similar to those described in Chapter 1, the main difference is the use of a flexible toothed timing belt instead of a chain.
2 Since the timing belt requires no lubrication, the front of the engine has been redesigned. The front of the rocker cover now locates over the camshaft oil seal housing on the front of the camshaft housing. The cylinder head is shortened, as there is no longer any need for a timing chest. The cylinder block is also shortened, as the crankshaft front oil seal is now located behind, instead of in front of, the crankshaft sprocket/gear, with changes also to the timing cover. The oil filter sealing face on the cylinder block now has a flat edge at the front, resulting in the need for an adaptor with the filter facing rearwards. The sump has also been shortened to align with the bottom face of the timing cover.

Timing belt – renewal

3 Remove the bonnet (Chapter 12) and the alternator/water pump drivebelt (Chapter 2).
4 Drain the cooling system (Chapter 2) then disconnect the top hose from the radiator and tie to the right-hand side of the engine compartment (photo).
5 Disconnect the battery negative lead.
6 Remove the fan shroud or electric cooling fan from the radiator, as applicable, with reference to Chapter 2 (photo).
7 Remove the air cleaner (Chapter 3).
8 Unscrew the upper nut and centre bolts and lift off the upper plastic timing belt cover (photos).
9 Unscrew the nuts and remove the rocker cover and gasket.
10 Turn the engine with a spanner on the crankshaft pulley dog nut until the detent on the rear of the camshaft sprocket is in line with the pointer on the camshaft bearing housing. The notch on the crankshaft drivebelt pulley must also be in line with the TDC mark on the timing belt cover. No 4 piston is now at TDC on its firing stroke.
11 Unbolt and remove the middle and lower timing belt covers (photo).
12 Loosen only the two bottom bolts of the tensioner bracket (photo), prise up the left-hand corner of the bracket to release the tension, then retighten the left-hand bolt to retain the tensioner in its released position.
13 Remove the distributor cap and mark the position of the contact end of the rotor on the rim of the distributor body.
14 Ease the timing belt from the camshaft sprocket, the tensioner, and the auxiliary and crankshaft sprocket and withdraw it upwards (photo). If the belt is to be refitted, mark it to indicate the direction of rotation as it must be fitted the same way round.
15 To refit the timing belt, first locate it on the crankshaft sprocket auxiliary drive shaft sprocket and the tensioner roller (photo).
16 Without moving the sprockets, position the belt on the two lower sprocket teeth so that it is taut then ease it onto the camshaft sprocket. Provided the sprockets have not been moved there should be no doubt as to which teeth to position the belt on.
17 Although the auxiliary shaft sprocket does not carry positioning marks, it should be positioned so that the distributor rotor arm is pointing at No 4 HT lead contact in the distributor cap. If the auxiliary

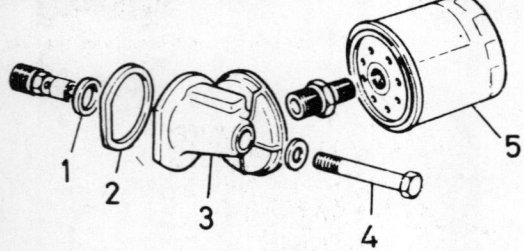

Fig. 13.1 Engine oil filter components (Sec 4)

1 Seal
2 Seal
3 Adaptor
4 Bolt
5 Filter cartridge

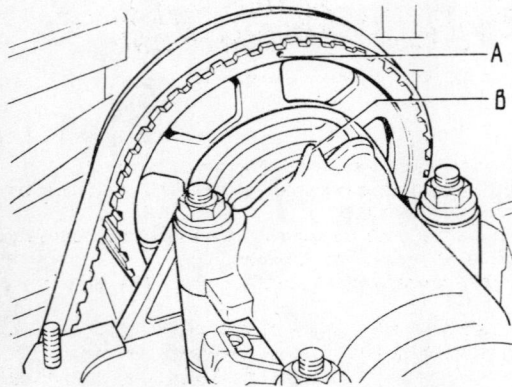

Fig. 13.2 Camshaft sprocket alignment marks (Sec 4)

A Dimple on rear face of sprocket
B Pointer on camshaft bearing housing

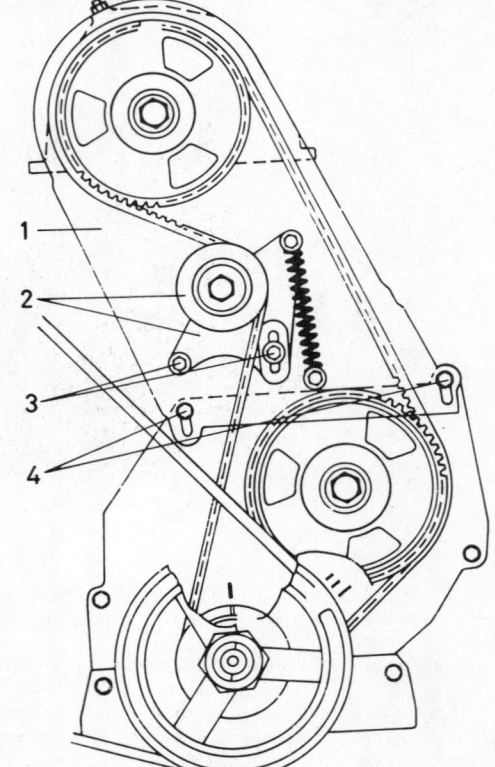

Fig. 13.3 Timing belt arrangement (Sec 4)

1 Belt cover upper section
2 Tensioner pulley and bracket
3 Tensioner bolts
4 Timing belt cover bolts

Chapter 13 Supplement

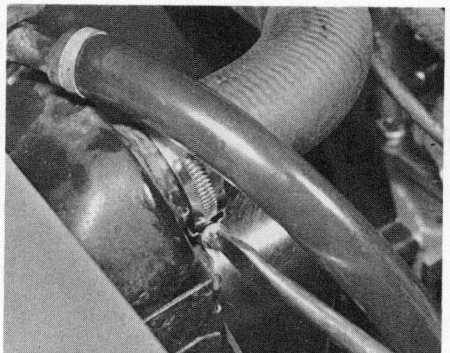
4.4 Disconnecting radiator top hose

4.6 Unbolting the radiator fan shroud

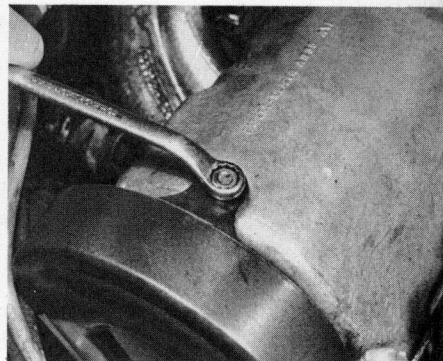
4.8A Unscrewing the timing belt cover upper nut

4.8B Unscrewing a timing belt cover centre bolt

4.11 A timing belt cover lower bolt (arrowed)

4.12 Loosening belt tensioner bracket bolt

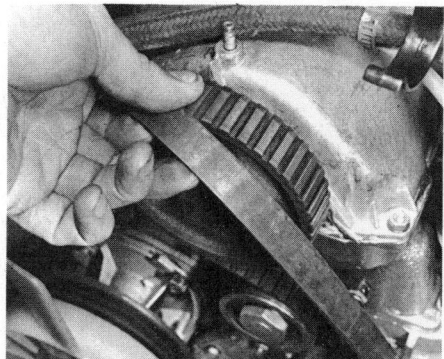
4.14 Removing the timing belt

4.15 Timing belt tensioner pulley and spring

shaft sprocket is not positioned in this way, then the distributor itself will have to be withdrawn and then refitted as described in Chapter 4, Section 7.
18 Loosen the tensioner bolt to allow its spring to tension the timing belt.
19 Using a spanner on the crankshaft pulley dog nut turn the engine two revolutions clockwise (ie normal engine rotation), then tighten the two bottom bolts of the tensioner bracket. Check that the detent on the rear of the camshaft sprocket is in line with the pointer on the camshaft housing.
20 Refit the middle and lower timing belt covers then check that the notch on the crankshaft drivebelt pulley is in line with the TDC mark on the timing belt cover.
21 Refit the rocker cover with a new gasket and tighten the nuts evenly in the sequence shown in Fig. 13.5.
22 Refit the upper timing belt cover, the air cleaner (Chapter 3), and

the fan shroud or electric cooling fan (Chapter 2).
23 Refit the alternator/water pump drivebelt (Chapter 2) and the bonnet (Chapter 12).
24 Reconnect the battery negative lead and the radiator top hose, then refill the cooling system (Chapter 2).
25 Finally check and, if necessary, adjust the ignition timing.

Timing cover oil seals – renewal
26 Remove the timing belt, as previously described.
27 Unscrew and remove the crankshaft dog nut while holding the crankshaft stationary, with reference to Chapter 1, Section 12. Remove the pulley.
28 Check that the mark on the end of the crankshaft timing belt sprocket is aligned with the TDC mark on the timing cover then remove the pulley wheel.
29 Unscrew the bolt from the auxiliary driveshaft while inserting a

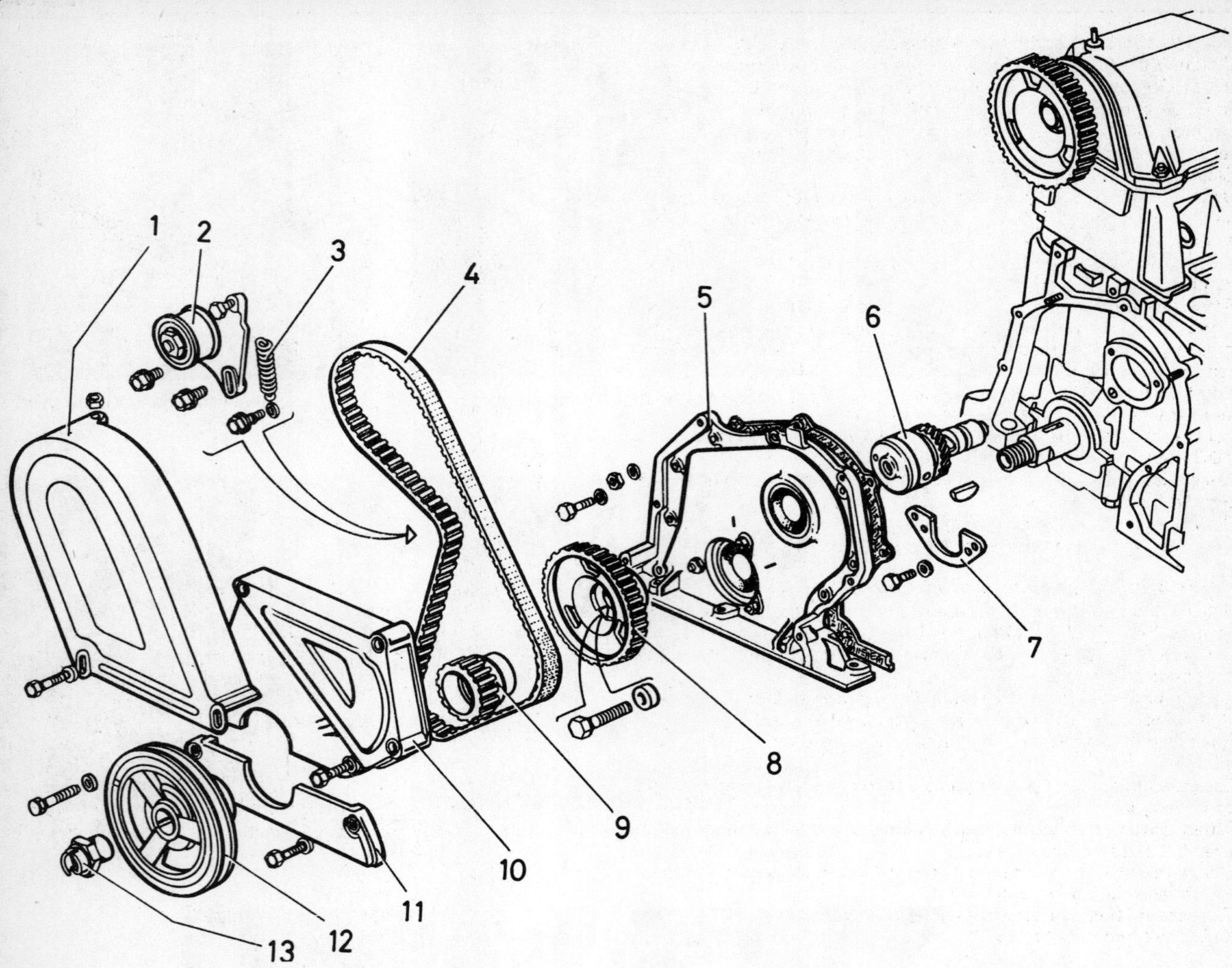

Fig. 13.4 Timing belt and associated components (Sec 4)

1 Belt cover upper section
2 Tensioner
3 Tensioner spring
4 Timing belt
5 Timing cover
6 Auxiliary driveshaft
7 Thrust plate
8 Auxiliary driveshaft sprocket
9 Crankshaft sprocket
10 Belt centre cover
11 Belt lower cover
12 Crankshaft pulley
13 Starting handle dog

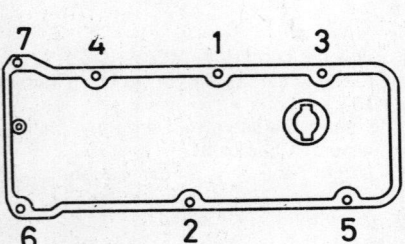

Fig. 13.5 Rocker cover nut tightening sequence (Sec 4)

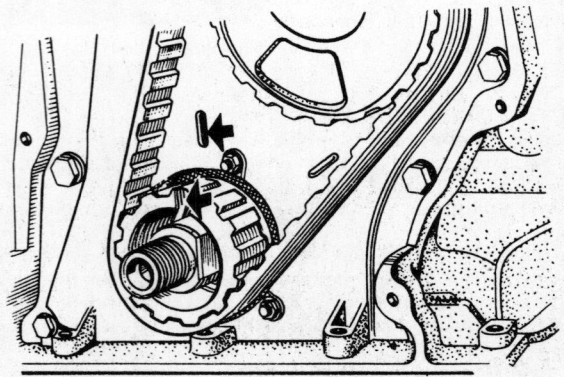

Fig. 13.6 Crankshaft sprocket TDC marks (Sec 4)

Chapter 13 Supplement

metal rod through one of the sprocket holes, then remove the sprocket.
30 Unscrew the front sump bolts and the timing cover retaining nuts and bolts. Remove the timing cover and gasket from the front of the cylinder block, taking care not to damage the sump gasket.
31 Unscrew the nuts and remove the crankshaft oil seal holder and gasket from the timing cover. Prise out the auxiliary shaft oil seal.
32 Clean all the components.
33 Drive in the new auxiliary shaft oil seal with a block of wood, then fit the crankshaft oil seal holder with a new gasket, leaving the nuts finger tight.
34 If necessary apply a little sealing compound on the front flange of the sump.
35 Smear the oil seal lips with a little engine oil then refit the timing cover with a new gasket. Check that the oil seal is located centrally on the auxiliary driveshaft then tighten the nuts and bolts evenly. Finally tighten the front sump bolts.
36 Apply sealing compound to the bolt threads then refit the pulley wheel to the auxiliary driveshaft and tighten the bolt.
37 Locate the timing belt sprocket on the front of the crankshaft with the TDC marks aligned.
38 Check that the oil seal is located centrally on the pulley wheel then tighten the holder nuts.
39 Refit the pulley to the crankshaft and tighten the dog nut.
40 Refit the timing belt, as previously described.

Cylinder head – removal and refitting
41 Follow the procedure in Chapter 1, Section 8, but instead of paragraphs 17 to 19 substitute the following paragraphs.
42 Unscrew the upper nut and centre bolts and lift off the upper plastic timing belt cover.
43 Loosen only the two bottom bolts of the tensioner bracket, prise up the left-hand corner of the bracket to release the tension, then retighten the left-hand bolt to retain the tensioner in its release position.
44 Ease the timing belt from the camshaft sprocket and tensioner, and tie it to one side.
45 When refitting the cylinder head, follow the procedure described in Chapter 1, Section 44 as applicable.
46 Check that the camshaft sprocket is at TDC with its marks aligned and re-engage the timing belt with it.
47 Tension the belt as described earlier in this Section.
48 Refit the timing belt cover.

Camshaft front oil seal – renewal
49 Remove the timing belt, as previously described.
50 Unscrew the bolt from the camshaft while inserting a metal rod through one of the sprocket holes, then remove the sprocket (photo).
51 Unscrew the nuts and remove the seal holder from the front of the camshaft housing (photo), then drive out the oil seal. Alternatively, the oil seal may be removed leaving the holder in position.
52 Drive in the new oil seal.
53 Where applicable, clean the mating faces of the seal holder and cylinder head, apply sealing compound, then refit the holder and tighten the nuts.
54 Refit the camshaft sprocket and tighten the bolt (photo).
55 Refit the timing belt, as previously described.

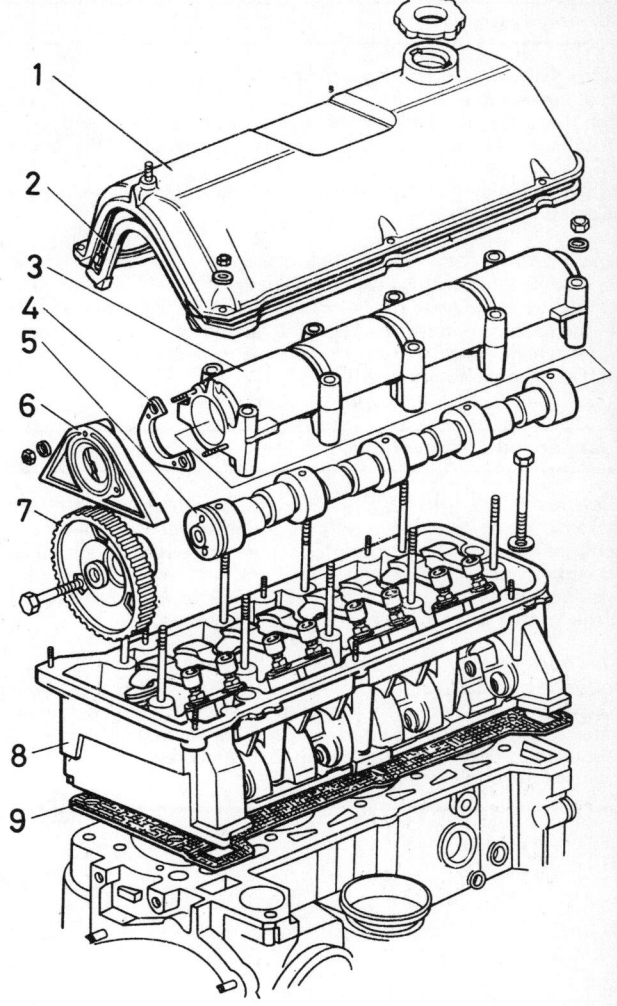

Fig. 13.7 Camshaft components (Sec 4)

1 Rocker cover
2 Gasket
3 Camshaft housing
4 Thrust plate
5 Camshaft
6 Oil seal retainer
7 Camshaft sprocket
8 Cylinder head
9 Gasket

Auxiliary driveshaft – removal and refitting
56 The procedure is similar to that described in Chapter 1, Section 13 after removal of the timing cover, as previously described in this Section.

4.50 Removing the camshaft sprocket

4.51 Camshaft front oil seal and holder

4.54 Tightening the camshaft sprocket bolt

5 Cooling system

Cooling system – draining and filling
1 As a conventional water valve is used to control the flow of water through the heater, the heater temperature control should be moved to the maximum heat position before draining and filling the cooling system.
2 Both the radiator and expansion tank caps should be removed when draining the system.

Electric cooling fan relay – description
3 On models fitted with an electric cooling fan, a relay is provided in the wiring circuit between the thermal switch (located in the left-hand side of the radiator bottom tank) and the fan motor. The relay is bolted to the left-hand front engine compartment panel and the wiring harness also incorporates an in-line fuse.

6 Fuel and exhaust systems

Air cleaner (Riva models) – description and adjustment
1 The air cleaner intake incorporates a unit which controls the air temperature entering the carburettor. There are two settings on the unit for summer or winter use and to reset the unit simply turn the lever on the side towards the appropriate symbol. This is particularly important in the winter when carburettor icing is possible.

Air cleaner (except Riva models) – adjustment
2 During the winter period it is important to position the air cleaner cover so that the red spot is in line with the arrow on the inlet. Failure to do this can result in icing up of the carburettor which is particularly noticeable after running the car for approximately ten miles from cold; icing causes the engine to cut out.

Ozone carburettor – description
3 Later models are fitted with what the manufacturers call the Ozone carburettor. Despite its name, the carburettor mixes fuel and air in much the same way as the twin-choke downdraught carburettor fitted to earlier models. The main functional difference in the Ozone carburettor is that the opening of the second choke throttle butterfly is achieved by vacuum, the amount of opening being determined by the engine speed and load. There is no mechanical link between the first and second choke butterflies.
4 The second choke actuating rod and throttle butterfly stop are preset by the manufacturers and should not be tampered with. Note that revving the engine when the car is stationary will not produce enough vacuum to open the butterfly.
5 On Riva models the carburettor has been further refined by incorporating an electronically controlled idle circuit. Basically, the idle circuit can be cut out by a needle valve which is operated by inlet manifold vacuum via a diaphragm valve. The diaphragm valve is activated both by a throttle microswitch and an electronic control box located on the left-hand side of the engine compartment (photos).
6 The electronic control box monitors the engine speed and only allows the idle circuit to operate at engine speeds below 1600 rpm on acceleration or 1200 rpm on deceleration. At all other times the idle circuit is inoperative, thus providing improved fuel economy.

Ozone carburettor – idle speed and mixture adjustment
7 Have the engine at normal operating temperature with the valve clearances and ignition correctly set. All electrical components should be switched off.
8 Connect a tachometer and exhaust gas analyser to the engine in accordance with the manufacturer's instructions.
9 Start the engine and run it at a fast idle for thirty seconds and then allow it to idle and check the idle speed. If this is not as specified, turn the idle speed screw.
10 Some carburettors are fitted with a limited-travel type of

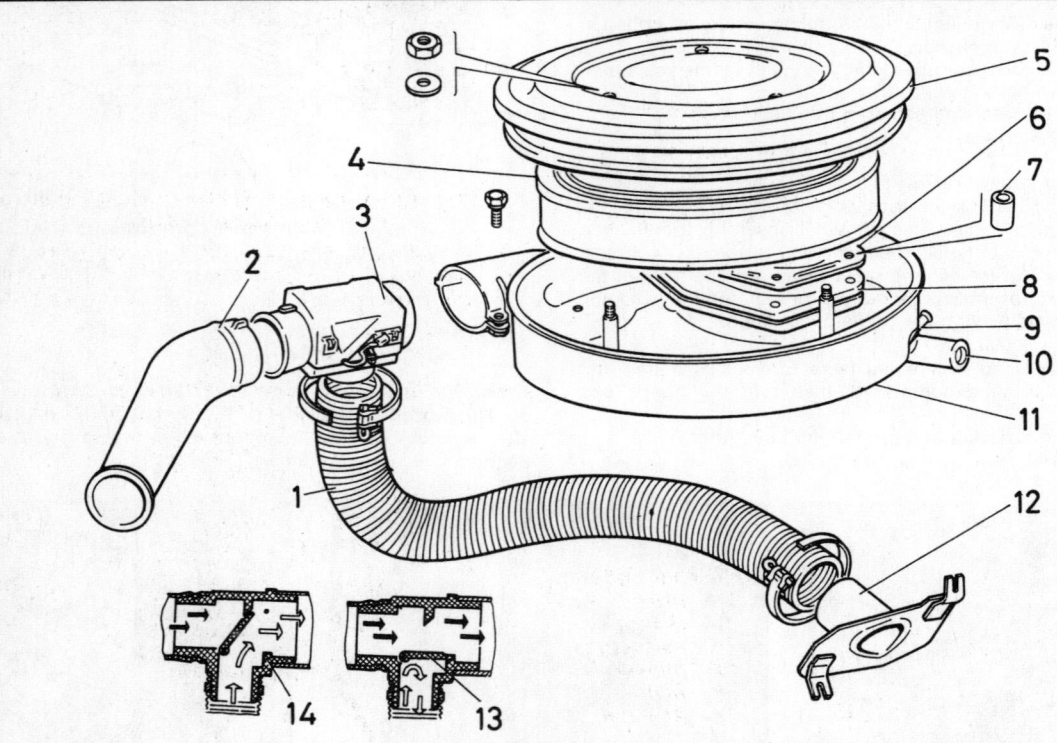

Fig. 13.8 Air cleaner, manual temperature control type (Sec 6)

1 Warm air intake
2 Cold air intake
3 Temperature control unit
4 Air cleaner element
5 Cover
6 Plate
7 Sleeve
8 Gasket
9 Crankcase ventilation pipe (to manifold)
10 Crankcase ventilation pipe (from crankcase)
11 Body
12 Warm air shroud
13 Air temperature control flap admitting cold air
14 Air temperature control flap admitting hot air

6.5A Throttle microswitch (arrowed)

6.5B Electronic control unit

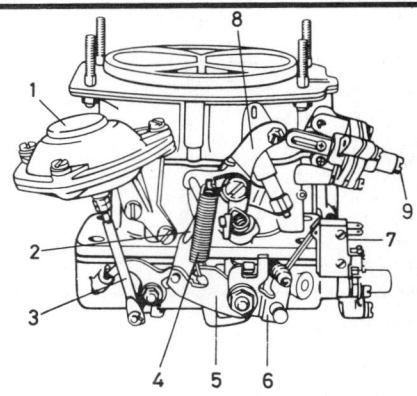

Fig. 13.9 Carburettor fitted to Riva models (Sec 6)

1 Secondary throttle vacuum unit
2 Choke/throttle link
3 Operating rod
4 Tension spring
5 Secondary throttle stop lever
6 Throttle control lever
7 Microswitch
8 Choke control lever
9 Choke control diaphragm unit

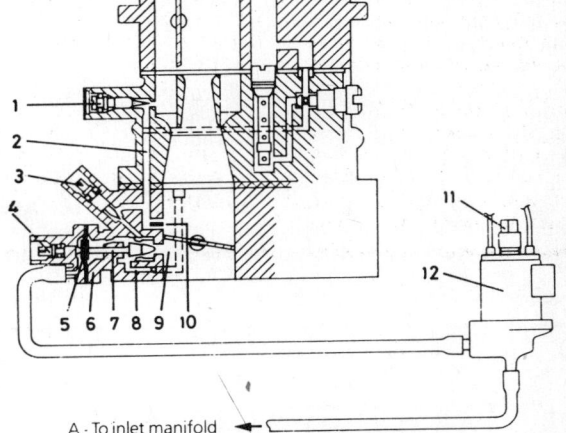

Fig. 13.10 Diagram of electronically controlled idle circuit (Sec 6)

1 Adjustment screw (plugged)
2 Idle circuit emulsion passage
3 Idle mixture screw
4 Idle speed screw
5 Economiser diaphragm
6 Economiser body

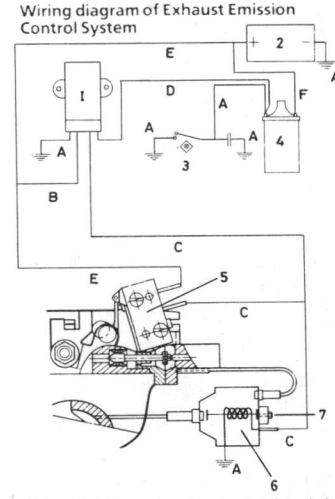

Fig. 13.11 Microswitch and wiring circuit for electronically controlled idle circuit (Sec 6)

1 Electronic control box
2 Battery
3 Contact breaker points
4 Coil
5 Microswitch
6 Vacuum valve
7 Vent to atmosphere

Colour code
A Black
B Orange
C Red/grey
D Brown
E Red
F Blue/red

Fig. 13.12 Ozone carburettor (Sec 6)

1 Idle speed screw
2 Distributor vacuum pipe connection
3 Mixture adjusting screw

tamperproof cap. This may have to be removed if abnormal adjustment is required.
11 The mixture is set during production and should only be adjusted if it is outside the specified range.
12 Adjust within the limits of the tamperproof cap. The cap will have to be removed if abnormal adjustment is required.
13 The tamperproof seals can be broken off by applying increased force to the screwdriver when unscrewing the adjustment screws.

Electronically controlled idle circuit – checking

14 Connect a tachometer to the engine then start the engine and let it idle.
15 Place a finger on the microswitch to prevent it switching on, then slowly open the throttle. When the engine speed reaches 1600 rpm the idle circuit should cut out, causing the engine to slow down, then at 1200 rpm it should cut in causing the engine speed to increase. The throttle must remain in the same position, and the engine speed should fluctuate continuously. If not, the electronic control box is proved faulty.
16 A further check is to let the engine idle then disconnect the wires from the diaphragm valve. If the engine does not stall immediately, check the vacuum hoses and economizer diaphragm for leaks. An audible click should also be heard when the diaphragm valve is switched on and off.
17 The microswitch contacts can be checked using an ohmmeter. If necessary, the position of the microswitch can be adjusted within the limits of the elongated hole.

Ozone carburettor – overhaul and adjustments

18 The limits of overhaul are as described in Chapter 3, Section 8.
19 As reassembly progresses, carry out the following adjustments

Choke valve plate setting
20 Refer to Fig. 13.13 and turn the lever (1) fully anti-clockwise. The choke valve plate should be completely closed. The link (3) should be located at the end of the slot in rod (4) but without any tendency to move it. Any adjustment required should be carried out by bending the link (3).
21 Close the choke valve plate with the fingers and then press the rod (4) fully into the choke housing. The choke valve plate should open to give a gap of between 5.0 and 5.5 mm (dimension C). Any adjustment required should be made using the screw (5).

Fast idle
22 When the choke valve plate is fully closed, the primary throttle valve plate should be open between 0.7 and 0.8 mm (dimension B). Check using a twist drill of suitable diameter. If adjustment is required, bend the link rod (7).
23 The correct fast idle speed at cold start should be between 1500 and 2000 rpm.

Throttle valve plate setting
24 If the secondary throttle vacuum diaphragm actuator has been removed, refit it in the following way.
25 Disconnect the diaphragm link rod (9 in Fig. 13.14) and push it fully into the actuator.
26 The hole in the link rod should be in exact alignment with the pivot pin on the coupling lever. If it is not, release the link rod locknut and turn the rod as necessary.
27 Tighten the locknut and reconnect the link rod.
28 The throttle valve plate should open, with a gap (C in Fig. 13.15) of 6.0 mm, when the upper lug of lever (3) contacts lever (2). Bend the upper lug if necessary to achieve this.
29 Turn the throttle valve control levers to fully open the valve plates and check the openings: primary 12.5 to 13.5 mm, secondary 14.5 to 15.5 mm. If adjustment is required for the primary valve plate, bend the lower lug of the lever (3). If adjustment is required for the secondary valve plate, turn the actuator rod after releasing the locknut and disconnecting the rod from its pivot pin.

Float level
30 The operations are as described in Chapter 3, Section 8, but use the dimensions given in the specifications at the beginning of this Supplement.

Accelerator pump
31 Fit a piece of thin bore flexible hose to the accelerator pump nozzle in the venturi and place the other end in a measuring glass.
32 Give several strokes of the pump by fully opening the throttle to prime the circuit and then discard the fuel.
33 Now open the throttle fully ten times and measure the volume of

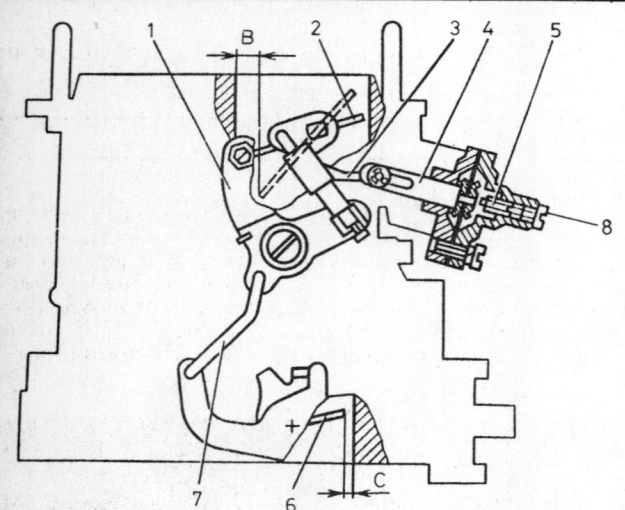

Fig. 13.13 Choke (cold start) valve plate and linkage (Sec 6)

1 Valve plate operating lever
2 Choke valve plate
3 Link
4 Actuating rod
5 Choke mechanism
6 Primary throttle valve
7 Throttle valve plate link rod
8 Choke pull-down adjustment screw
B = 0.7 to 0.8 mm
C = 6.0 mm

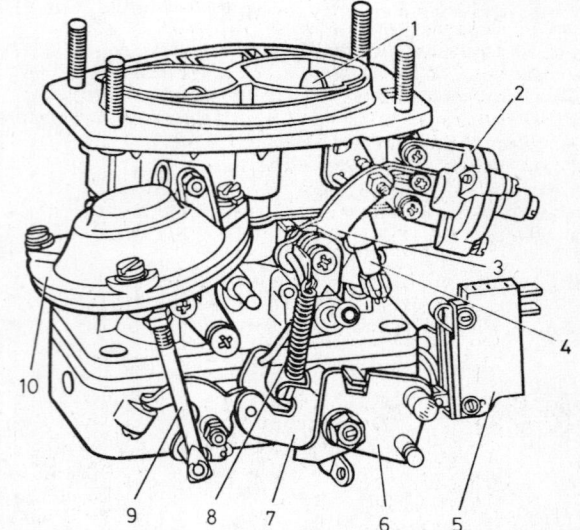

Fig. 13.14 Carburettor linkage arrangement (Sec 6)

1 Choke valve plate
2 Choke mechanism
3 Choke valve lever
4 Telescopic link
5 Microswitch
6 Throttle valve operating lever
7 Secondary throttle opening limiter
8 Tension spring
9 Actuating rod
10 Secondary throttle vacuum control unit

Chapter 13 Supplement

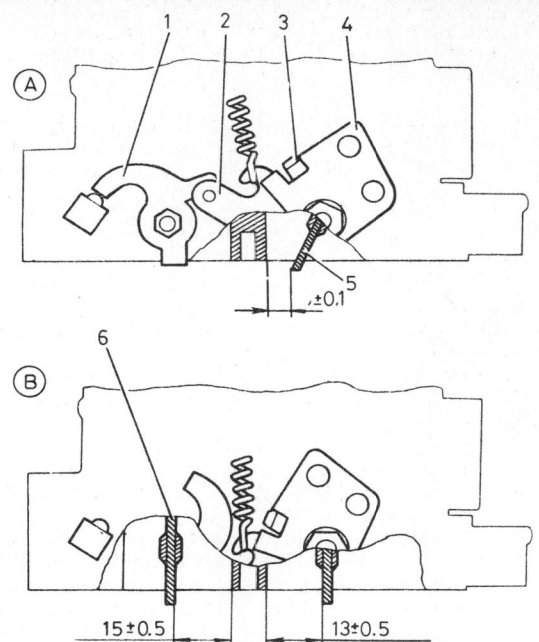

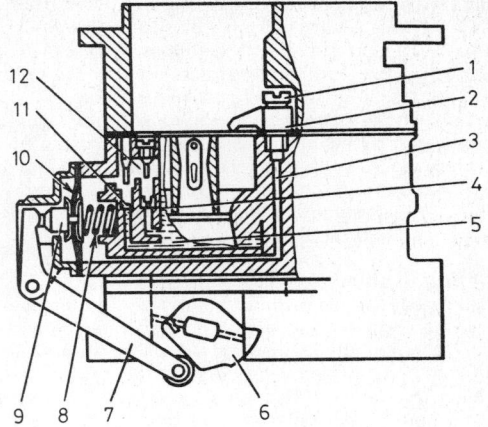

Fig. 13.16 Diagrammatic view of accelerator pump (Sec 6)

1. Delivery ball valve
2. Atomizer
3. Fuel channel
4. Bypass jet
5. Float chamber
6. Pump operating cam
7. Operating lever
8. Return spring
9. Diaphragm cup
10. Diaphragm
11. Fuel inlet ball valve
12. Vapour chamber

Fig. 13.15 Throttle valve adjustment diagrams (Sec 6)

- A Primary throttle valve partially open
- B Throttle valve plates fully open
- 1 Secondary throttle shaft lever
- 2 Secondary throttle stop lever
- 3 Lever linked with primary throttle shaft
- 4 Throttle control lever
- 5 Primary throttle valve plate
- 6 Secondary throttle valve plate

Dimensions in mm

the fuel collected. This should be 7.0 cc otherwise suspect blocked passages or split diaphragm.

Accelerator linkage – removal and refitting
34 The control linkage is of rod and ball socket type.
35 The accelerator pedal rod passes through the engine compartment rear bulkhead and connects with the link rods and pivots (photo).
36 The link rods are adjustable for length after releasing the locknut and prising the socket from the ball-pin (photo).
37 Set the linkage so that there is just a slight amount of slackness before the throttle valve plate is actuated (photo).
38 To remove the pedal, first take off the cover from under the facia panel, disconnect the return spring and the appropriate balljoint.

Clutch (cold start) control cable – renewal
39 The choke control knob is located on a bracket below the facia panel and secured by a locknut.
40 The inner and outer cables are secured at the carburettor end by pinch screws.
41 To remove the cable, release the screws, undo the locknut and withdraw the cable through the bulkhead grommet.
42 When fitting the cable, pull out the knob slightly, hold the choke valve plate fully open and tighten the pinch screws.

Unleaded fuel
43 All Lada models will run on unleaded fuel, and require no adjustment.

7 Ignition system

Distributor – description
1 As from the introduction of the Ozone carburettor in 1980 the distributor has been modified to incorporate a vacuum advance unit. The contact breaker points baseplate is redesigned to accommodate the vacuum advance unit operating arm, and the octane selector (vernier adjuster) is no longer fitted. With the exception of these items the other distributor components remain the same.
2 The inspection and overhaul instructions as given in Chapter 4

6.35 Accelerator pedal rod and bulkhead grommet

6.36 Throttle control rods at carburettor

6.37 Accelerator pedal return spring at bulkhead

apply also to the new distributor, but in addition the vacuum advance unit diaphragm should be checked only by sucking on the outlet tube then placing a finger over the tube. If the operating arm quickly returns to its initial position the diaphragm is fractured; this can be confirmed by checking with a timing light, as described in paragraph 4.

3 The rotor arm (more accurately referred to as a rotor) incorporates a suppressor between the centre and outer HT contacts in order to counteract interference with radio reception (photo). Early suppressors were prone to burning out and an improved version was fitted as from late 1982. The resistance of the suppressor can be checked with an ohmmeter – it should be between 5000 and 6000 ohm.

Ignition timing – checking and adjustment

4 On models fitted with the distributor having a vacuum advance unit (ie 1980 on), the vacuum hose should be disconnected and plugged when checking the timing with a stroke lamp. Correct operation of the vacuum unit may be checked by carrying out the advance mechanism check described in Chapter 4, Section 7, paragraph 10, both with and without the vacuum hose connected. The advance should be greater when the vacuum hose is connected, especially at part throttle openings.

Ignition switch (with steering lock) – description, removal and refitting

5 A common fault with the ignition switch is that the internal starter solenoid control contacts fail due to the contacts becoming dirty or simply not touching each other. Lada have modified the switch and the starter motor solenoid, and it is also possible to obtain a relay switch for fitting into the starter circuit to overcome the problem, although most models manufactured after late 1985 should already have the relay switch fitted. To remove the ignition switch proceed as follows.
6 Disconnect the battery negative lead.
7 Remove the steering column shrouds.
8 Disconnect the multi-plug from the ignition switch (photo).
9 Remove the screws securing the switch to the bracket. Up to three screws are located on the circumference of the switch (photo).
10 Using a small screwdriver through the access hole, depress the lockplate on the left-hand side of the switch then withdraw the switch from the bracket (photos). Note that the steering lock must be in the unlocked position otherwise it will be impossible to remove the switch.
11 If necessary, the switch contact unit may be removed from the body by prising out the circlip. Take care not to lose the small plate which locates on the key barrel, as the key number is stamped on it (photos).
12 Refitting is a reversal of removal, but when inserting the switch contact unit make sure that the circlip is seated correctly in its groove.

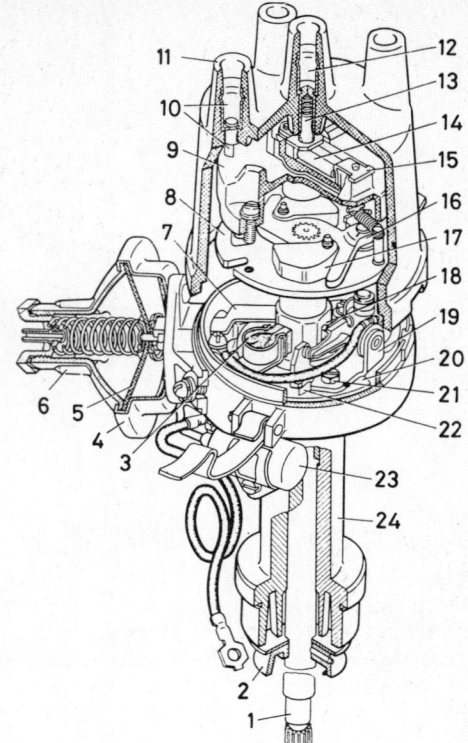

Fig. 13.17 Cutaway view of later type distributor with vacuum advance unit (Sec 7)

1 Shaft
2 Oil deflecting sleeve
3 Lubrication wick
4 Vacuum unit body
5 Diaphragm
6 Vacuum unit cover
7 Operating rod
8 Centrifugal weight baseplate
9 Rotor arm (rotor)
10 HT segment/terminal
11 Cap
12 Central HT terminal
13 Carbon brush
14 Suppressor
15 Rotor outer contact
16 Shaft driveplate
17 Centrifugal weight
18 Cam
19 Contact points
20 Contact points baseplate
21 Screw
22 Adjusting notch
23 Condenser
24 Distributor body

7.3 Distributor rotor arm with suppressor arrowed

7.8 Ignition switch multi-plug

Chapter 13 Supplement

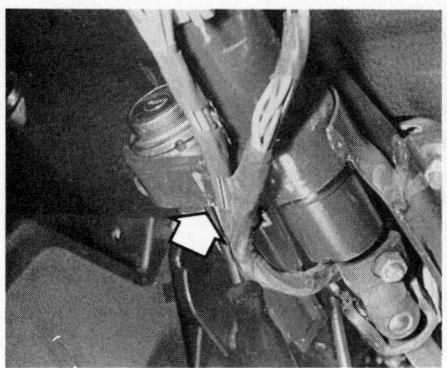
7.9 Ignition switch fixing screw (arrowed)

7.10A Depressing ignition switch lockplate

7.10B Withdrawing the ignition switch

7.11A Prising out the ignition switch circlip

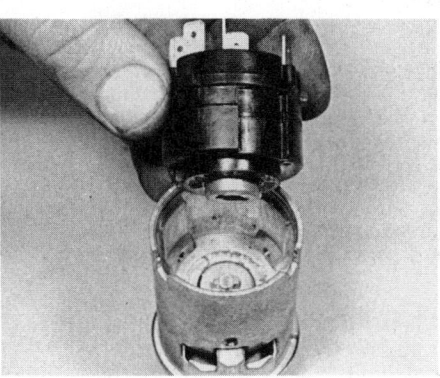
7.11B Removing ignition switch contact unit

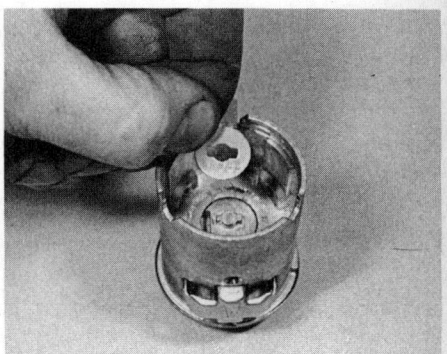
7.11C Removing ignition switch key number plate

8 Clutch

Clutch – inspection and renovation

1 If on inspection the clutch disc friction linings have worn down to or near 0.008 in (0.2 mm) of the rivets the disc assembly should be renewed.

2 The run-out of the friction linings must not exceed 0.02 in (0.5 mm). To check this, fit the disc on the gearbox input shaft splines and use a dial gauge or fixed block and feeler gauges at several points. Make sure that only the run-out of the linings is being measured and not any movement due to wear of the splines.

3 When removing the clutch it is always worthwhile checking the condition of the ball-bearing in the end of the crankshaft. Excessive wear of this bearing could result in unnecessary wear of the friction linings and gearbox input shaft bearing.

4 A suitable puller will be required to remove the bearings, or it may be possible to remove it by filling the inner cavity behind the bearing with thick grease and driving a close fitting metal rod through the bearing. The pressure of the grease should force the bearing out. Drive in the new bearing using a metal tube.

Release lever – removal and refitting

5 On early models the release lever is retained on its ball pivot by a spring plate riveted to the lever. To remove the lever it must be first withdrawn from the side of the clutch bellhousing to disengage the spring plate. On later models the spring plate has been replaced by a spring clip retained by formed tabs to the lever. To remove this type, hold the outer end of the lever with one hand then, with the other hand, position the fingers under the spring clip and pull it sharply to expand it over the ball pivot.

6 When refitting the lever, lubricate the ball pivot with a little molybdenum disulphide grease then position the spring clip on it and press on the lever so that the clip snaps onto the ball.

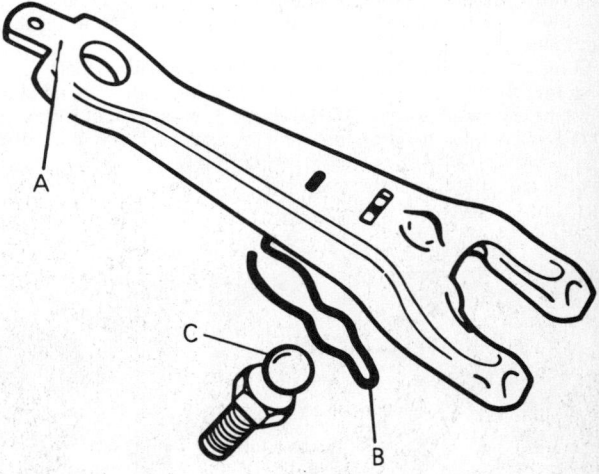

Fig. 13.18 Later type of clutch release lever (Sec 8)

A Lever
B Spring clip
C Ball pivot

9 Gearbox (five-speed)

Description
1 This type of gearbox, which is fitted to later 1500 and 1600 models, is basically a 4-speed gearbox which has been modified to incorporate a 5th (overdrive) gear.
2 The mainshaft has been extended to carry 5th gear and the synchro unit, while the layshaft has a bolt-on extension, a 5th gear and a shaft support bearing located in the rear extension housing cover.
3 The reverse selector shaft has been extended to carry the 5th/reverse selector fork.

Removal and refitting
4 The removal and refitting operations are very similar to those described in Chapter 6, Section 2.
5 All five speed models are equipped with a centre console which must be removed in order to give better access for prising out the plastic slotted ring with a small screwdriver (photos).

Overhaul
Dismantling
6 Refer to Chapter 6, Section 3 and carry out the operations described in paragraphs 1 to 7 inclusive.
7 Unbolt the gear lever turret and remove the assembly. Do not unscrew the inner ring of nuts as their bolts will drop into the gearbox interior (photos). This is very important if the turret is being removed with the gearbox in the car. Only unscrew the outer ring of nuts. With the turret removed, discard the gasket.
8 Unscrew the nuts and take off the rear extension housing. Note the nut inside the main casing, also the one hidden by the bracket (photos).
9 Discard the gasket.
10 Pull off the mainshaft rear bearing.
11 Withdraw the speedometer drivegear and take out the locking ball.
12 Remove the oil slinger.
13 Unscrew the lockbolt from the 5th/reverse selector fork and slide the fork down the shaft whilst holding the shaft stationary.
14 Pull either one of the two other selector shafts upwards by gripping the shaft dog. Two gears will now be locked up to prevent the geartrains turning.
15 Unscrew the layshaft bolt.
16 Return the selector shaft and 5th/reverse fork to neutral. Screw in the fork lockbolt finger tight.
17 Pull 5th/reverse selector fork upwards to engage 5th gear.
18 Lift the layshaft 5th gear assembly and remove it. Note the blind washer below it.
19 Pull off 5th gear with its spacer/sleeve from the mainshaft.
20 Unscrew the finger tight 5th/reverse fork lockbolt and remove it. Remove the 5th gear selector shaft dog lockbolt and remove the dog.
21 Remove 5th gear synchromesh unit, the selector fork and reverse idler gear with the spacer washer as an assembly.
22 Take reverse gear from the mainshaft.
23 Continue dismantling by carrying out the operations described in Chapter 6, Section 3, paragraphs 8 and 15 to 23.

Mainshaft – dismantling
24 Refer to Chapter 6, Section 4. The 5th gear synchromesh unit is similar to the other units but the hub-to-sleeve relationship should be marked before separating the components.

Inspection and renewal of components
Refer to Chapter 6, Section 5.

Mainshaft – reassembly
Refer to Chapter 6, Section 6.

Reassembly
25 Refer to Chapter 6, Section 7 and carry out the operations described in paragraphs 1 to 23 (photo).
26 Fit the Woodruff key, if removed, to the mainshaft (photo).
27 Fit the reverse gear to the mainshaft so that its extended hub is towards the end of the shaft (photo).
28 Refit the reverse idler gear, 5th/reverse selector fork and 5th speed synchro unit as an assembly (photo).
29 Screw in the 5th/reverse selector fork lockbolt finger tight. Locate the mainshaft thrust washer (photos).
30 Fit 5th gear with its spacer/sleeve to the mainshaft (photo).
31 Fit the dog (chamfered edge towards the bellhousing) to the 5th/reverse selector shaft. This dog operates the reversing lamp switch. Tighten the dog lockbolt (photo).
32 Pull 5th/reverse selector fork upwards to engage 5th gear.
33 Fit the shaft bearing blind washer and 5th gear assembly to the layshaft (photos).
34 Return the 5th/reverse selector fork and shaft to neutral and remove the finger tight fork lockbolt.
35 Push the 5th/reverse selector fork down the shaft while holding the shaft stationary.
36 Now pull either one of the other selector shafts upwards so locking two gears to prevent the geartrains turning.
37 Screw in the layshaft bolt and tighten to the specified torque (photos).
38 Return the selector fork and shaft to neutral and screw in the 5th/reverse fork lockbolt. Tighten it to the specified torque.
39 Refit the oil slinger, convex side towards the end of the mainshaft (photo).
40 Locate the locking ball and fit the speedometer drivegear with its shank towards the end of the mainshaft (photos).
41 Fit the mainshaft rear bearing so that the engraved marks can be read from the end of the shaft (photo).
42 Using a new gasket, fit the extension housing and tighten nuts to the specified torque (photo).
43 Using a new gasket, bolt on the gearlever turret.
44 Carry out the operations described in Chapter 6, Section 7, paragraphs 40 to 47, noting that the output drive flange should be rotated in order to be able to insert the three connecting bolts which are otherwise obstructed by the extension housing casing (photo).

9.5A Centre console

9.5B Gearchange lever slotted ring

9.7A Removing gear lever turret

9.7B Nuts securing turret to casing (arrowed)

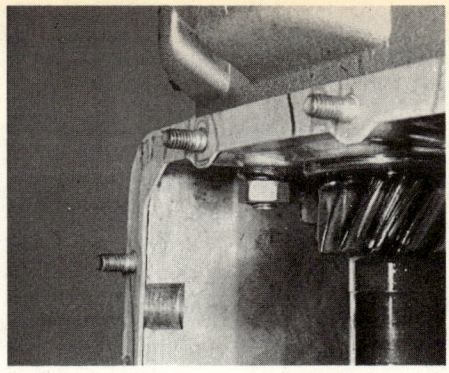
9.8A Rear extension housing nut inside main casing

9.8B Rear extension housing nut masked by bracket

9.8C Removing rear extension housing

9.25 The gearbox ready for assembling 5th/reverse components

9.26 Woodruff key in mainshaft

9.27 Reverse gear on mainshaft

9.28 Fitting 5th/reverse components to mainshaft as an assembly

9.29A 5th/reverse selector fork lockbolt

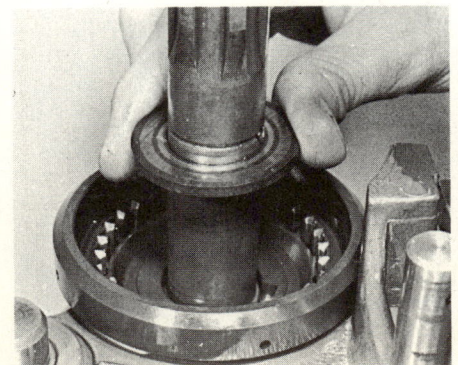
9.29B Mainshaft thrust washer

9.30 Mainshaft 5th gear with spacer/sleeve

9.31 5th/reverse selector shaft dog (chamfered edge arrowed)

9.33A Shaft bearing blind washer

9.33B Fitting layshaft 5th gear assembly

9.37A Screwing in the layshaft bolt

9.37B Tightening the layshaft bolt

9.39 Fitting the oil slinger

9.40A Speedo drive gear locking ball

Chapter 13 Supplement

9.40B Fitting speedo drive gear

9.41 Mainshaft rear bearing

9.42 Rear extension housing gasket in position

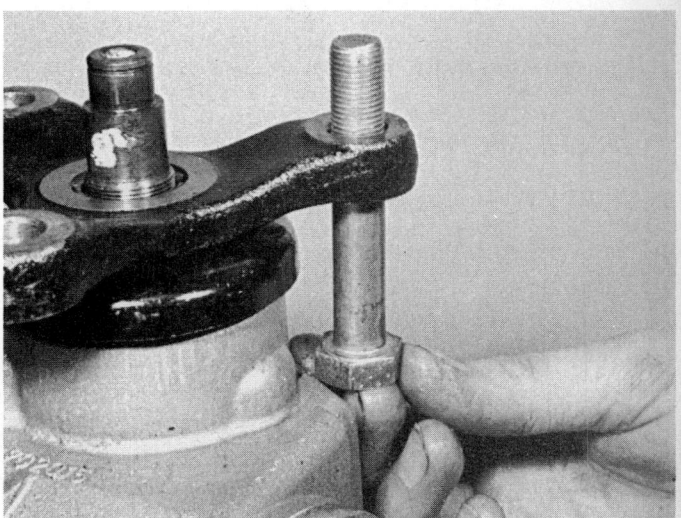

9.44 Inserting drive flange coupling bolt

10 Propeller shaft

Universal joints – overhaul

1 When reassembling the universal joints it is important to ensure that the spider has the specified axial clearance between the bearing cups. To achieve this the circlips are available in five different thicknesses.

2 To check the clearance, assemble the joint as instructed in Chapter 7 up to the point where the circlips are to be fitted. Working on each opposite pair of bearing cups, fit a 0.061 in (1.56 mm) circlip to one cup then press on the opposite cup to eliminate all clearance.

3 Using a feeler blade, determine the clearance between the cup face and the top of the circlip groove, then deduct the nominal clearance of 0.001 in (0.025 mm) to give the thickness of the circlip to fit.

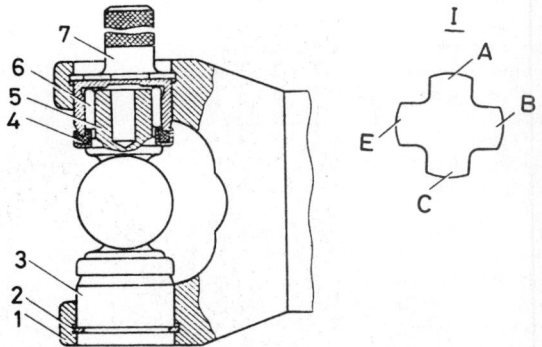

Fig. 13.19 Cross-sectional view of propeller shaft universal joint (Sec 10)

1	Yoke	7 and 1 Special Lada gauge
2	Circlip	(feeler blades can be used instead)
3	Cup	A = 0.060 in (1.53 mm)
4	Seal	B = 0.061 in (1.56 mm)
5	Spider	C = 0.063 in (1.59 mm)
6	Needle bearing	E = 0.064 in (1.62 mm)

11 Rear axle

Axleshaft bearings – checking

1 Wear in the axleshaft bearings may be checked by measuring the

endplay of the axleshaft. To do this, jack up the rear of the car and support on axle stands. Chock the front wheels and release the handbrake.
2 The check may be made without removing the wheels, although it is preferable to remove both wheels and brake drums. If the former method is used remove just one wheel bolt from each wheel.
3 If available, attach a dial gauge to the wheel with an extension probe through the wheel bolt hole and in contact with the backplate. Alternatively use a suitable bolt and feeler gauge, using the underside of the bolt head as a datum face against the wheel or drive flange.
4 Move the wheel or drive flange fully in and out to measure the total amount of endplay present in the bearing. If this exceeds the amount given in the Specifications the bearing must be renewed, with reference to Chapter 8.

Pinion oil seal – renewal
5 The method of tightening the pinion nut is now modified to disregard the sleeve wall thickness of the collapsible spacer. The specified torque for the nut is now between 87 and 188 lbf ft (12 and 28 kgf m) and the preload must be checked as soon as the lower limit is reached.

12 Braking system

Braking fluid warning switch – checking
1 On later models, a brake fluid low level warning switch is incorporated in the fluid reservoir cap (photo). To check that the switch and circuit is functioning correctly switch on the ignition then depress the plunger on the top of the reservoir cap. The warning lamp on the instrument panel must remain on while the plunger is depressed. If not, check the wiring, bulb and switch.

Rear brake drums – removal
2 If difficulty is experienced when removing the rear brake drums, the drum retaining bolts (Chapter 9 photo 4.2a) should be screwed into the two other holes provided and tightened evenly until the drum is forced from the drive flange (photo).

Rear brake shoes (except Riva 1500 and 1600 models) – renewal
3 All non-Riva 1500 and 1600 models are fitted with a self-adjusting mechanism for the rear brake shoes. The components of the system are shown in Chapter 9, Fig. 9.4 and consist of spring tensioned friction pads attached to the shoe web within an elongated hole. The internal sleeves are located on a pin welded to the backplate. The friction pads are tensioned to prevent movement of the shoes by the return spring. Outward movement of the shoes due to hydraulic pressure from the footbrake pedal relocates the shoes on the friction pads. The bushes on the backplate pins have an inbuilt clearance to allow the normal shoe-to-drum clearance. Wear of the linings causes the shoes to be repositioned on the backplate pins.
4 The renewal procedure is similar to that for other models except that the shoes must be lifted from the backplate pins. Transfer the friction pads to the new shoes. To do this, Lada use their special fixture A.72259 (Fig. 13.22); a suitable tool can be made out of metal tubing. The upper section of the tool should have a crossbar so that the spring can be compressed in order to screw the two sleeves together.

Rear brakes (Riva and later models) – description
5 All Riva and later models are fitted with rear brake wheel cylinders incorporating self-adjusting pistons (photo). The retracted position of the pistons is governed by thrust rings which are a tight fit in the cylinder bore but which can be moved by hydraulic pressure when the brake shoes are expanded. Normal movement of each piston is between the stop screw and retainers which are located either side of the thrust ring. Movement of the thrust rings, therefore, only occurs when wear of the linings causes the pistons to move further out of the cylinder.

Rear brake shoes (Riva and later models) – renewal
6 The procedure is as described in Chapter 9. However, before fitting the new shoes, it will be necessary to force the pistons fully back into the wheel cylinder using a G-clamp. On completion set the shoes with reference to paragraph 18.

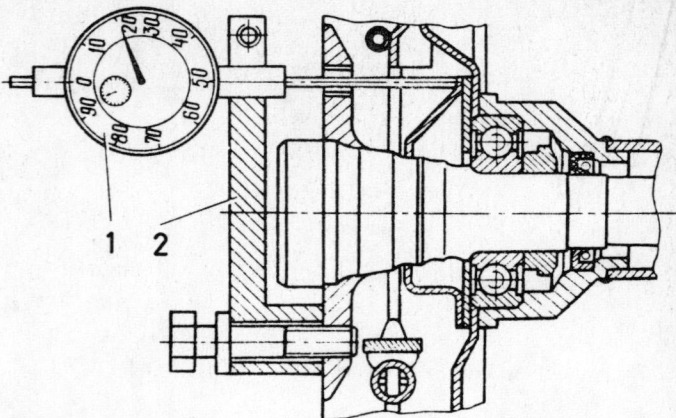

Fig. 13.20 Measuring axleshaft endfloat with wheel and brake drum removed (Sec 11)

1 Dial gauge
2 Typical dial gauge mounting bracket

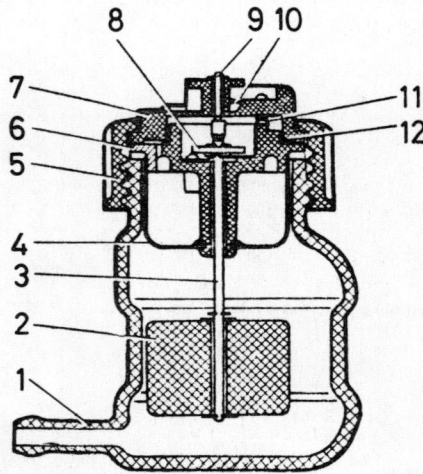

Fig. 13.21 Cross-sectional view of brake fluid warning switch (Sec 12)

1 Reservoir
2 Float
3 Pushrod
4 Deflector
5 Cap
6 Contact unit
7 Terminal body
8 Contact (moving)
9 Plunger
10 Terminal
11 Contact (fixed)
12 Seal

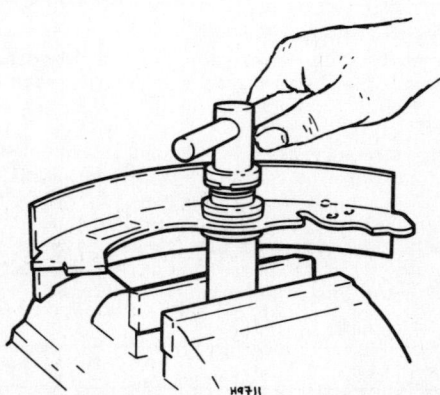

Fig. 13.22 Using Lada special tool to remove self-adjuster friction pads from a rear brake shoe (Sec 12)

Chapter 13 Supplement

12.1 Brake fluid level warning switch

12.2 Removing a tight brake drum

12.5 Rear brake on later Riva models

Rear brake wheel cylinder (Riva and later models) – removal, refitting and overhaul

7 The removal procedure is as described in Chapter 9 and the overhaul procedure is as follows.
8 Clean the exterior of the unit then mount it in a vice.
9 Prise the dust covers from each end of the cylinder.
10 Pull each piston from the cylinder, together with the thrust rings.
11 Examine the pistons and cylinder bore for signs of wear or scoring, and if evident renew the complete cylinder. If the surfaces of the pistons and bore are in good condition, only the seals need to be renewed. The thrust rings will need to be renewed if the assembly force on the pistons is less than 77 lbf (35 kgf) (see paragraph 16).
12 To dismantle the pistons proceed as follows. Grip the piston in a soft-jawed vice then unscrew the inner stop screw from it and remove the seal, seat, spring, retainers and thrust ring.
13 Clean all the components with hydraulic fluid or methylated spirit.
14 Locate the thrust ring on the stop screw followed by the retainer halves, spring, seat and seal then tighten the stop screw in the piston.
15 Immerse each piston assembly in hydraulic fluid then insert it into the cylinder, making sure that the slot of the thrust ring faces upwards towards the bleed screw outlet. This will facilitate bleeding air from the hydraulic system.
16 Press each piston fully into the cylinder, initially using a spring balance and suitable hook to check that the assembly force of the thrust rings is at least 77 lbf (35 kgf). If less than this, the thrust ring must be renewed otherwise its self-adjusting action will be impaired.
17 With both pistons fitted, check the dimensions shown in Fig. 13.23 then fit the dust covers. Check that the free movement of each piston (ie without moving the thrust rings) is between 0.049 and 0.065 in (1.25 and 1.65 mm).
18 Refitting is as described in Chapter 9, but after refitting the brake drum and roadwheel, depress the footbrake pedal firmly several times in order to set the self-adjusting thrust rings in their correct positions.

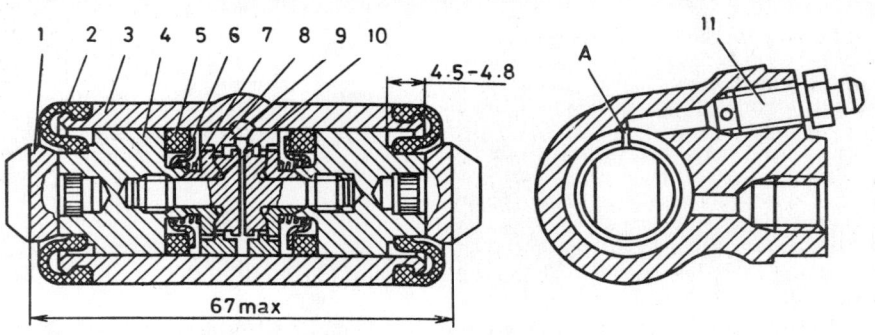

Fig. 13.23 Cross-sectional view of self-adjusting type rear wheel cylinder (Sec 12)

1 Shoe stop	5 Seal	8 Retainers	11 Bleed screw
2 Dust cover	6 Seat	9 Thrust ring	A Slot in thrust ring
3 Cylinder	7 Spring	10 Inner stop screw	Dimensions in mm
4 Piston			

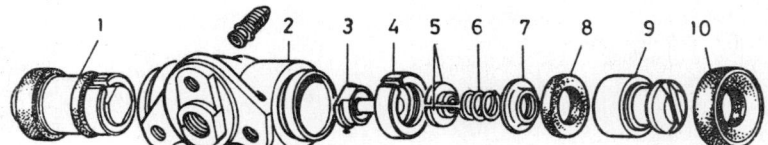

Fig. 13.24 Exploded view of self-adjusting type rear wheel cylinder (Sec 12)

1 Piston assembly	5 Retainers	8 Seal
2 Cylinder	6 Spring	9 Piston
3 Inner stop screw	7 Seat	10 Dust cover
4 Thrust ring		

Rear brake pressure compensator valve – adjustment
19 As from mid 1980 the operating bar adjustment dimension described in Chapter 9, Section 13 is reduced from 7.87 in (200.0 mm) to 5.31 to 5.71 in (135.0 to 145.0 mm).

Hydraulic system – additional bleeding methods
20 If the master cylinder has been disconnected and reconnected then the complete system (both circuits) must be bled.
21 If a component of one circuit has been disturbed then only that particular circuit need be bled.
22 Unless the pressure bleeding method is being used, do not forget to keep the fluid level in the master cylinder reservoir topped up to prevent air from being drawn into the system which would make any work done worthless.
23 Before commencing operations, check that all system hoses and pipes are in good condition with all unions tight and free from leaks.
24 Take great care not to allow hydraulic fluid to come into contact with the vehicle paintwork as it is an effective paint stripper. Wash off any spilled fluid immediately with cold water.
25 As the system incorporates a vacuum servo, destroy the vacuum by giving several applications of the brake pedal in quick succession.
26 If the complete system is to be bled, start with the rear circuit.

Bleeding – using one-way valve kit
27 There is a number of one-man, one-way brake bleeding kits available from motor accessory shops. It is recommended that one of these kits is used wherever possible as it will greatly simplify the bleeding operation and also reduce the risk of air or fluid being drawn back into the system quite apart from enabling the work to be done without the help of an assistant.
28 To use the kit, connect the tube to the bleed screw and open the screw one half a turn.
29 Depress the brake pedal fully and slowly release it. The one-way valve in the kit will prevent expelled air from returning at the end of each pedal downstroke. Repeat this operation several times to be sure of ejecting all air from the system. Some kits include a translucent container which can be positioned so that the air bubbles can actually be seen being ejected from the system.
30 Tighten the bleed screw, remove the tube and repeat the operations on the remaining brakes.
31 On completion, depress the brake pedal. If it still feels spongy repeat the bleeding operations as air must still be trapped in the system.

Bleeding – using a pressure bleeding kit
32 These kits too are available from motor accessory shops and are usually operated by air pressure from the spare tyre.
33 By connecting a pressurized container to the master cylinder fluid reservoir, bleeding is then carried out by simply opening each bleed screw in turn and allowing the fluid to run out, rather like turning on a tap, until no air is visible in the expelled fluid.
34 By using this method, the large reserve of hydraulic fluid provides a safeguard against air being drawn into the master cylinder during bleeding which often occurs if the fluid level in the reservoir is not maintained.
35 Pressure bleeding is particularly effective when bleeding 'difficult' systems or when bleeding the complete system at time of routine fluid renewal.

All methods
36 When bleeding is completed, check and top up the fluid level in the master cylinder reservoir.
37 Check the feel of the brake pedal. If it feels at all spongy, air must still be present in the system and further bleeding is indicated. Failure to bleed satisfactorily after a reasonable period of the bleeding operation may be due to worn master cylinder seals.
38 Discard brake fluid which has been expelled. It is almost certain to be contaminated with moisture, air and dirt making it unsuitable for further use. Clean fluid should always be stored in an airtight container as it absorbs moisture readily (hygroscopic) which lowers its boiling point and could affect braking performance under severe conditions.

Handbrake warning switch – removal and refitting
39 A warning lamp is provided on the instrument panel to indicate when the handbrake is applied, and the operating switch is located on a bracket in front of the handbrake lever. The warning lamp doubles as a front brake disc pad wear warning lamp or low brake fluid warning lamp on some models.
40 To remove the switch, undo the screws and withdraw the cover from the base of the handbrake lever (photo). Use an angled screwdriver to remove the side screw and push the cover forwards to release it from the metal lip.
41 Unscrew the nut and release the switch from the bracket then withdraw it, pull back the boot and disconnect the wire (photo).
42 Refitting is a reversal of removal.

Brake pedal free movement – adjustment
43 On 1500 models manufactured before October 1975 the brake pedal free movement should be between 0.276 and 0.394 in (7.0 and 10.0 mm). On all other models the dimension should be as given in Chapter 9, Section 18.

Brake servo unit – adjustment
44 Whenever the master cylinder or brake servo unit is removed, the servo unit pushrod protrusion should be checked and if necessary adjusted. The tip of the pushrod should protrude 0.041 to 0.049 in (1.05 to 1.25 mm) from the master cylinder mounting face on the servo unit. Check the dimension using either vernier calipers or a dial gauge, and if necessary make the adjustment by holding the pushrod and turning the domed nut.

Brake servo unit and non-return valve – checking
45 To check the operation of the brake servo unit first depress the brake pedal several times with the engine stopped to destroy the vacuum in the unit. Hold the pedal depressed with moderate pressure then start the engine and check the pedal moves down slightly as a result of the assistance from the servo unit.

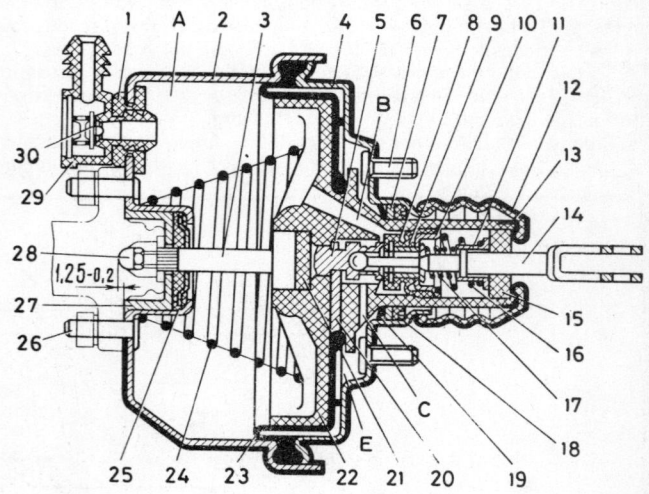

Fig. 13.25 Longitudinal sectional view of brake vacuum servo unit (Sec 12)

1 Grommet
2 Servo body
3 Pushrod
4 Cover
5 Piston
6 Mounting stud
7 Distance ring
8 Valve spring seat
9 Valve
10 Valve seat
11 Return spring seat
12 Boot
13 Holder
14 Pushrod
15 Air filter
16 Valve return spring
17 Valve spring
18 Cover seal
19 Seal lockring
20 Thrust plate
21 Buffer
22 Valve body
23 Diaphragm
24 Valve body return spring
25 Pushrod seal
26 Master cylinder mounting stud
27 Pushrod seal holder
28 Adjusting bolt
29 Non-return valve body
30 Non-return valve
A Vacuum chamber
B Channel (vacuum)
C Channel (atmospheric)
E Atmospheric chamber

Pushrod protrusion is in mm

Chapter 13 Supplement

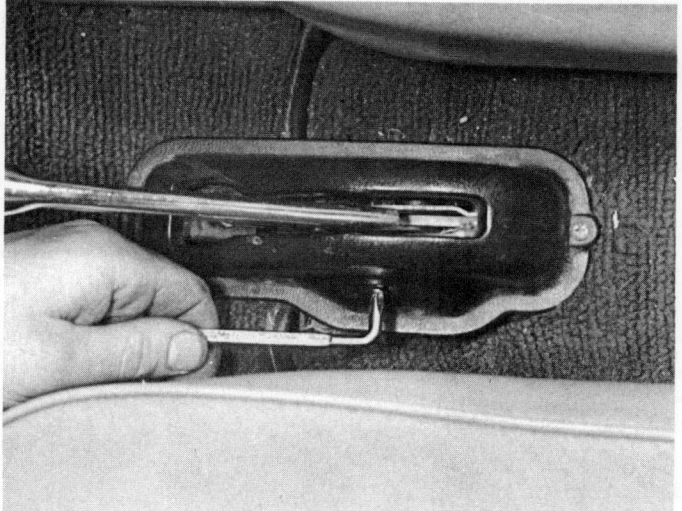

12.40 Removing the handbrake lever cover

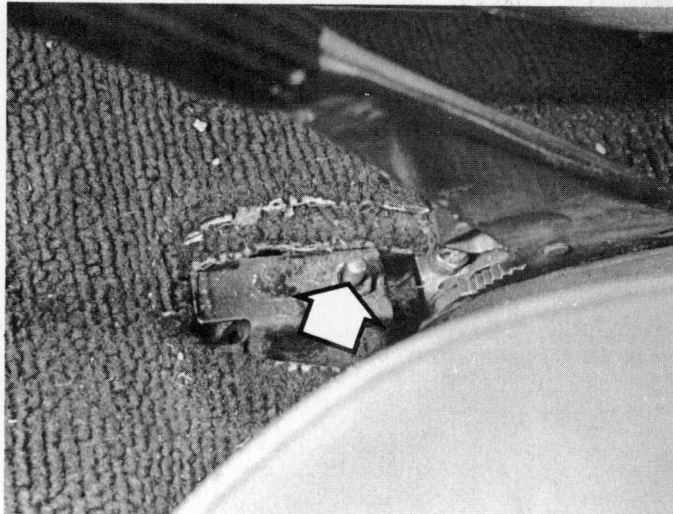

12.41 Handbrake 'on' warning switch (arrowed)

46 If the unit fails to operate as described check the condition and security of the vacuum hose. Incidents have been reported of sticking non-return valves, therefore this should also be checked. The fault occurs as a result of lapping fluid remaining on the valve seats, and the remedy is to immerse the non-return valve in boiling water for approximately one minute then blow through the unit with an air line. The valve can be removed by disconnecting the vacuum hose and carefully prising it from the servo unit.

13 Electrical system

Alternator (Riva models) – description

1 The new alternator incorporates a transistor voltage regulator instead of the remote vibrating contact type previously fitted in the charging circuit. It is located over the brush holder on the rear of the alternator and is removed by extracting the screws retaining the small cover. The brush holder and a condenser are removed at the same time.

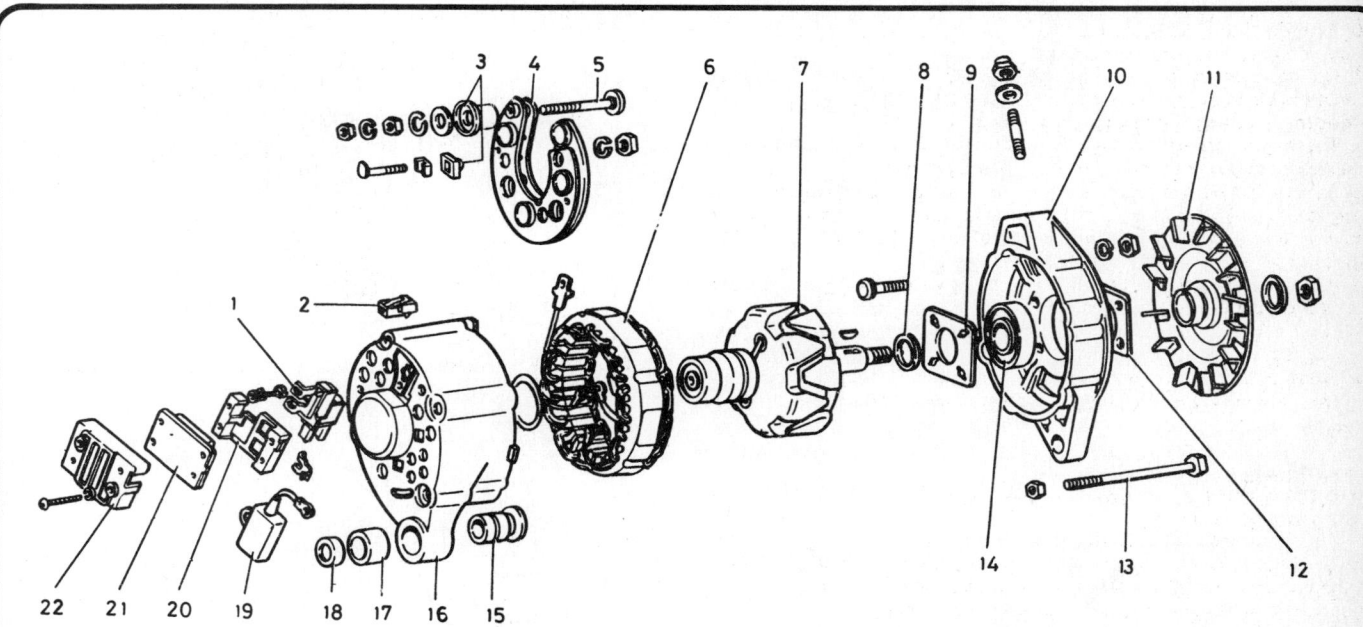

Fig. 13.26 Exploded view of integral voltage regulator type alternator (Sec 13)

1 Brush holder	7 Rotor	13 Through-bolt	18 Bush
2 Neutral wire plug socket	8 Washer	14 Bearing	19 Condenser
3 Insulating bushes	9 Bearing retaining plate	15 Bush	20 Base
4 Rectifier	10 Drive end bracket	16 Slip ring end bracket	21 Voltage regulator
5 Terminal bolt	11 Pulley	17 Sleeve	22 Cover
6 Stator	12 Bearing retaining plate		

Chapter 13 Supplement

Fuses and relays (early Riva models)
2 On Riva models the fusebox is located in the engine compartment on the left-hand side of the bulkhead (photo).
3 The main relays (Fig. 13.27) are mounted on the fusebox, but the handbrake warning lamp relay is located on the right-hand side of the bulkhead, and the no-charge warning lamp relay is located on the right-hand side of the engine compartment.

Fuses and relays (later Riva models)
4 On later models, the fusebox with relays is as shown in the photograph.
5 The small relay on the right-hand wing valance within the engine compartment is the starter motor relay (photo).
6 The flasher unit and main relay are located behind the instrument panel on the right-hand side (photo).
7 Access to these relays is obtained by withdrawing the instrument panel as described in paragraphs 45 to 49 of this Section.

Ignition warning lamp
8 On later models with 1500 or 1600 cc engines, the ignition lamp (charge relay) has been deleted. Reference should be made to the battery charge indicator (voltmeter) for confirmation of the alternator charging.

Headlamp wiper motor – description, removal and refitting
9 Some models are equipped with a headlamp wash/wipe system. The control switch is mounted on the steering column and the wiper motors are located on the inner sides of the headlamps.
10 To remove a motor, first remove the wiper arms and blades by lifting the caps and unscrewing the spindle nuts (photo).
11 Remove the radiator grille then disconnect the wiring and unbolt the motor unit (photo).
12 If faulty, the unit should be renewed as it is of sealed construction.
13 Refitting is a reversal of removal, but before fitting the wiper arms switch the wipers on and off so that the spindles are at the parked position.

Washer pump (electric) – description, removal and refitting
14 Both the windscreen and headlamp washer pump are located in the washer reservoir on the left-hand side of the engine compartment. They are incorporated in the filler caps with tubing connecting them to their respective jets.
15 To remove a pump, unscrew it from the reservoir, then disconnect the tubing and wiring (photo).
16 Undo the screws, remove the motor and cover, and disconnect the tube.
17 Prise out the bottom rim and filter.
18 Remove the coupling, then tap the shaft to drive out the support. Withdraw the shaft with the rotor.
19 Refitting is a reversal of removal.

Washer pump (foot-operated) – description
20 Some early models are equipped with a foot-operated washer pump located next to the clutch pedal. The pump incorporates a switch for the windscreen wipers, and is designed to spray the windscreen with water before the switch contacts close.
21 The unit is secure by studs to the scuttle. The larger outlet should be connected to the reservoir and the other outlet to the jets.
22 It is not possible to dismantle the pump so, if it is proved faulty, it should be renewed.

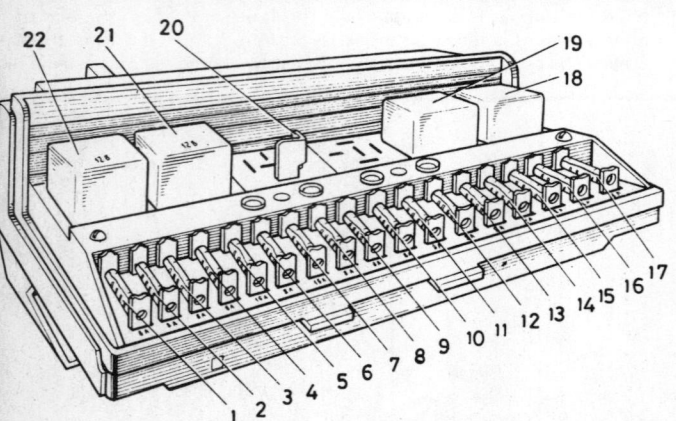

Fig. 13.27 Riva fusebox and relays (Sec 13)

1 to 17 Fuses
18 Headlamp dipped beam relay
19 Headlamp main beam relay
20 Bridge connector
21 Headlamp wash/wipe relay
22 Heated rear window relay

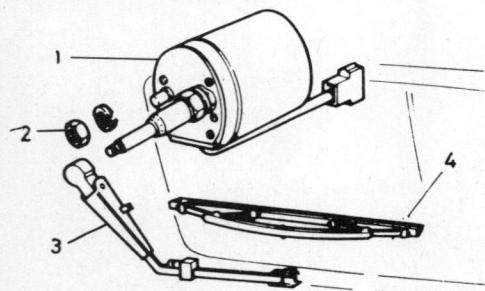

Fig. 13.28 Headlamp wiper motor components (Sec 13)

1 Motor
2 Nut
3 Wiper arm
4 Wiper blade

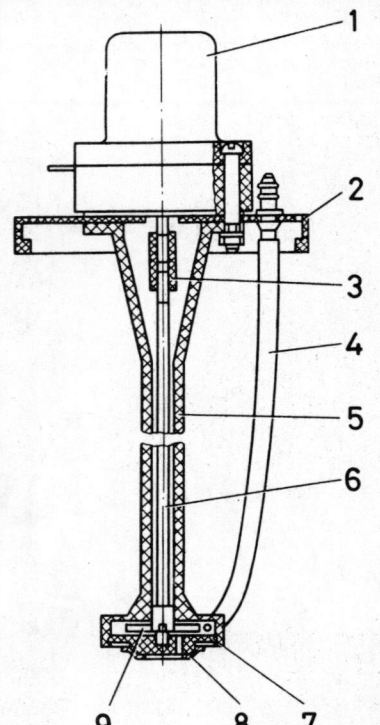

Fig. 13.29 Sectional view of washer pump (Sec 13)

1 Electric motor
2 Cover
3 Coupling
4 Tube
5 Casing
6 Rotor shaft
7 Support
8 Cap with filter screen
9 Rotor

Chapter 13 Supplement

Headlight bulb (Riva models) – removal and refitting
23 Remove the cover from the rear of the headlamp by turning it anti-clockwise (photo).
24 Pull the connector from the bulbholder.
25 Unhook the spring clips and withdraw the bulbholder and bulb (photo).
26 Refitting is a reversal of removal, but do not touch the bulb glass with the fingers. Clean the glass with methylated spirit if touched.

Front parking bulb (Riva models) – removal and refitting
27 Remove the cover from the rear of the headlamp by turning it anti-clockwise.
28 Pull out the bulbholder then depress and twist the bulb to remove it (photo).
29 Refitting is a reversal of removal.

Front direction indicator bulb (Riva models) – removal and refitting
30 Turn the bulbholder anti-clockwise and remove it from the rear of the headlamp (photo).
31 Depress and twist the bulb to remove it.
32 Refitting is a reversal of removal.

Rear lamp cluster bulbs (Riva models) – removal and refitting
33 Working in the boot space, unscrew the knurled nuts and remove the lamp inner cover (photo).
34 Prise open the plastic clips and withdraw the printed circuit board with the bulbs (photo).
35 Depress and twist the bulbs to remove them.
36 Refitting is a reversal of removal.

Number plate lamp (Riva models) – removal and refitting
37 Undo the screws, remove the lamp and take off the lens.
38 Remove the festoon type bulb from the spring contacts.
39 Refitting is a reversal of removal.

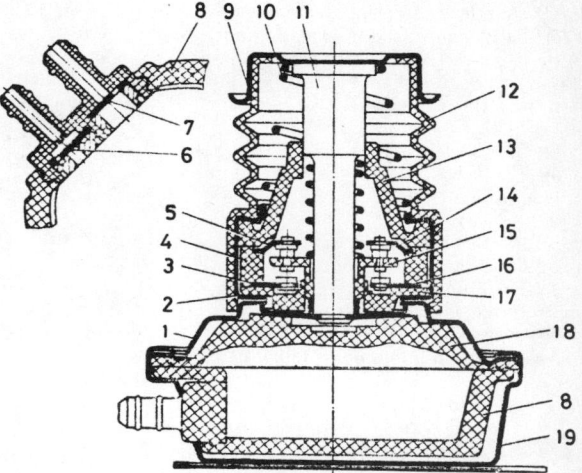

Fig. 13.30 Sectional view of foot-operated washer pump (Sec 13)

1 Cover
2 Switch holder
3 Lower contact
4 Contact block
5 Upper contact
6 Outlet valve
7 Inlet valve
8 Casing
9 Push button
10 Return spring
11 Pushrod
12 Boot
13 Guide
14 Upper contact
15 Movable contact holder
16 Lower contact
17 Guide
18 Diaphragm
19 Housing

13.2 Fuses and relays – early Riva models

13.4 Fuses and relays – later Riva models

1 Heated rear window
2 Headlamp wiper
3 Horn
4 Engine fan
5 Headlamp
6 Headlamp

13.5 Starter motor relay

13.6 Flasher unit (relay)

13.10 Removing headlamp wiper arm

13.11 Headlamp wiper motor

13.15 Washer pump integral with filler cap

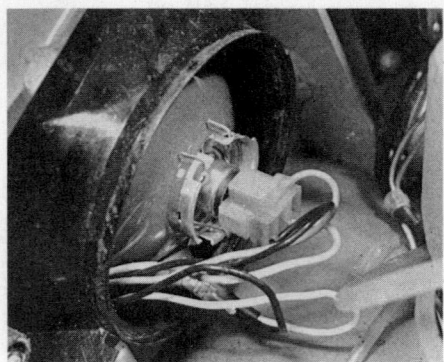
13.23 Headlamp wiring plug, cover removed

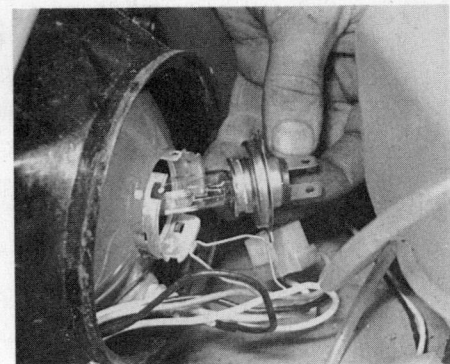
13.25 Withdrawing headlamp bulb

13.28 Front parking lamp bulbholder

13.30 Front direction indicator bulbholder

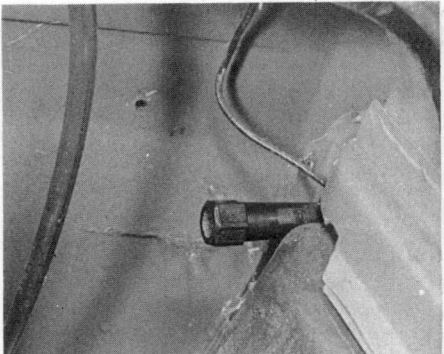
13.33 Rear lamp cluster knurled nut and cover

Hydraulic headlamp adjuster (Riva models) – description
40 The hydraulic headlamp adjuster is located on the right-hand side of the facia panel and has four settings – corresponding to the load being carried.
41 The system consists of the control unit and two tubes connected to servo units on the headlamps. Refer also to paragraphs 50 to 56.

Battery (Riva models)
42 Later Riva models are equipped with a low-maintenance type battery.
43 Topping up is not normally required provided the electrolyte level is seen to be correct when viewed through the translucent case (photo).
44 The need for frequent topping up will probably be due to overcharging, which should be investigated.

Instrument panel (Riva models) – removal and refitting
45 Disconnect the battery.
46 Remove the heater symbol blanking plate to expose two screws. Extract the upper screw (photos).
47 Pull the left-hand side of the instrument panel towards you against the tension of the retaining spring clips. Have an assistant push the speedometer drive cable from within the engine compartment. The cable passes through a hole in the transmission tunnel (photos).
48 Reach around behind the left-hand end of the panel and disconnect the switches and plugs, and unscrew the speedometer cable knurled ring. Release the trip cable (photos).
49 Withdraw the panel noting the retaining nibs at the right-hand end (photo).

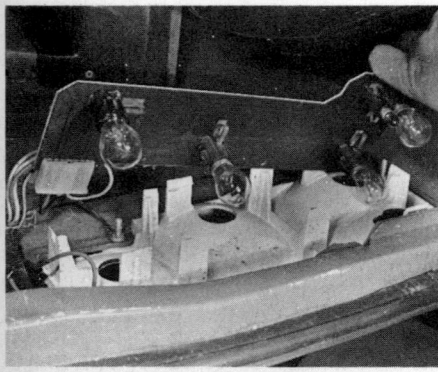
13.34 Rear lamp cluster printed circuit board

13.43 Battery showing electrolyte level

13.46A Removing heater symbol plate

Chapter 13 Supplement

13.46B Upper fixing screw unscrewed

13.47A Withdrawing the instrument panel

13.47B Instrument panel fixing clips

13.48A Instrument panel switch

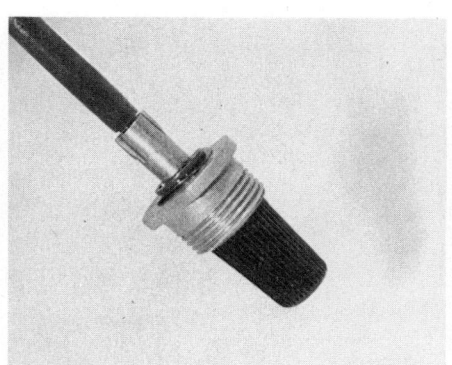

13.48B Speedometer trip recorder knob and cable

13.49 Instrument panel right-hand nibs

Headlamp (Riva models) – removal and refitting

50 Open the bonnet, disconnect the battery.
51 Disconnect the wiring from the rear of the lamp. Remove the headlamp wiper arm (photo).
52 From the front of the lamp, extract the self-tapping fixing screws, one at the top and two at the bottom (photo).
53 Pull the lamp assembly forward and disconnect the hydraulic beam adjusting device. To do this, depress the retaining tab, grip the device and pull it sharply from its balljoint. Take care not to break or disconnect the hydraulic tube (photos).
54 Remove the lamp assembly.
55 Refit by reversing the removal operations, but when reconnecting the hydraulic beam adjuster to its balljoint, grease the ball and hold up the socket by inserting a screwdriver as shown (photo), but take care not to allow the screwdriver to press against the headlamp glass reflector.
56 Adjust the headlamp beam on completion.

Headlamp dim-dip system

57 This system is fitted to post 1987 models to comply with current regulations. The system prevents the car being driven with parking lamps only.
58 If the parking lights are switched on with the ignition on, the headlamps come on at reduced intensity by means of a modified wiring circuit.

13.51 Headlamp wiper arm attachment to drive spindle

13.52 Headlamp fixing screw

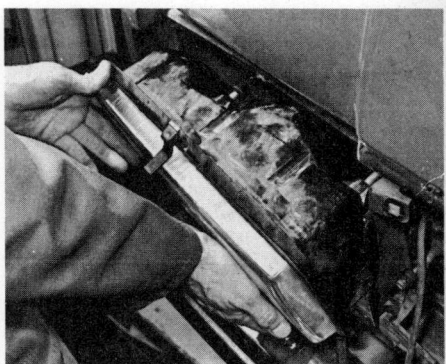

13.53A Withdrawing headlamp

13.53B Headlamp beam adjuster withdrawn

13.53C Headlamp beam adjuster. Retaining clip arrowed

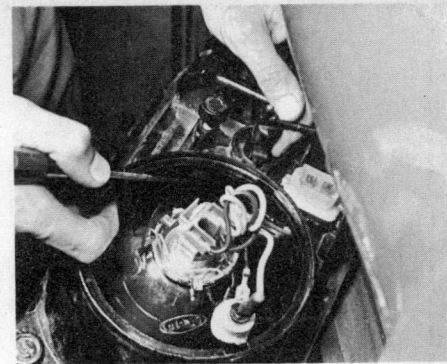
13.55 Supporting beam adjusting device socket with a screwdriver in order to engage ball

Key to wiring diagram for later 1200 models

1 Side direction indicators
2 Sidelights
3 Headlights
4 Horns
5 Voltage regulator
6 Ignition coil
7 Ignition distributor
8 Spark plugs
9 Windshield wiper motor
10 Brake fluid level sender
11 Oil pressure gauge sender
12 Coolant temperature sender
13 Bonnet lamp
14 Generator
15 Starter
16 Battery
17 Battery no-charge warning relay
18 Windshield wiper switch
19 Windshield wiper relay
20 Handbrake or lamp flasher
21 Fuse block
22 Handbrake on lamp switch
23 Reversing light switch
24 Direction indicator flasher
25 Stop-light switch
26 Inspection lamp socket
27 Heater motor
28 Heater motor series resistor
29 Glovebox lamp
30 Windshield wiper switch
31 Instrument lighting switch
32 External lighting switch
33 Direction indicator switch
34 Headlight dip switch
35 Horn switch
36 Ignition switch
37 Cigarette lighter
38 Heater switch
39 Rear door interior lamp switches
40 Front door interior lamp switches
41 Interior lamps
42 Instrument board
43 Fuel level and reserve signal lamp
44 Instrument board lamp
45 Handbrake and brake fluid level lamp
46 Oil pressure warning lamp
47 Battery no-charge warning lamp
48 External lighting signal lamp
49 Direction indicator signal lamp
50 Headlight main beam signal lamp
51 Boot lamp
52 Fuel level and reserve sender
53 Tail lights
54 Reversing lamp
55 Number plate lamp
56 Rear wire bundle
57 Rear body lamp

Not all items are fitted to all models

Colour Codes

W White
Bl Blue
Y Yellow
G Green
Br Brown
R Red
O Orange
P Pink
Gy Grey
B Black

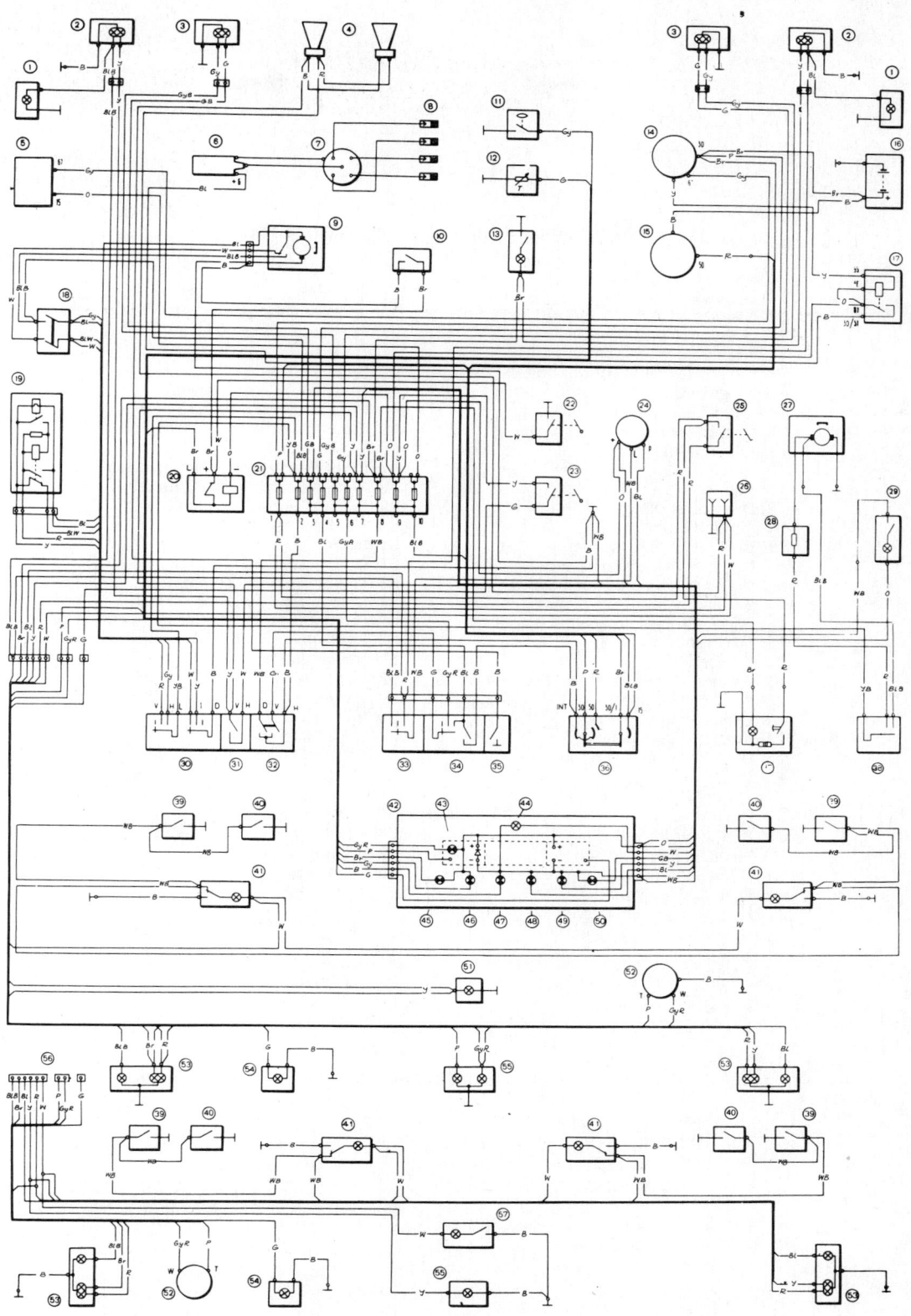

Fig. 13.31 Wiring diagram for later 1200 models

Key to wiring diagram for Riva 1300 models

1 Direction indicators
2 Headlamps
3 Headlamp wiper motors
4 Horns
5 Pneumatic valve control unit
6 Headlamp washer motor
7 Windshield washer motor
8 Distributor
9 Spark plugs
10 Generator
11 Battery
12 Diagnostic panel
13 Ignition coil
14 Brake fluid level sender
15 Coolant temperature gauge sender
16 Oil pressure signal lamp sender
17 Bonnet lamp
18 Carburettor microswitch
19 TDC sender
20 Starter
21 Battery no-charge signal lamp relay
22 Pneumatic valve
23 Windscreen wiper relay
24 Inspection lamp socket
25 Hazard warning and indicator relay
26 Stop-light switch
27 Windshield wiper motor
28 Heater motor
29 Heater motor series resistor
30 Reversing light switch
31 Handbrake on warning switch
32 Fuse and relay unit
33 Headlamp lower beam relay
34 Headlamp upper beam relay
35 Headlamp wiper and washer relay
36 Heated rear window relay
37 Instrument lighting switch
38 Ignition switch
39 Windshield wiper switch
40 Windshield washer switch
41 Horn switch
42 Headlamp dip switch
43 Direction indicator switch
44 Hazard warning switch
45 Cigarette lighter
46 Glovebox lamp
47 Front door interior light switches
48 Rear door interior light switches
49 External lighting switch
50 Tail lamp foglamp switch
51 Instrument cluster
52 Oil pressure signal lamp
53 Fuel reserve signal lamp
54 Battery no-charge signal lamp
55 Handbrake on signal lamp flasher
56 Signal lamp unit
57 Brake fluid level signal lamp
58 Tail light foglamp signal lamp
59 Handbrake on signal lamp
60 Voltmeter
61 Speedometer
62 External lighting signal lamp
63 Direction indicator signal lamp
64 Headlamp main beam signal lamp
65 Heater motor switch
66 Heated rear window switch
67 Interior lamp
68 Tail lamps
69 Heated rear window
70 Number plate lamps
71 Fuel level and reserve sender

Not all items are fitted to all models

Colour Codes

W   White   Br Brown   P   Pink
Bl Blue   R   Red   Gy Grey
Y   Yellow   O   Orange   B   Black
G   Green

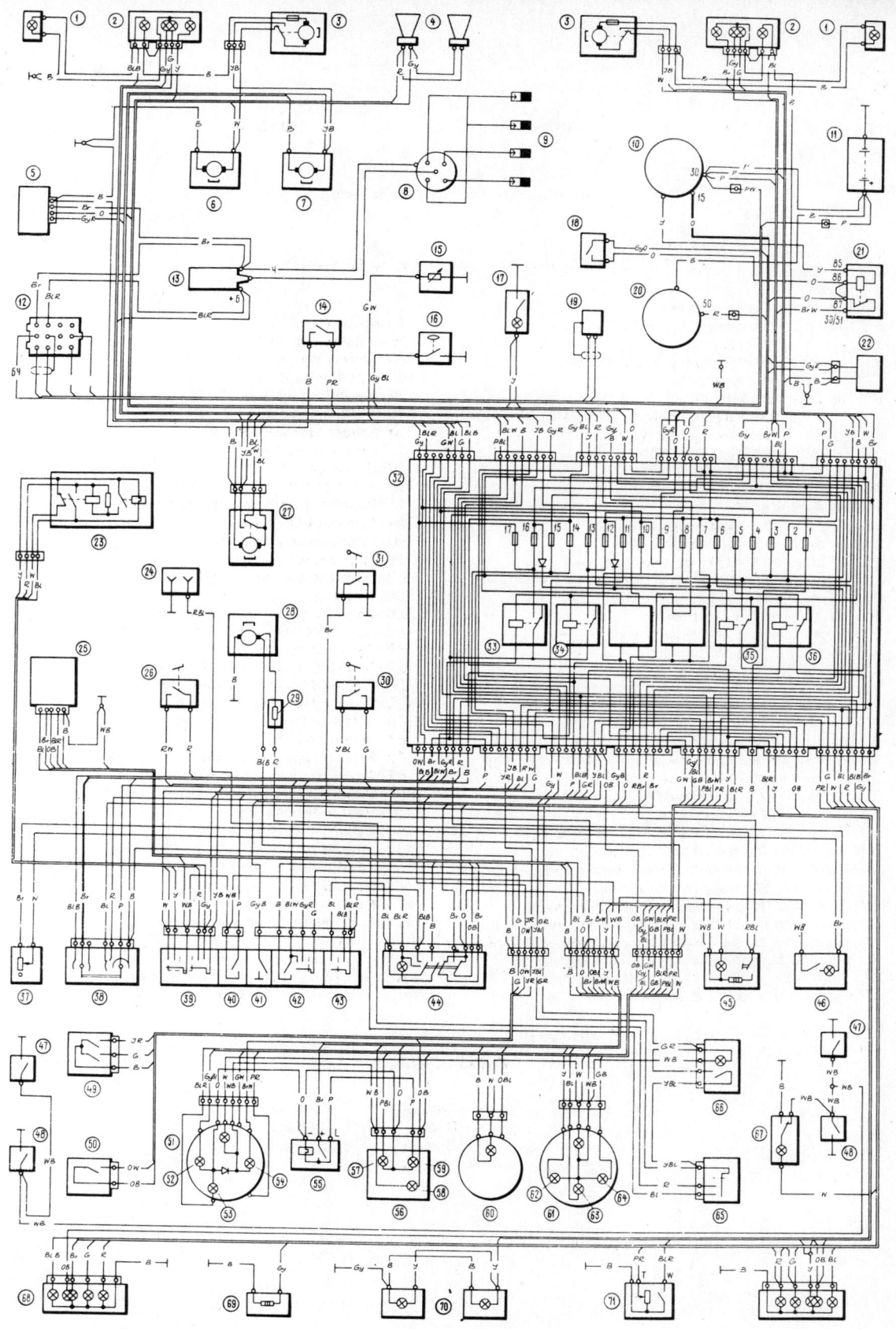

Fig. 13.32 Wiring diagram for Riva 1300 models

Key to wiring diagram for Riva 1500 models up to 1986

1 Headlamps
2 Headlamp wiper motors
3 Horns
4 Fan motor cut-in sender
5 Cooling fan motor
6 Direction indicators
7 Headlamp washer motor
8 Pneumatic valve control unit
9 Diagnostic panel
10 Windshield washer motor
11 Ignition coil
12 Brake fluid level signal lamp sender
13 Brake shoe wear indicator sender
14 Distributor
15 Spark plugs
16 Coolant temperature sender
17 Oil pressure signal lamp sender
18 Oil pressure sender
19 Carburettor microswitch
20 Bonnet lamp
21 TDC sender
22 Generator
23 Starter
24 Brake shoe wear indicator sender
25 Battery
26 Battery no-charge signal lamp relay
27 Pneumatic valve
28 Windshield wiper relay
29 Windshield wiper motor
30 Heater motor
31 Handbrake on lamp switch
32 Fuse and relay unit
33 Direction indicator and hazard warning relay
34 Inspection lamp socket
35 Heater motor series resistor
36 Stop-light switch
37 Reversing lamp switch
38 Headlamp lower beam relay
39 Headlamp upper beam relay
40 Cooling fan motor relay
41 Horn relay
42 Headlamp wiper and washer relay
43 Heated rear window relay
44 Ignition switch
45 Windshield wiper switch
46 Windshield washer switch
47 Horn switch
48 Headlamp dip switch
49 Direction indicator switch
50 Hazard warning switch
51 Heater motor switch
52 Glovebox lamp
53 Instrument lighting switch
54 Clock
55 Cigarette lighter
56 Front door open signal lamps
57 Front door open signal lamp switches
58 Instrument cluster
59 Oil pressure gauge with signal lamp
60 Coolant temperature gauge
61 Handbrake on signal lamp flasher
62 Tachometer
63 Direction indicator signal lamp
64 Brake fluid level signal lamp
65 Battery no-charge signal lamp
66 Tail light foglamp signal lamp
67 Headlamp main beam signal lamp
68 External lighting signal lamp
69 Brake shoe wear lamp (constant) and handbrake on (flashing)
70 Fuel level gauge with reserve signal lamp
71 Voltmeter
72 External lighting switch
73 Tail lamp foglamp switch
74 Heated rear window switch
75 Front door interior lamp switches
76 Rear door interior lamp switches
77 Interior lamp
78 Tail lamps
79 Heated rear window
80 Number plate lamp
81 Fuel level and reserve sender

Not all items are fitted to all models

Colour Codes

W	White	Br	Brown	P	Pink
Bl	Blue	R	Red	Gy	Grey
Y	Yellow	O	Orange	B	Black
G	Green				

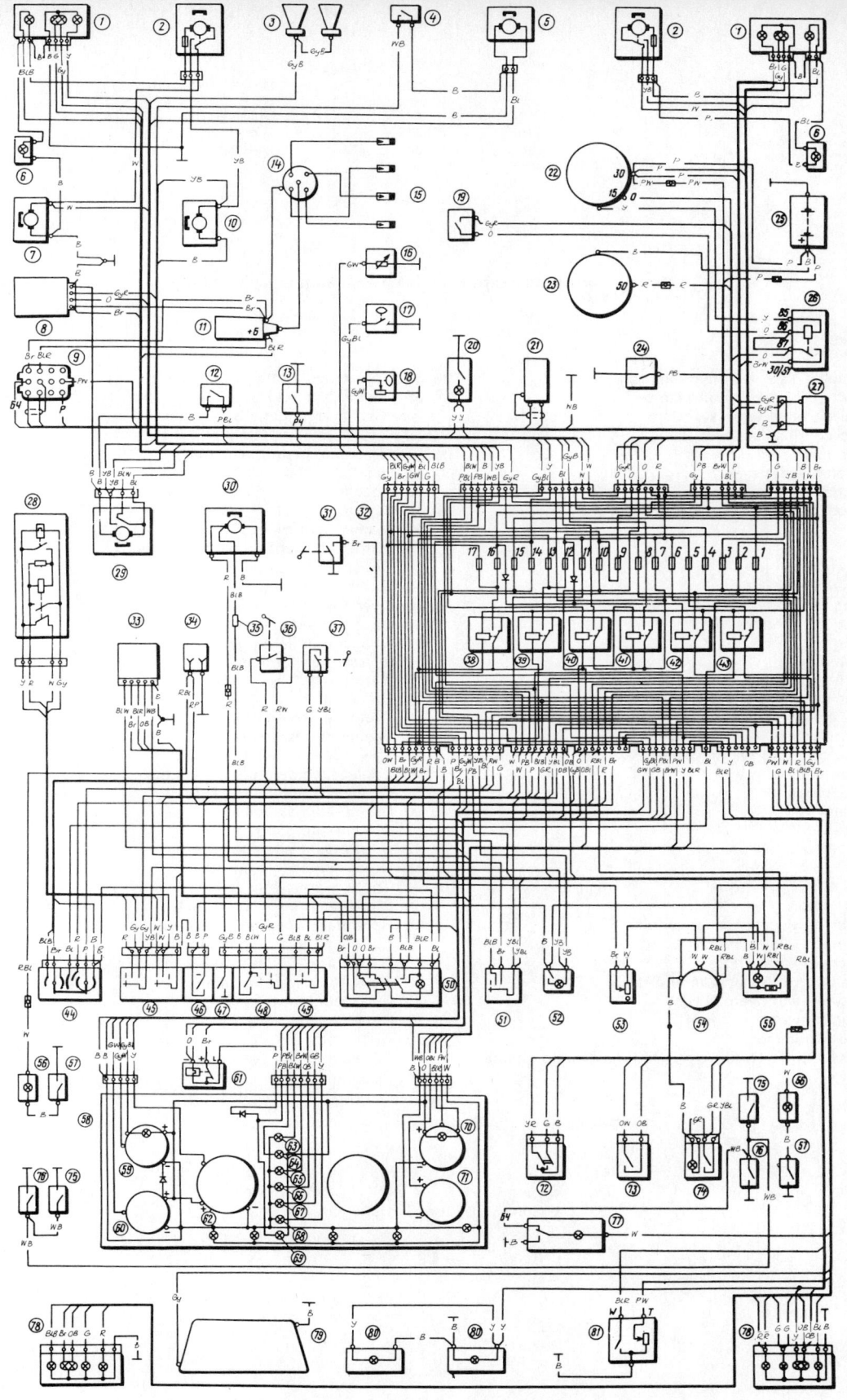

Fig. 13.33 Wiring diagram for Riva 1500 models up to 1986

Key to wiring diagram for Riva 1600 models up to 1986

1. Front door 'open' warning lamp switches
2. Front door 'open' warning lamps
3. Interior courtesy lamp switches
4. Interior courtesy lamps
5. Stop-lamps
6. Stop-lamp switch
7. Inspection lamp socket
8. Cigar lighter
9. Clock
10. Horns
11. Horn relay
12. Horn switch
13. Cooling fan
14. Cooling fan thermo switch
15. Cooling fan motor relay
16. Spark plugs
17. Ignition distributor
18. Ignition coil
19. Starter motor
20. Ignition switch
21. Ignition (charge) warning lamp
22. Ignition lamp relay
23. Voltage regulator
24. Alternator
25. Battery
26. Windscreen washer pump
27. Heater motor switch
28. Heater motor resistor
29. Heater motor
30. Windscreen wiper/washer switch
31. Windscreen wiper relay
32. Windscreen wiper motor
33. Headlamp main beam warning lamp
34. Headlamp main beam (LH)
35. Headlamp main beam (RH)
36. Dipped headlamps
37. Marker lamp (RH), tail lamp (LH) and number plate lamp (LH)
38. Engine compartment lamp
39. Marker lamp (LH), tail lamp (RH), number plate lamp (RH) and luggage boot lamp
40. Cigar lighter illumination
41. Headlamp main beam relay
42. Headlamp dip switch
43. Headlamp dipped beam relay
44. Lighting switch
45. Marker light switch
46. Instrument lighting with rheostat
47. Direction indicator relay
48. Direction indicator warning lamp
49. Direction indicator switch
50. Direction indicator (LH)
51. Direction indicator (RH)
52. Instrument illumination
53. Hazard warning relay (flasher unit)
54. Tachometer
55. Fuel level transmitter
56. Fuel contents gauge
57. Low fuel level warning lamp
58. Coolant temperature transmitter
59. Coolant temperature gauge
60. Oil pressure switch
61. Oil pressure gauge
62. Oil pressure warning lamp
63. Oil pressure lamp switch
64. Choke warning lamp
65. Choke warning lamp switch
66. Handbrake warning lamp switch
67. Handbrake warning lamp relay
68. Handbrake 'ON' warning lamp
69. Brake fluid warning lamp
70. Brake fluid level sensor
71. Glovebox lamp
72. Carburettor anti-run on valve
73. Reversing lamp switch
74. Reversing lamps

Colour code (included on diagram where available)

B	Black	Gy	Grey	V	Violet
Bl	Blue	O	Orange	W	White
Br	Brown	P	Pink	Y	Yellow
G	Green	R	Red		

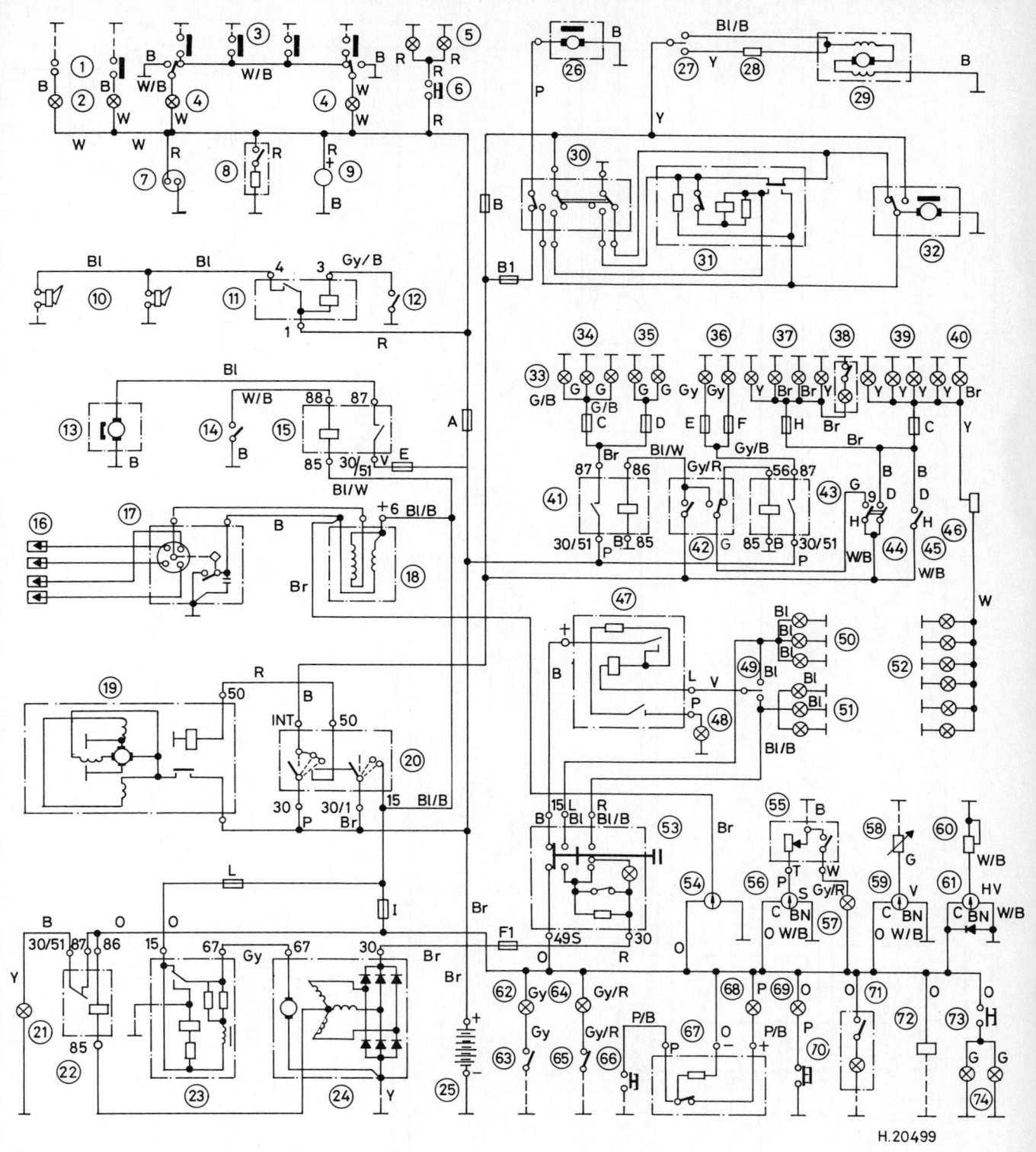

Fig. 13.34 Wiring diagram for Riva 1600 models up to 1986

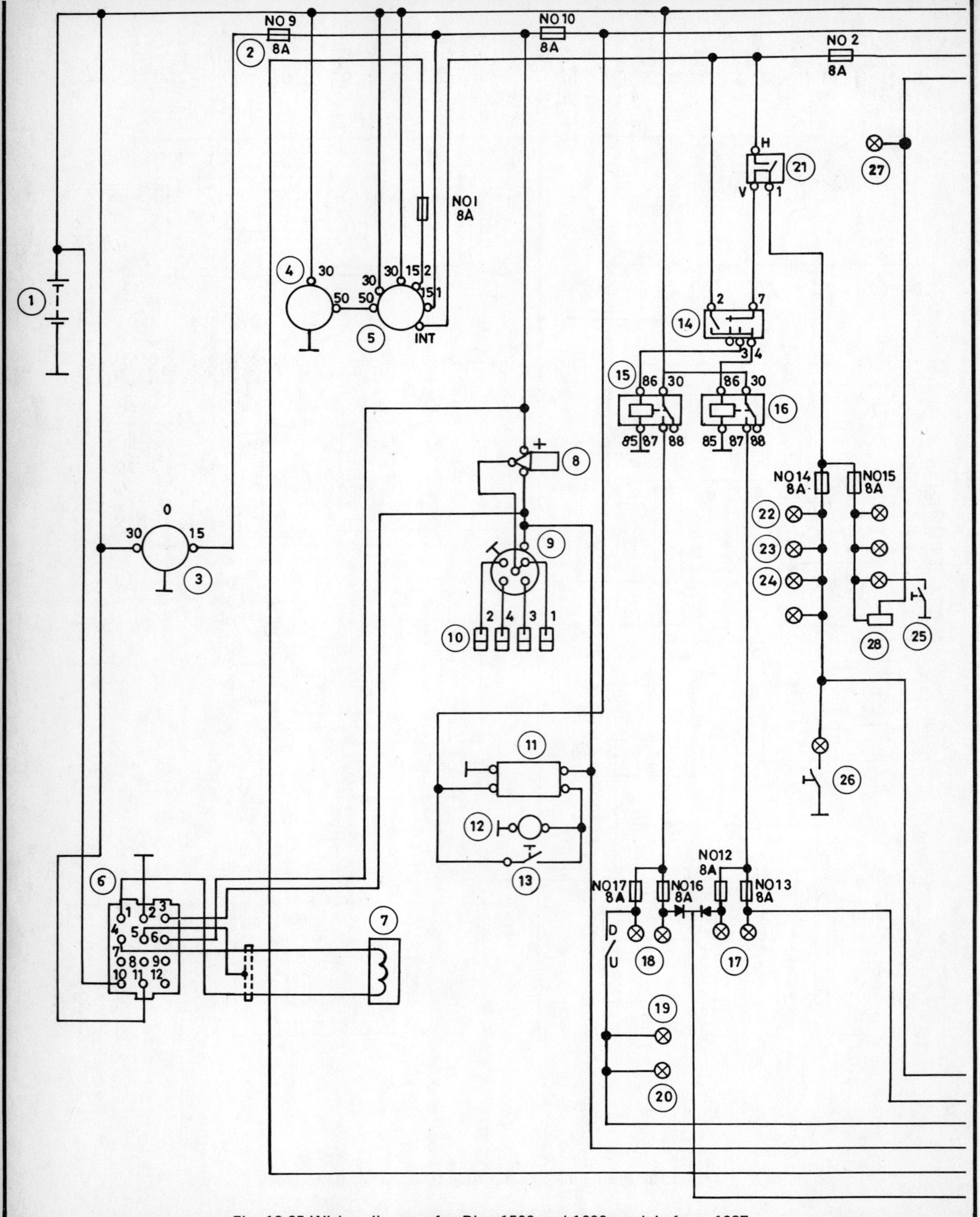

Fig. 13.35 Wiring diagram for Riva 1500 and 1600 models from 1987

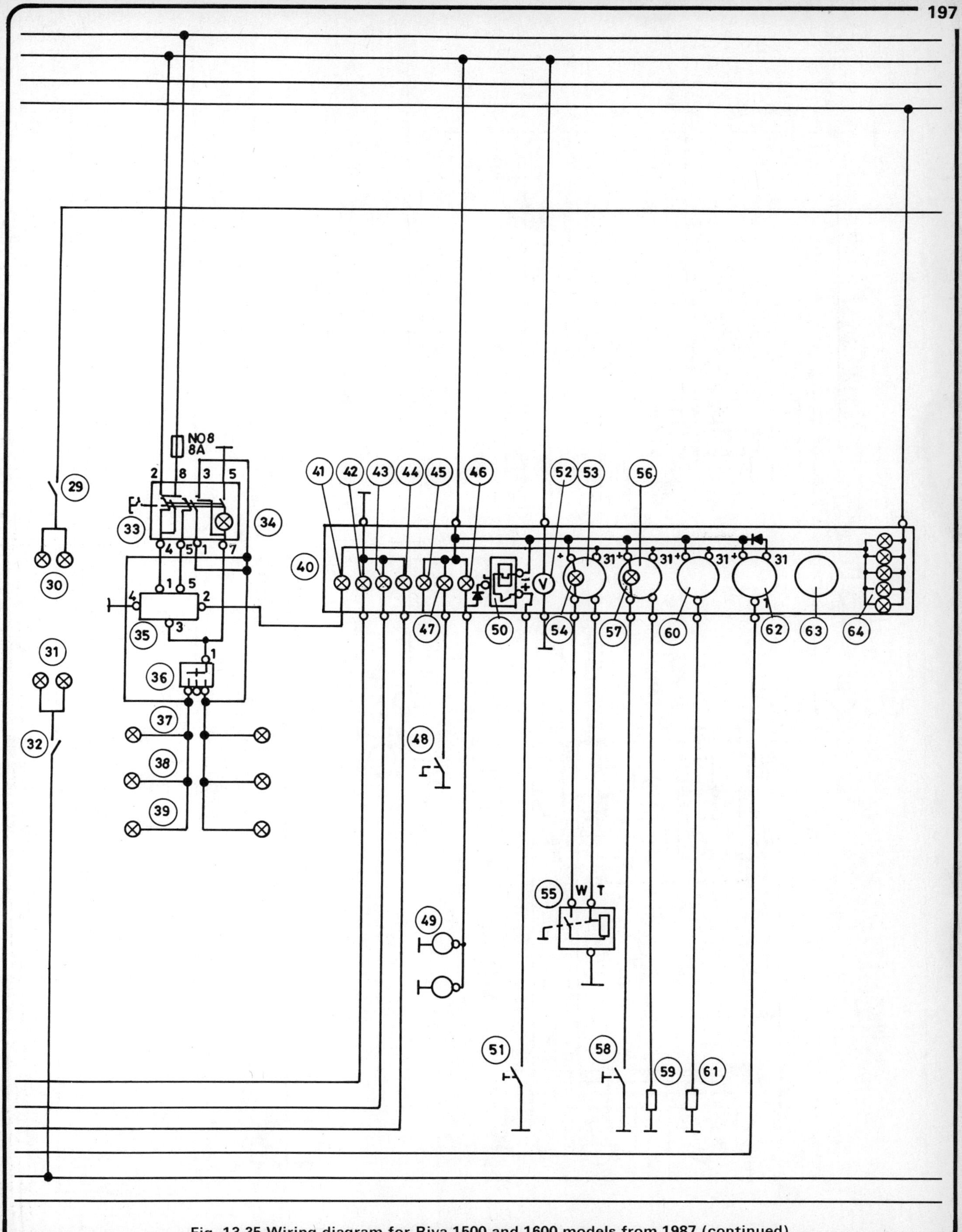

197

Fig. 13.35 Wiring diagram for Riva 1500 and 1600 models from 1987 (continued)

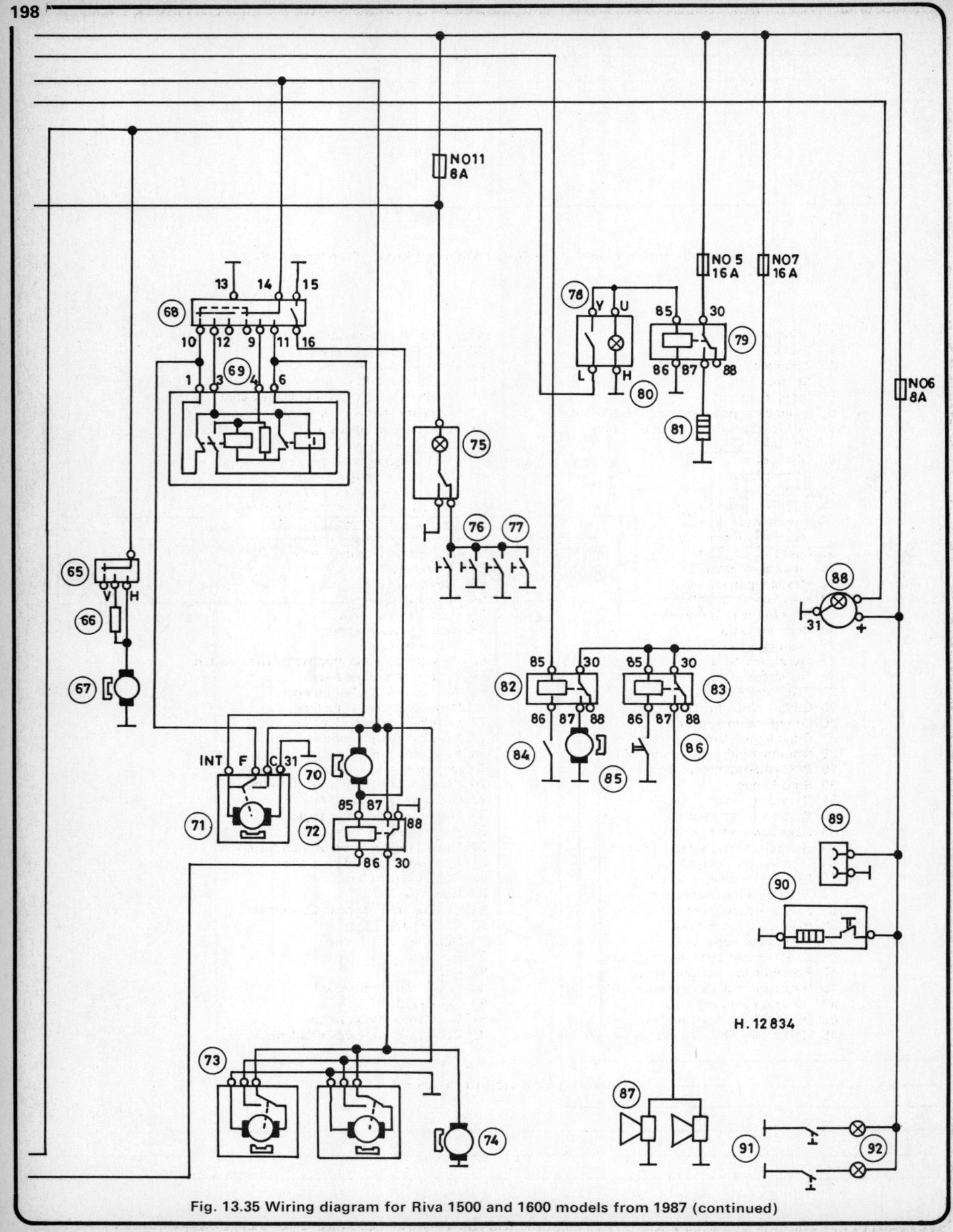

Fig. 13.35 Wiring diagram for Riva 1500 and 1600 models from 1987 (continued)

Key to wiring diagram for Riva 1500 and 1600 models from 1987

1 Battery
2 Fuses
3 Alternator
4 Starter
5 Ignition switch
6 Diagnostic panel (certain models only)
7 TDC sensor (certain models only)
8 Ignition coil
9 Distributor
10 Spark plugs
11 Control unit
12 Electropneumatic valve
13 Microswitch
14 Headlamp dipper switch
15 Dipped beam relay
16 Main beam relay
17 Headlamp main beam
18 Headlamp dipped beam
19 Rear fog lamp switch
20 Rear fog lamp
21 Lighting switch
22 Front parking lamps
23 Tail lamps
24 Rear number plate lamps
25 Glove compartment lamp
26 Engine compartment lamp
27 Cigar lighter lamp
28 Instrument illumination switch
29 Brake stop-lamp switch
30 Stop-lamps
31 Reversing lamps
32 Reversing lamp switch
33 Hazard warning switch
34 Hazard warning system indicator lamp
35 Flasher unit (relay)
36 Direction indicator switch
37 Front direction indicators
38 Side direction indicator repeaters
39 Rear direction indicators
40 Instrument panel
41 Direction indicator warning lamp
42 Parking lamp warning lamp
43 Headlamp main beam warning lamp
44 Tail lamp warning lamp
45 Reverse lamp warning lamp
46 Handbrake 'ON' and pad sensor warning lamp
47 Low brake fluid warning lamp
48 Brake fluid level sensor
49 Disc pad wear sensor
50 Handbrake 'ON' relay
51 Handbrake 'ON' warning lamp switch
52 Voltmeter (battery condition indicator)
53 Fuel contents gauge
54 Low fuel level warning lamp
55 Fuel level sender unit
56 Oil pressure gauge
57 Oil pressure warning lamp
58 Low oil pressure warning lamp switch
59 Oil pressure sender unit
60 Coolant temperature gauge
61 Coolant temperature sender unit
62 Tachometer
63 Speedometer
64 Instrument lamps
65 Heater blower switch
66 Heater blower resistor
67 Heater blower
68 Windscreen/headlamp wash/wipe switch
69 Windscreen wiper relay
70 Windscreen washer pump
71 Windscreen wiper motor
72 Headlamp wiper relay
73 Headlamp wiper motors
74 Headlamp washer pump
75 Interior courtesy lamp
76 Courtesy lamp switches (front door)
77 Courtesy lamp switches (rear doors)
78 Heated rear window switch
79 Heated rear window relay
80 Heated rear window warning lamp
81 Rear window heating element
82 Cooling fan relay
83 Horn relay
84 Cooling fan thermostatic switch
85 Cooling fan motor
86 Horn switch
87 Horns
88 Clock
89 Inspection lamp socket
90 Cigar lighter
91 Door open lamp switch
92 Door open edge warning lamps

No wiring colours information available

14 Suspension and steering

Steering column and shaft (Riva type) – description
1 On Riva models the main steering column has been shortened and an intermediate shaft added between the main column and the steering box. The intermediate shaft which incorporates two universal joints provides a safety feature by allowing the steering box to move sideways in the event of a frontal impact rather than forcing the steering column into the car interior (photo).

Steering column (Riva type) – removal and refitting
2 Disconnect the battery.
3 Extract the screws and take off the shrouds from the upper end of the column (photos).
4 Peel back the carpet and, with the front roadwheels in the straight-ahead attitude, mark the alignment of the universal joint coupling to the pinion shaft of the steering box.
5 Unscrew and remove the coupling pinch-bolt.
6 Unscrew the column upper bracket mounting bolts. The two lower bolts are of shear type and may not have had their heads broken off. If their heads have been broken off then the bolts will have to be drilled out in order to remove them (photo).
7 Lower the column so that the steering wheel rests on the front seat cushion and then disconnect the steering column wiring plugs.
8 Withdraw the column; it may require gently tapping upwards to release the lower coupling or, alternatively, slightly prise open the coupling slot with a large screwdriver.
9 Refitting is a reversal of removal.

Steering column and shaft (Riva models) – overhaul
10 With the assembly removed from the car, remove the steering wheel, ignition switch and column switches with reference to Chapters 11 and 10 respectively.
11 Mark the intermediate shaft in relation to the upper shaft then unscrew the clamp bolt and pull the intermediate shaft from the splines.

Type A (Fig. 13.36)
12 Lever out the staking points on the bracket tube then pull out the upper shaft complete with bearings.
13 Drive the bearings from each end of the upper shaft.

Type B (Fig. 13.37)
14 Remove the lower bearing coil spring, followed by the expander ring and lower bearing.
15 Extract the circlip from the top of the shaft followed by the washer, coil spring, expander ring and upper bearing.
16 Withdraw the shaft from the bracket tube.

All types
17 Check the bearings for wear and renew them if necessary. Clean all the components and check them for wear and damage.
18 Reassembly is a reversal of dismantling, but, if required, pack the bearings with a suitable grease. On the type A column, stake the bracket tube on each end in two places to retain the bearings. On the type B column, make sure that the clamp bolt engages the groove in the upper shaft.

Steering box (Riva models) – removal and refitting
19 Due to the length of the pinion shaft on these later steering boxes, the removal procedure differs from earlier units and is as follows (photo).

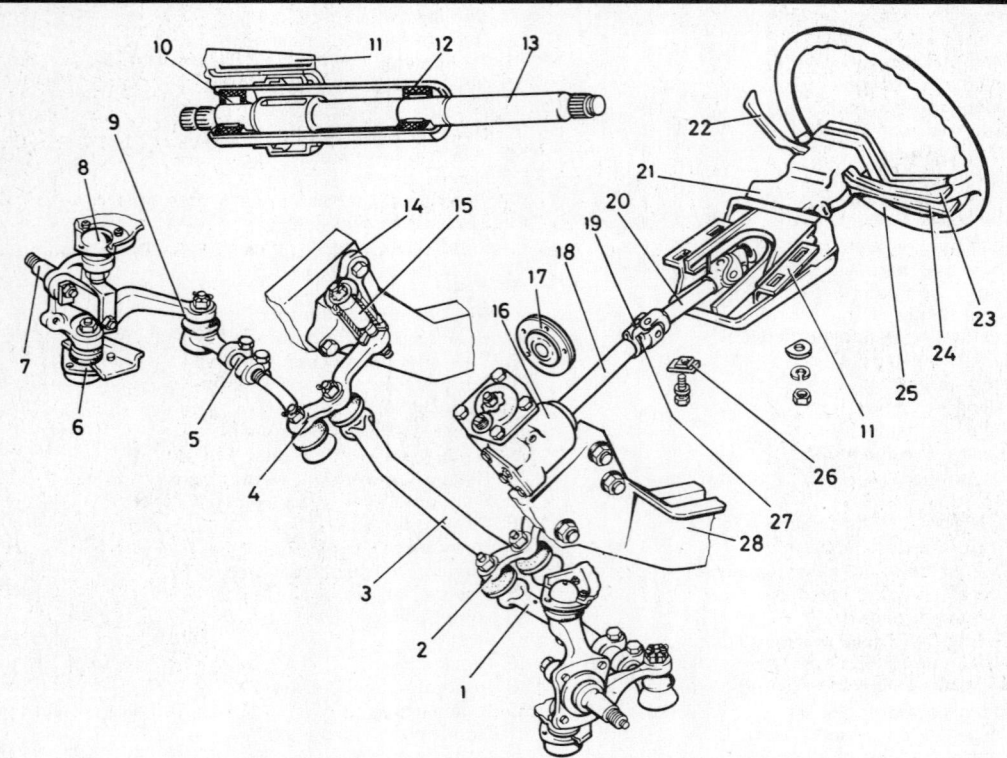

Fig. 13.36 Riva steering column and gear – LHD shown (Sec 14)

1 Track rod	11 Bracket tube	21 Shroud
2 Drop arm	12 Column upper bearing	22 Wiper/washer switch
3 Relay rod	13 Upper shaft	23 Headlamp switch
4 Idler arm	14 Idler	24 Direction indicator switch
5 Adjustment	15 Shaft	25 Steering wheel
6 Lower balljoint	16 Steering box	26 Plate
7 Stub axle	17 Grommet	27 Clamp bolt
8 Upper balljoint	18 Worm shaft	28 Body side member
9 Track rod end	19 Universal joint	
10 Column lower bearing	20 Intermediate shaft	

Chapter 13 Supplement

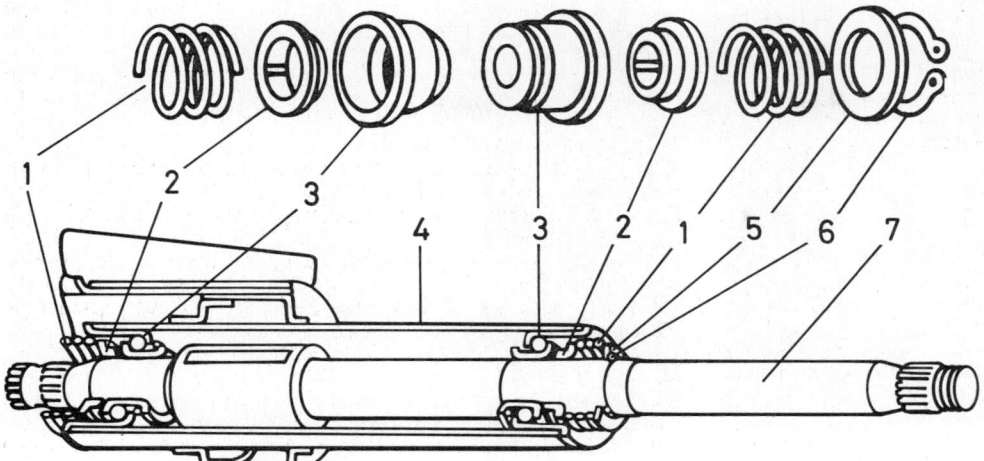

Fig. 13.37 Cut-away view of Riva steering column (Sec 14)

1. Coil spring
2. Expander ring
3. Bearing
4. Bracket tube
5. Washer
6. Circlip
7. Upper shaft

14.1 Steering column intermediate shaft (arrowed)

14.3A Extracting steering column shroud screws

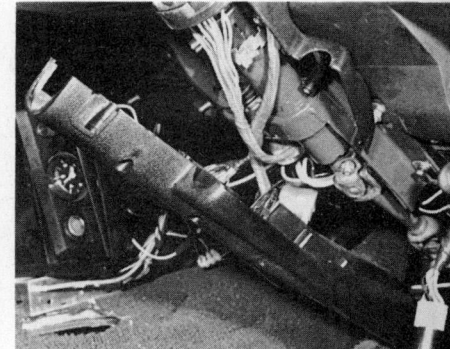

14.3B Steering column shrouds released

14.6 Steering column bracket bolts. Shear type bolt arrowed

14.19 Long pinion shaft type steering box

14.25 Balljoint nut split pin withdrawn

20 Disconnect the battery, and remove the air cleaner.
21 Remove the engine compartment undershield.
22 Disconnect the exhaust downpipes from the manifold. Remove the starter motor shield.
23 Working underneath the vehicle, disconnect the exhaust pipe bracket from the gearbox and then remove the clamp and separate the front section of the exhaust from the rest of the system. If rusted, the pipe socket joint will require a good soaking in freeing fluid to release it.
24 Before the front section of the exhaust pipe can be removed, the engine right-hand mounting must be removed to provide clearance for the pipe flange. To do this, refer to the note in Chapter 3, Section 16.
25 Remove the split pins and unscrew the two castellated nuts from the balljoints on the steering box drop arm (photo).
26 Using a balljoint splitter tool, disconnect the balljoints from the drop arm (photo).
27 Working under the front wing, unscrew the steering box mounting bolts (photo).
28 Working inside the vehicle, peel back the carpet and unscrew and remove the steering column shaft coupling pinch-bolt. Prise the

14.26 Splitting a steering balljoint

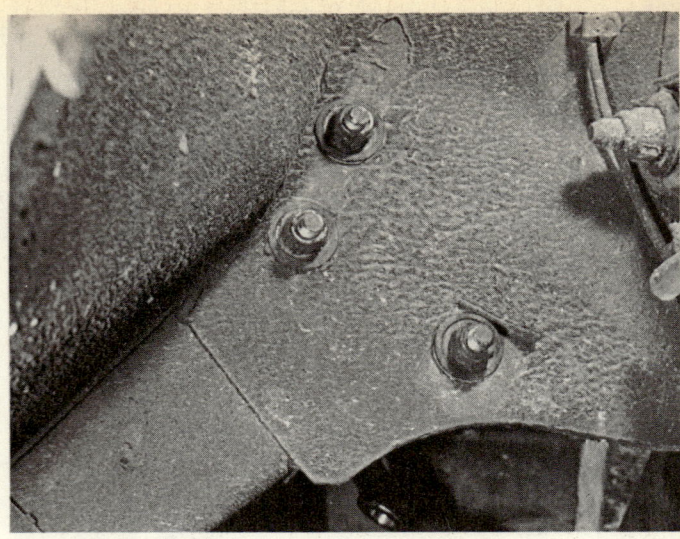
14.27 Steering box mounting nuts under front wing

14.28 Steering shaft lower coupling pinch-bolt

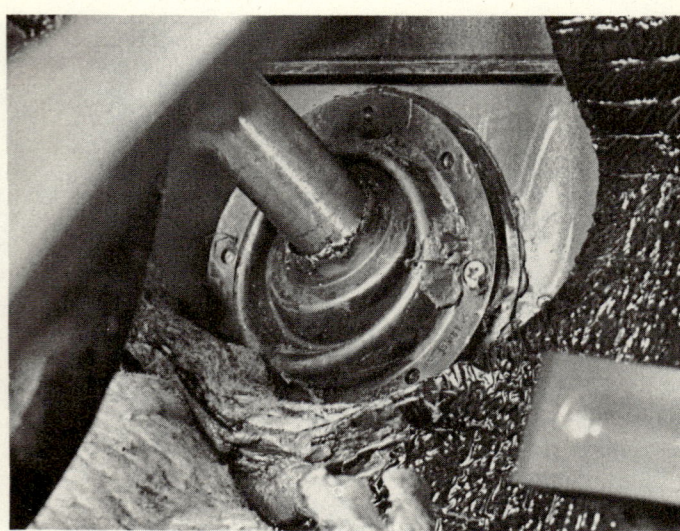
14.29 Steering shaft sealing plate

14.30 Unscrewing brake master cylinder fixing bolts

14.31A Steering box being removed from bulkhead (gearbox removed for clarity)

Chapter 13 Supplement

14.31B Withdrawing steering box upwards from engine compartment

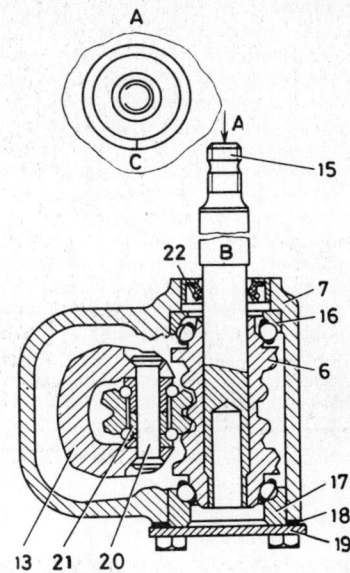

Fig. 13.38 Riva ball type steering box (Sec 14)

- 6 Worm
- 7 Steering box
- 13 Sector shaft
- 15 Wormshaft
- 16 Upper bearing
- 17 Lower bearing
- 18 Shim
- 19 End cover
- 20 Roller spindle
- 21 Ball-bearing
- B and C Alignment marks

coupling jaws open slightly so that the steering box pinion shaft will release easily (photo).
29 Extract the self-tapping screws and release the shaft sealing plate (photo).
30 Unscrew the nuts which secure the brake hydraulic master cylinder to the servo unit (photo).
31 Working under the vehicle, pull the steering box forward until the pinion shaft clears the bulkhead. The steering box should now be removed upwards from the engine compartment, but the brake master cylinder which was previously released must be moved carefully against the tension of the hydraulic pipes to provide the necessary clearance for the steering box (photos).
32 Refitting is a reversal of removal, but make sure that the engine is still well supported while the engine right-hand mounting is still removed.
33 Set the front roadwheels and the steering wheel in the straight-ahead attitude before connecting the shaft coupling.
34 Tighten all nuts and bolts to the specified torque.

Steering box (Riva models) – overhaul
35 The procedure is basically as described in Chapter 11; the only design difference being that the roller runs on ball-bearings instead of needle bearings.
36 The initial turning torque of the wormshaft should be between 1.7 and 4.3 lbf in (2.0 and 5.0 kgf cm).

Steering idler and arm – checking
37 When checking the steering idler there should be no noticeable endplay of the shaft. This can be ascertained by gripping the idler arm and attempting to move it up and down. If necessary adjust the idler as follows.
38 Disconnect the track rod and relay rod from the drop arm.
39 Tighten the shaft nut until the shaft only moves with a force of 2.2 to 4.4 lbf (1.0 to 2.0 kgf) applied at the end of the drop arm with a spring balance. If the idler has been removed, and is held with the drop arm horizontal, an alternative method is to tighten the nut until the arm is just retained in the horizontal position. An approximate tightening torque is given in the Specifications in Chapter 11 under 'Idler arm self-locking nut'.

Wheels and tyres – general care and maintenance
40 Wheels and tyres should give no real problems in use provided that a close eye is kept on them with regard to excessive wear or damage. To this end, the following points should be noted.
41 Ensure that tyre pressures are checked regularly and maintained correctly. Checking should be carried out with the tyres cold and not immediately after the vehicle has been in use. If the pressures are checked with the tyres hot, an apparently high reading will be obtained owing to heat expansion. Under no circumstances should an attempt be made to reduce the pressures to the quoted cold reading in this instance, or effective underinflation will result.
42 Underinflation will cause overheating of the tyre owing to excessive flexing of the casing, and the tread will not sit correctly on the road surface. This will cause a consequent loss of adhesion and excessive wear, not to mention the danger of sudden tyre failure due to heat build-up.
43 Overinflation will cause rapid wear of the centre part of the tyre tread coupled with reduced adhesion, harsher ride, and the danger of shock damage occurring in the tyre casing.
44 Regularly check the tyres for damage in the form of cuts or bulges, especially in the sidewalls. Remove any nails or stones embedded in the tread before they penetrate the tyre to cause deflation. If removal of a nail *does* reveal that the tyre has been punctured, refit the nail so that its point of penetration is marked: Then immediately change the wheel and have the tyre repaired by a tyre dealer. Do *not* drive on a tyre in such a condition. In many cases a puncture can be simply repaired by the use of an inner tube of the correct size and type. If in any doubt as to the possible consequences of any damage found, consult your local tyre dealer for advice.
45 Periodically remove the wheels and clean any dirt or mud from the inside and outside surfaces. Examine the wheel rims for signs of rusting, corrosion or other damage. Light alloy wheels are easily damaged by 'kerbing' whilst parking, and similarly steel wheels may become dented or buckled. Renewal of the wheel is very often the only course of remedial action possible.
46 The balance of each wheel and tyre assembly should be maintained to avoid excessive wear, not only to the tyres but also to the steering and suspension components. Wheel imbalance is normally signified by vibration through the vehicle's bodyshell, although in many cases it is particularly noticeable through the steering wheel. Conversely, it should be noted that wear or damage in suspension or steering components may cause excessive tyre wear. Out-of-round or out-of-true tyres, damaged wheels and wheel bearing wear/maladjustment also fall into this category. Balancing will not usually cure vibration caused by such wear.
47 Wheel balancing may be carried out with the wheel either on or off the vehicle. If balanced on the vehicle, ensure that the wheel-to-hub relationship is marked in some way prior to subsequent wheel removal so that it may be refitted in its original position.
48 General tyre wear is influenced to a large degree by driving style –

harsh braking and acceleration or fast cornering will all produce more rapid tyre wear. Interchanging of tyres may result in more even wear, but this should only be carried out where there is no mix of tyre types on the vehicle. However, it is worth bearing in mind that if this is completely effective, the added expense of replacing a complete set of tyres simultaneously is incurred, which may prove financially restrictive for many owners.

49 Front tyres may wear unevenly as a result of wheel misalignment. The front wheels should always be correctly aligned according to the settings specified by the vehicle manufacturer.

50 Legal restrictions apply to the mixing of tyre types on a vehicle. Basically this means that a vehicle must not have tyres of differing construction on the same axle. Although it is not recommended to mix tyre types between front axle and rear axle, the only legally permissible combination is crossply at the front and radial at the rear. When mixing radial ply tyres, textile braced radials must always go on the front axle, with steel braced radials at the rear. An obvious disadvantage of such mixing is the necessity to carry two spare tyres to avoid contravening the law in the event of a puncture.

51 In the UK, the Motor Vehicles Construction and Use Regulations apply to many aspects of tyre fitting and usage. It is suggested that a copy of these regulations is obtained from your local police if in doubt as to the current legal requirements with regard to tyre condition, minimum tread depth, etc.

15 Bodywork and underframe

Front and rear bumpers (Riva models) – removal and refitting

1 On Riva models the bumpers are mounted on tubular brackets bolted to the underframe. Towing eyes are welded to the brackets. The removal and refitting procedure is similar to that described in Chapter 12 – remember to disconnect the wiring from the rear number plate lamp when removing the rear bumper on saloon models.

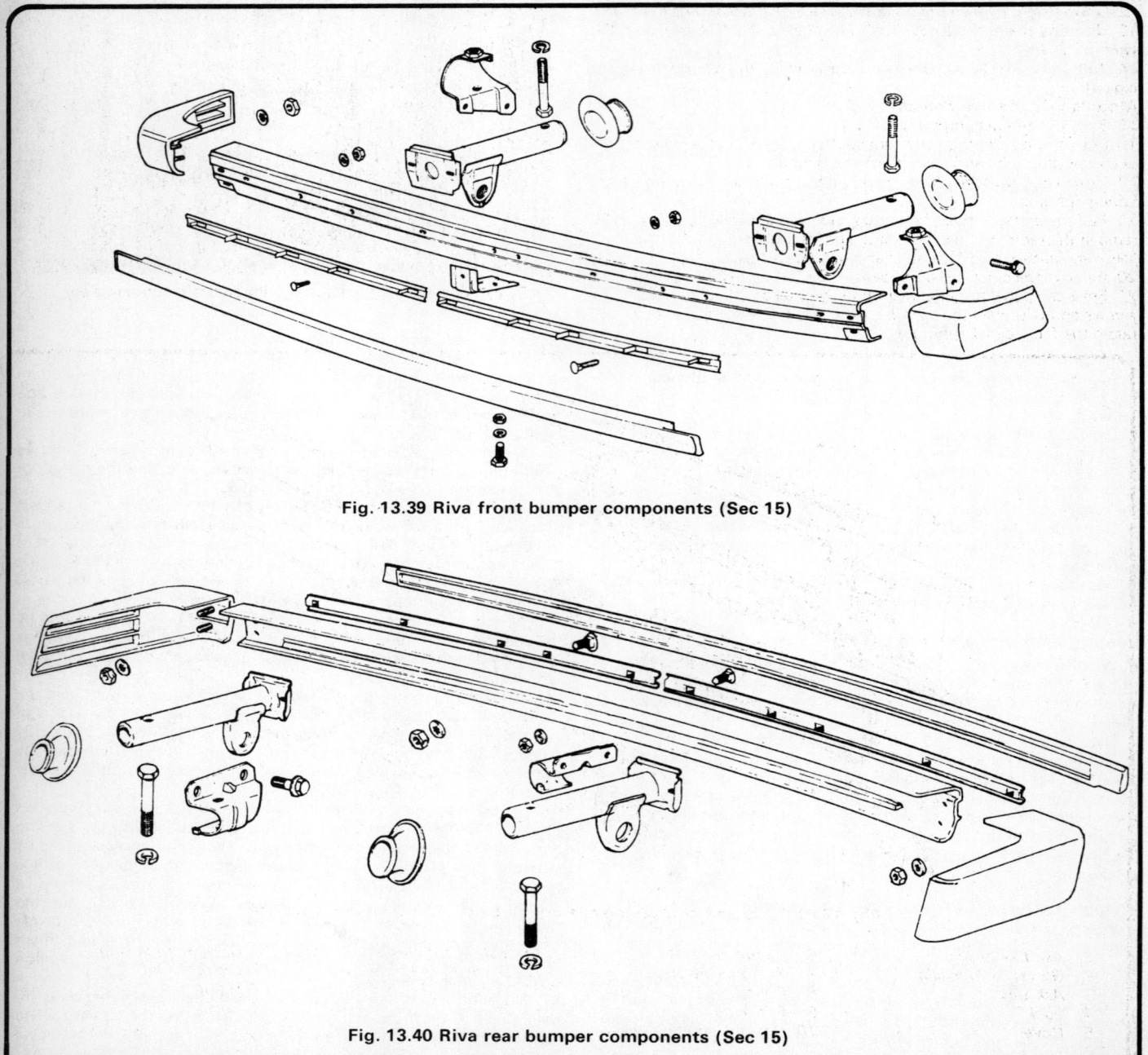

Fig. 13.39 Riva front bumper components (Sec 15)

Fig. 13.40 Riva rear bumper components (Sec 15)

Chapter 13 Supplement

Facia panel (Riva models) – removal and refitting
2 Disconnect the battery negative lead.
3 Undo the screws and remove the steering column cowls.
4 Remove the instrument cluster, as described in Section 13 of this Supplement.
5 Where a hydraulic headlamp adjuster is fitted, pull off the control knob, unscrew the mounting nut, and push the adjuster from the facia panel (photo).
6 Prise the side vents from each end of the facia panel and disconnect the ducts.
7 Remove the glovebox and shelf by undoing the retaining screws.
Models without centre console
8 Pull off the heater control knobs then remove the surround.
9 Undo the screws and withdraw the radio housing, then disconnect the aerial and wiring.
10 Undo the facia lower mounting screws.
11 Working through the instrument panel and glovebox aperture, undo the facia upper mounting nuts.
12 Undo the screws and remove the insert from the top of the facia.
13 Unscrew the nuts, remove the loudspeaker(s) and disconnect the wiring.
14 Withdraw the facia panel and disconnect the wiring from the cigarette lighter.
Models with centre console
15 Prise out the heating symbol plug.
16 Prise the covers from the loudspeakers, then remove the screws, withdraw the units and disconnect the wiring.
17 Extract the switches from the centre console and disconnect the wiring (photos).
18 Pull the control knobs from the radio then unscrew the fastenings, remove the radio and disconnect the aerial and wiring (photos).
19 Remove the radio baseplate and oddments tray.
20 Remove the clock surround panel from the centre console (photo).
21 Press the panel from the rear of the centre console then remove the screws and withdraw the console. Note that the two upper screws also retain the facia panel (photos).
22 Remove the remaining facia lower mounting screws then, working through the instrument panel and glovebox apertures, unscrew the upper mounting nuts.
23 Withdraw the facia panel from the car.

Doors (Riva models) – dismantling and reassembly
24 Prise out the plug then undo the screw securing the door pull to the armrest. Unhook the top of the door pull from the trim panel (photos).

15.5 Hydraulic headlamp adjuster with control knob removed

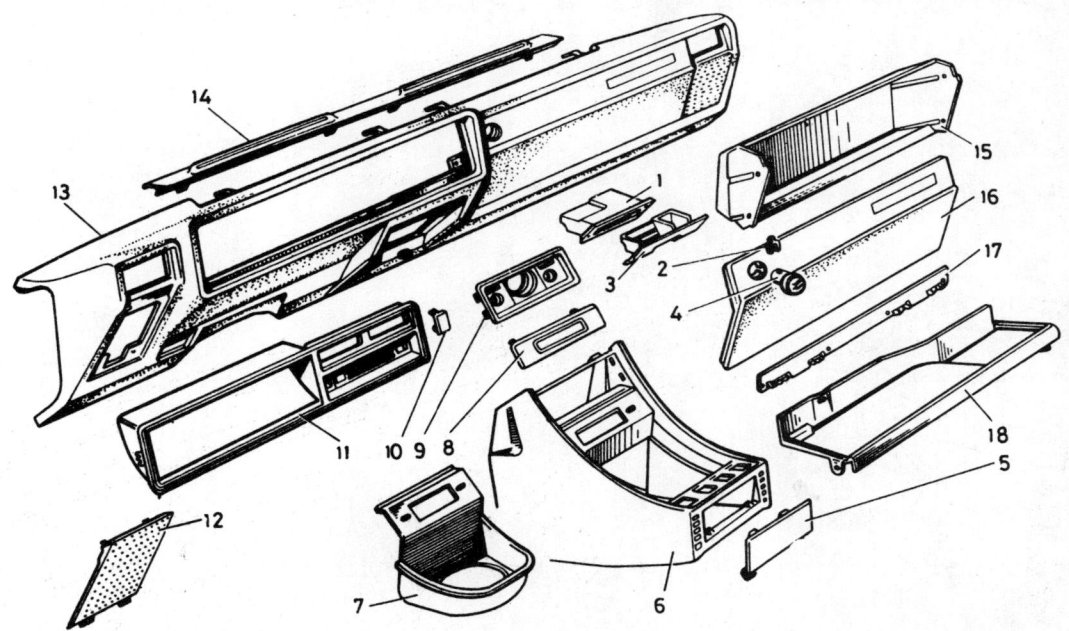

Fig. 13.41 Facia and centre console components – LHD shown (Sec 15)

1 Ash tray casing	6 Centre console	11 Instrument cluster surround	15 Glovebox housing
2 Glovebox lock clip	7 Radio baseplate	12 Loudspeaker cover	16 Glovebox lid
3 Ash tray	8 Cover	13 Facia panel	17 Hinge
4 Lock	9 Surround	14 Top insert	18 Shelf
5 Cover	10 Heating symbol plug		

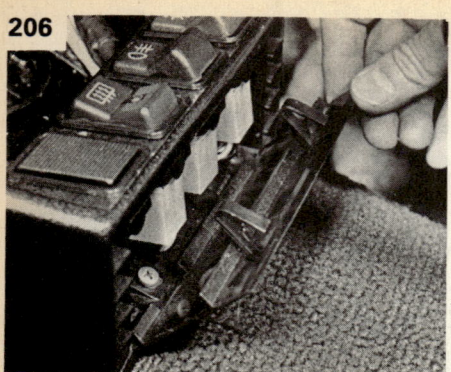
15.17A Removing centre console end plate to expose switch wiring plugs

15.17B Centre console switches

15.17C Withdrawing a centre console switch/plug assembly

15.18A Radio control knobs removed

15.18B Clock panel and radio released from centre console

15.20 Clock panel withdrawn

15.21A Centre console rear fixing screws

15.21B Centre console front fixing screw

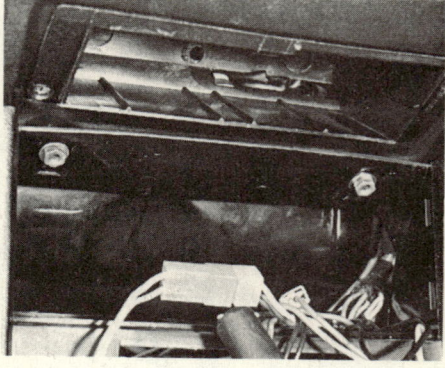
15.21C Centre console upper fixing screws/nuts

15.24A Prising out door pull plug

15.24B Extracting door pull screw

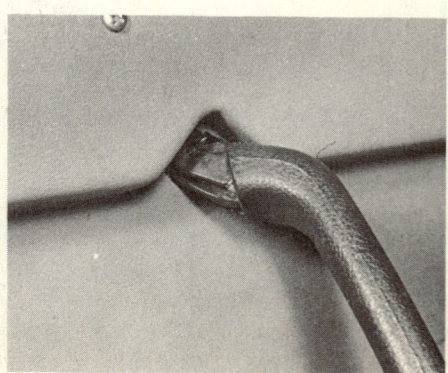
15.24C Unhooking door pull

Chapter 13 Supplement

25 Fully close the window and note the position of the regulator handle.
26 Slightly depress the escutcheon then push the two shoulders of the locking clip along the window regulator handle away from the end knob. This will release the clip from the groove in the regulator spindle and enable the handle and escutcheon to be moved (photos).
27 Prise the small plastic cover from the lower half of the door locking knob, then extract the small clip and withdraw the knob from the top of the rod (photos).
28 Prise out and remove the door inner handle surround (photo).
29 Using a wide-bladed screwdriver inserted between the trim panel and door, lever out the retaining clips and remove the trim panel (photo).
30 If working on the rear door, lower the window then undo the screw from the top of the rear channel and pull out the fixed quarter light, together with the seals.
31 Where necessary, remove the window lower sealing strips.
32 Unscrew the nut and screw and remove the front guide channel.
33 Undo the screws and remove the rear guide channel.
34 Undo the screws and remove the front corner plates on the front door only.
35 Loosen the bolt(s) of the regulator cable tensioner roller (photo).
36 Support the window then unscrew the clamp bolts and detach the cable from the window bracket(s).
37 Lift the window glass upwards from the door.
38 Unscrew the nuts and withdraw the window regulator from the door, at the same time releasing the cable from the rollers (photos).
39 Undo the screws and remove the inner door handle, at the same time disconnecting it from the rod (photo).
40 Working through the inner panel aperture, unscrew the outer handle retaining nuts, remove the handle, and disconnect it from the rod (photo).
41 Undo the lock retaining screws, disconnect the control rods as necessary and withdraw the lock from the door (photos).
42 Pull out the spring clip, disconnect the control rod and extract the private lock from the outside of the door (photo).

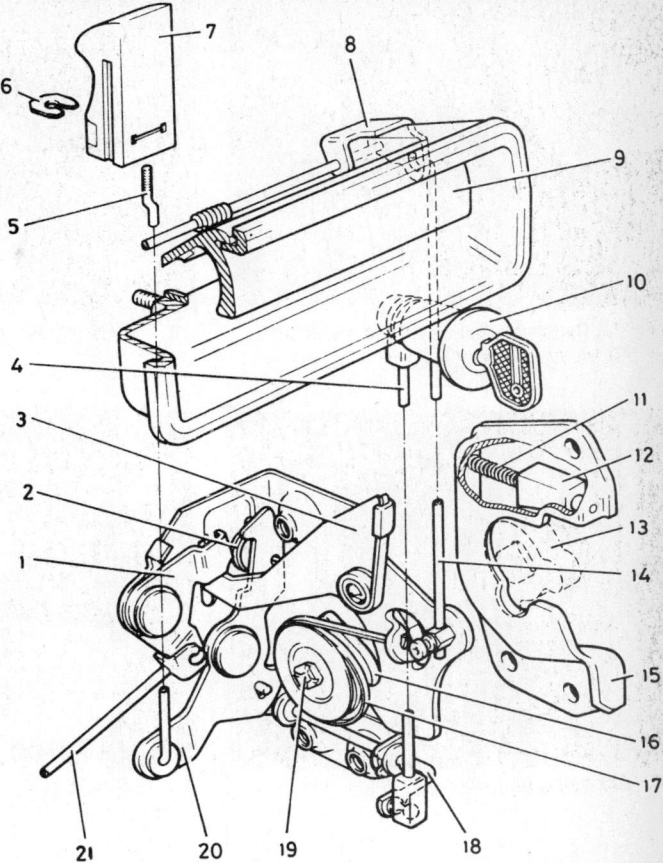

Fig. 13.42 Riva door lock components (Sec 15)

1 Inner control lever
2 Closing lever spring
3 Outer control lever
4 Private lock control rod
5 Locking knob rod
6 Clip
7 Locking knob
8 Exterior handle link
9 Exterior handle
10 Private lock
11 Spring
12 Thrust block
13 Rotor
14 Exterior handle control rod
15 Body
16 Ratchet
17 Central shaft spring
18 Lock release shaft
19 Central shaft
20 Closing lever
21 Inner handle control rod

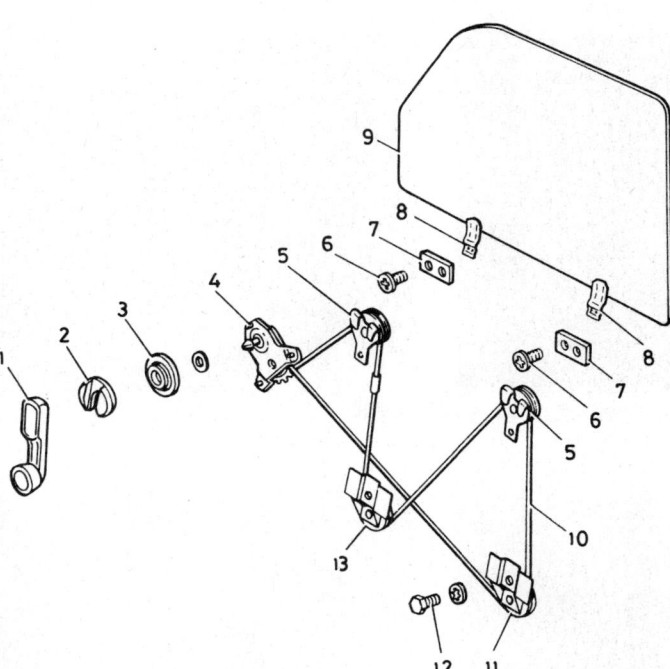

Fig. 13.43 Riva front door glass and regulator (Sec 15)

1 Handle
2 Locking clip
3 Escutcheon
4 Regulator mechanism
5 Roller
6 Screw
7 Clamp
8 Bracket
9 Glass
10 Cable
11 Roller
12 Bolt
13 Tensioner roller

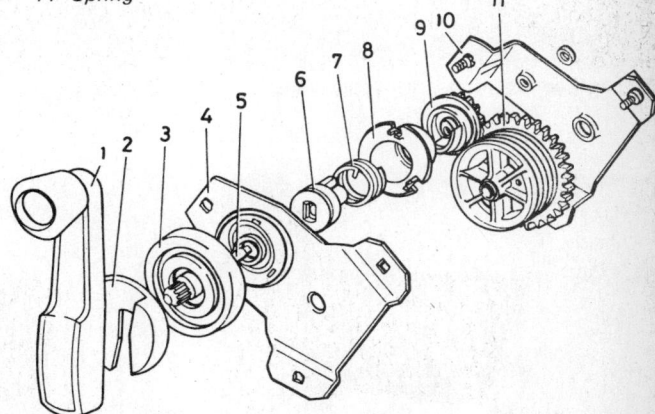

Fig. 13.44 Window regulator mechanism (Sec 15)

1 Handle
2 Locking clip
3 Escutcheon
4 Cover
5 Shaft
6 Spring brake drive link
7 Spring brake
8 Support
9 Pinion
10 Body
11 Drum

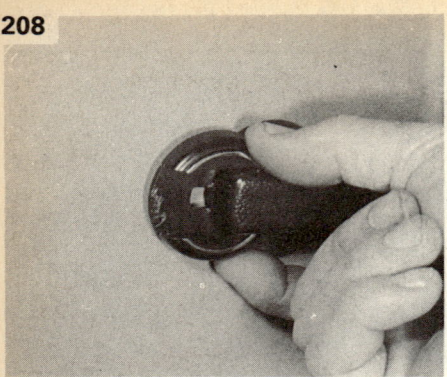

15.26A Releasing door window regulator handle

15.26B Removing regulator handle escutcheon

15.27A Prising out door lock knob plastic cover

15.27B Extracting door lock knob clip

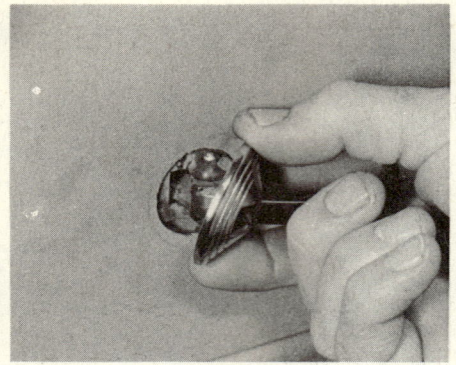

15.28 Door inner handle surround

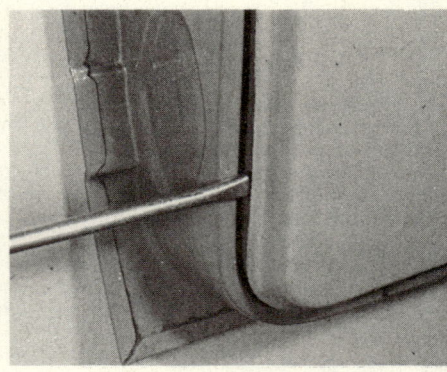

15.29 Prising off a door trim panel

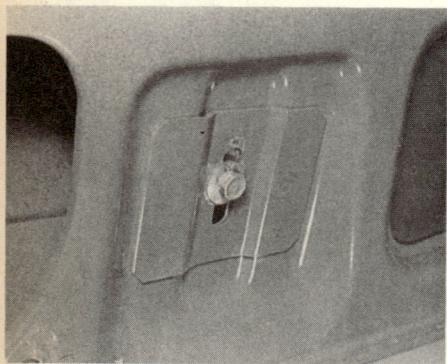

15.35 Window regulator cable tensioner adjusting bolt

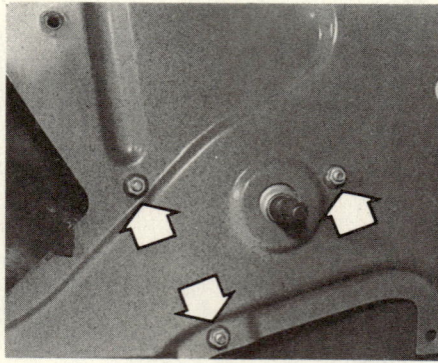

15.38A Window regulator securing nuts (arrowed)

15.38B Window regulator cable and roller

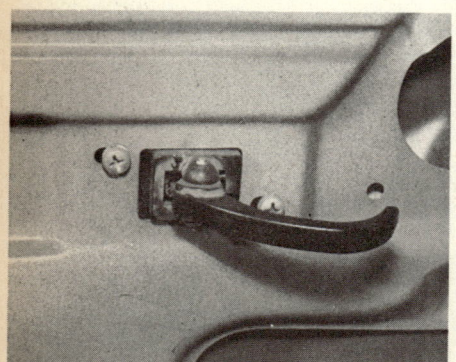

15.39 Door interior handle

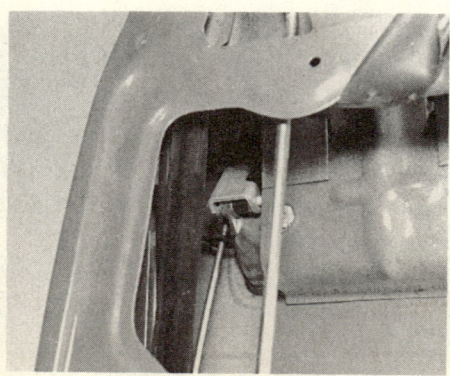

15.40 Door exterior handle operating link

15.41A Door lock retaining screws

Chapter 13 Supplement

15.41B Door lock and control rod arrangement

15.42 Door private lock with retaining clip

43 Reassembly is a reversal of the dismantling procedure with reference to the following points. Check that the cable turns do not overlap each other on the regulator drum. Lubricate the cable rollers with a little grease, and tension the cable as required by adjusting the tensioner roller.

Centre console – removal and refitting
The operations are described earlier in this Section as part of the facia removal procedure.

Seats – removal and refitting
Front seat
44 Push the seat fully rearwards and remove the bolts from the front ends of the seat runners (photo).
45 Now push the seat fully forwards and remove the bolts from the rear ends of the runners (photo).
46 Remove the seat from the vehicle.
Rear seat
47 Grip the front edge of the seat cushion and lift the cushion off its locating pegs (photo).
48 The seat back lower fixing tabs will now be exposed. Flatten the tabs and lift the seat back to disengage it from its upper fixings (photos).
All seats
49 Refitting is a reversal of removal.

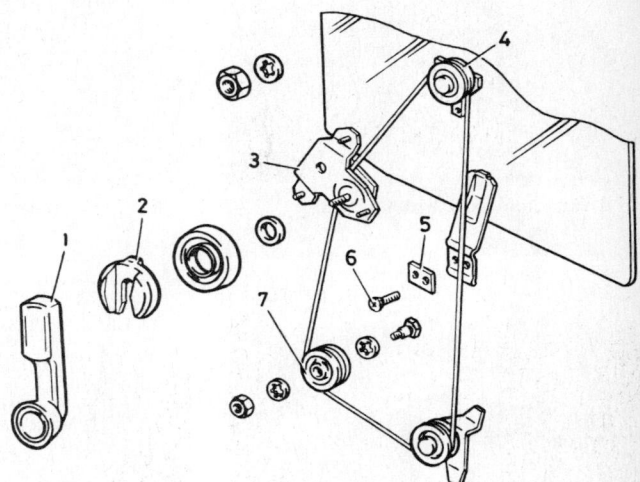

Fig. 13.45 Riva rear door glass and regulator (Sec 15)

1 Handle
2 Locking clip
3 Regulator mechanism
4 Roller
5 Clamp
6 Screw
7 Tensioner roller

Fig. 13.46 Front seat runners (Sec 15)

1 Seat track
2 Seat slide
3 Roller
4 Roller rubber ring
5 Stop
6 Rail catch
7 Inner track retainer
8 Reinforcement plate
9 Rod
10 Spring
11 Cotter pin
12 Seat back rake adjuster
13 Fore and aft adjuster
14 Bracket

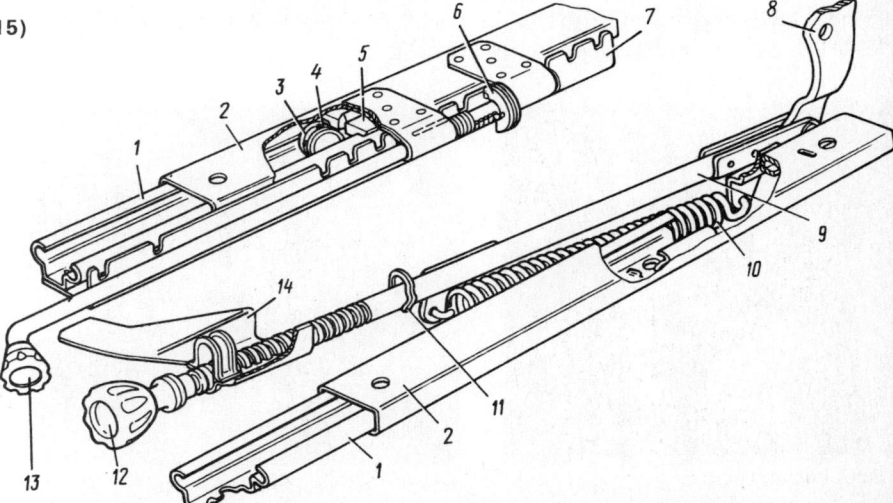

Seat belts

50 Front seat belts are fitted to all models and rear belts to later vehicles.
51 Regularly inspect the belts for damage or fraying. Dirt or stains should be removed using warm water and liquid detergent. Allow the belt to dry before it is retracted into its reel.
52 If a belt anchorage is disconnected, make sure that the original sequence of washers and spacers is retained (photos).
53 The front seat belt reel is secured to the base of the centre pillar by a single bolt (photo).
54 The rear seat belt reels are accessible after having removed the rear seat and rear upper corner trim panels (photo).
55 Bend up the tabs and withdraw the rear parcel shelf (photo).
56 Unscrew the belt retaining bolt (photo).

Radiator grille (1300 and 1600 Riva) – removal and refitting

57 The grille is secured by self-tapping screws. Three long screws hold the top while three shorter ones are located in cut-outs at the bottom (photos).

Grab rails – removal and refitting

58 Pull off the end covers to expose the securing screws. Remove the screws and the rail (photo).

Remote control exterior mirror (except Riva models) – removal and refitting

59 Prise the plug from the control handle.
60 Undo the screw and withdraw the handle and washers.
61 Remove the mirror from the outside of the door.
62 Refitting is a reversal of removal.

Remote control exterior mirror (Riva models) – removal and refitting

63 Open the door and extract the screw from the door edge level with the mirror (photo).
64 Prise away the triangular trim plate and move the mirror upwards to remove it (photo).
65 If the reason for removal was due to a cracked glass, remove the glass by prising it from its adhesive backing.

15.44 Front seat front mounting bolt

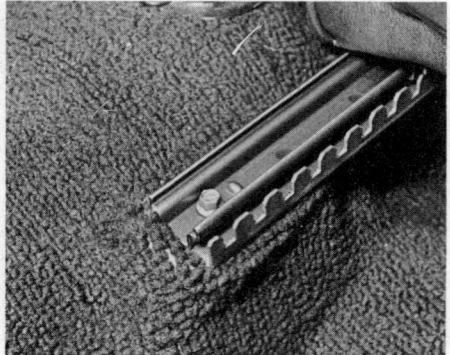
15.45 Front seat rear mounting bolt

15.47 Rear seat cushion locating peg

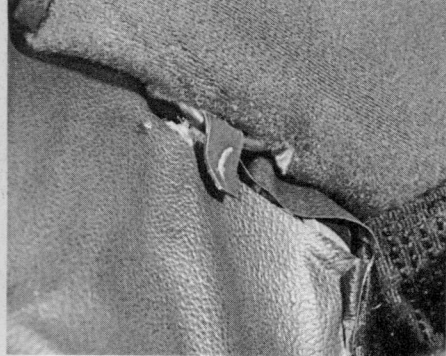
15.48A Rear seat back lower fixing tab

15.48B Rear seat back upper fixing

15.52A Seat belt upper pillar bolt

15.52B Seat belt anchor bolt components

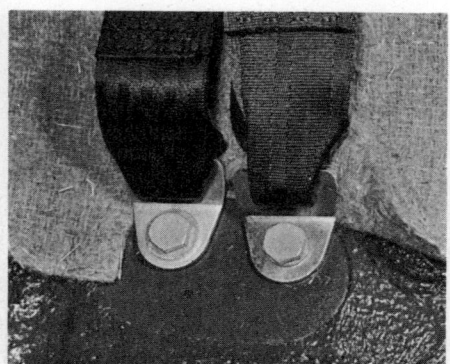
15.52C Rear seat belt floor bolts

15.52D Front seat belt stalk

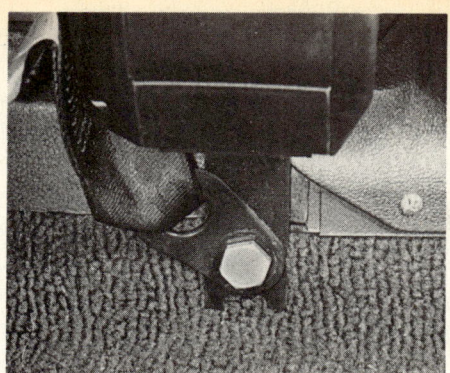
15.53 Front seat belt reel

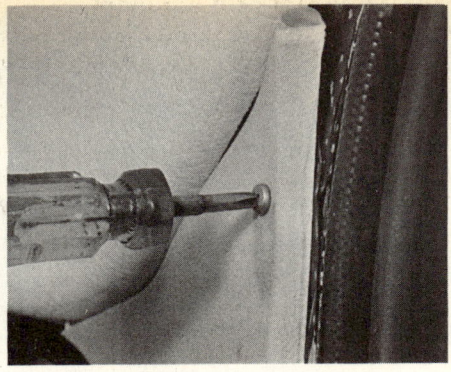
15.54 Removing rear quarter panel screw

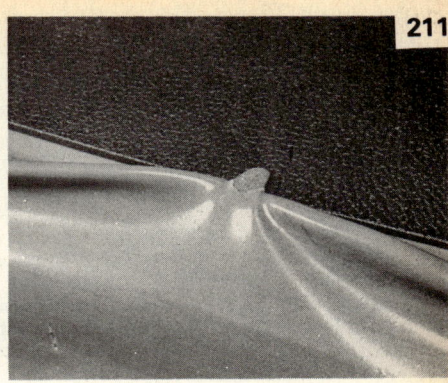
15.55 Rear parcel shelf retaining tab

15.56 Rear seat belt reel

15.57A Removing a radiator grille upper centre screw

15.57B Radiator grille upper side screw

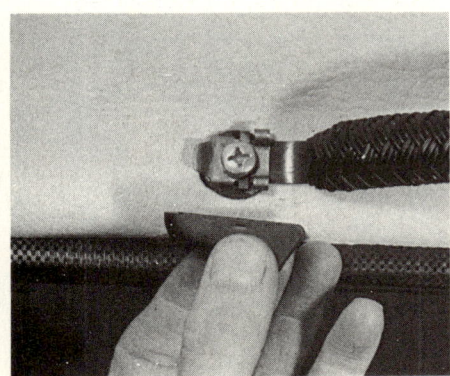
15.58 Grab rail and screw, cover removed

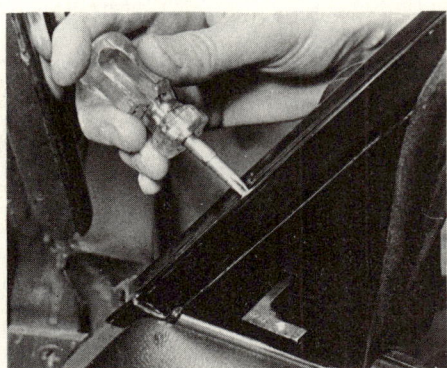
15.63 Removing mirror screw from door edge

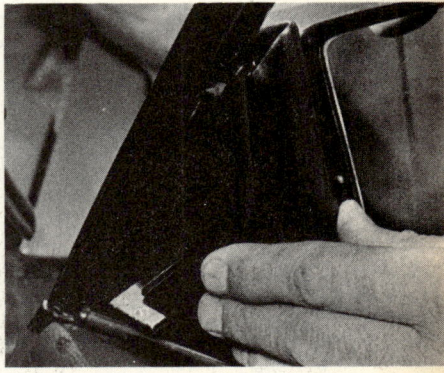
15.64A Removing mirror trim plate

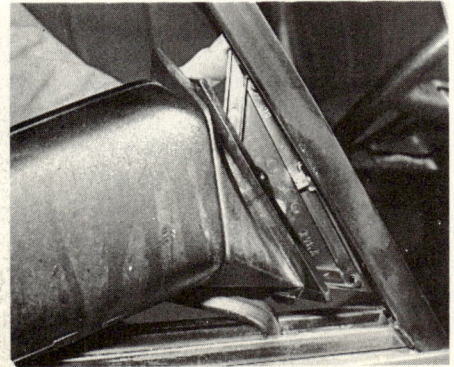
15.64B Removing mirror

15.66 Removing mirror handle screw

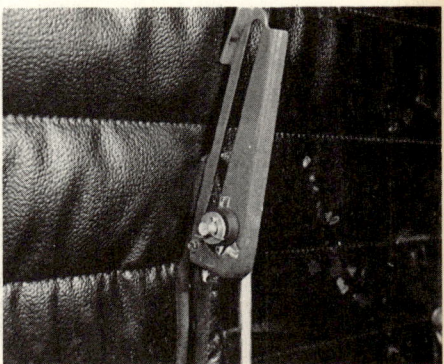
15.69 Bonnet strut and slide bracket

66 Extract the screw and remove the handle (photo).
67 Withdraw the handle from the mirror housing to disconnect the slide and balljoint.
68 Stick the new mirror glass in position, reassemble and refit the mirror by reversing the removal and dismantling operations.

Bonnet strut (later models)
69 Later models have a solid rod-type strut to support the bonnet in its open position instead of the spring type used previously (photo).
70 To remove the strut, simply release the lower pivot, but make sure that an assistant is supporting the weight of the bonnet.

Tailgate struts (later models)
71 Later models have two gas-filled struts to support the tailgate instead of the torsion springs used on earlier models.
72 To remove, unscrew the upper and lower pivot bolts.

Heating and ventilation system (Riva models) – description
73 The heating and ventilation system components are shown in Fig. 13.42 and include side vents, windscreen vents and footwell vents. Certain models have two additional vents mounted centrally on the facia panel. Stale air is exhausted through flap valves located behind the rear quarter trim panels.

Heating and ventilation unit (Riva models) – removal and refitting
74 Disconnect the battery negative lead.
75 Remove the instrument panel, glovebox and casing, and the centre console and radio housing (as applicable).
76 Drain the cooling system, as described in Chapter 2.
77 Disconnect the heater hoses from the matrix pipes on the bulkhead and unbolt the seal.
78 Unbolt the heater control bracket then identify the control cables for position and disconnect them by loosening the clamp bolts.
79 Undo the screws and remove the footwell air duct.
80 Prise out the side vents (and centre vents where fitted) and remove the seals.
81 Remove the left and right-hand air ducts by unscrewing the mounting nuts and prising them from the heater casing.
82 Prise off the spring clips from the fan shroud.
83 Unscrew the matrix housing nuts, noting that the earth wire is located beneath the left-hand side nut. Remove the fan housing and windscreen vent.
84 Refitting is a reversal of removal, but make sure that the gasket on the bulkhead is in good condition. Fill the cooling system with reference to Chapter 2 and adjust the control cables, as follows, before refitting the instrument panel. Fully close the heater valve and air intake shutter, and fully open the windscreen vent shutters. Adjust the position of the cables to give the dimensions shown in Fig. 13.44, then tighten the clamp bolts.

Heating and ventilation unit (Riva models) – dismantling and reassembly
85 Prise off the spring clips and lift the fan and motor from the shroud. If necessary unscrew the nut and separate the fan from the motor.
86 Unscrew the nuts and separate the air distribution cover from the fan housing.
87 Extract the side vent shutters and disconnect the control rods.
88 Loosen the clamp bolts and remove the control cables.
89 Unclip the matrix and lift it from the air distribution cover.
90 Unscrew the nuts and remove the pipes and valve from the matrix.
91 Loosen the clamp bolt and remove the control rod.
92 If necessary, unscrew the nuts and remove the air intake cover from the bulkhead.
93 Reassembly is a reversal of dismantling.

Plastic components
94 With the use of more and more plastic body components by the vehicle manufacturers (eg bumpers, spoilers, and in some cases major body panels), rectification of more serious damage to such items has become a matter of either entrusting repair work to a specialist in this field, or renewing complete components. Repair of such damage by the DIY owner is not really feasible owing to the cost of the equipment and materials required for effecting such repairs. The basic technique involves making a groove along the line of the crack in the plastic using a rotary burr in a power drill. The damaged part is then welded back together by using a hot air gun to heat up and fuse a plastic filler rod into the groove. Any excess plastic is then removed and the area rubbed down to a smooth finish. It is important that a filler rod of the correct plastic is used, as body components can be made of a variety of different types (eg polycarbonate, ABS, polypropylene).
95 Damage of a less serious nature (abrasions, minor cracks etc) can be repaired by the DIY owner using a two-part epoxy filler repair material, like Holts Body + Plus or Holts No Mix which can be used directly from the tube. Once mixed in equal proportions (or applied direct from the tube in the case of Holts No Mix), this is used in similar fashion to the bodywork filler used on metal panels. The filler is usually cured in twenty to thirty minutes, ready for sanding and painting.
96 If the owner is renewing a complete component himself, or if he has repaired it with epoxy filler, he will be left with the problem of finding a suitable paint for finishing which is compatible with the type of plastic used. At one time the use of a universal paint was not possible owing to the complex range of plastics encountered in body component applications. Standard paints, generally speaking, will not bond to plastic or rubber satisfactorily, but Holts Professional Spraymatch paints to match any plastic or rubber finish can be obtained from dealers. However, it is now possible to obtain a plastic body parts finishing kit which consists of a pre-primer treatment, a primer and coloured top coat. Full instructions are normally supplied with a kit, but basically the method of use is to first apply the pre-primer to the component concerned and allow it to dry for up to 30 minutes. Then the primer is applied and left to dry for about an hour before finally applying the special coloured top coat. The result is a correctly coloured component where the paint will flex with the plastic or rubber, a property that standard paint does not normally possess.

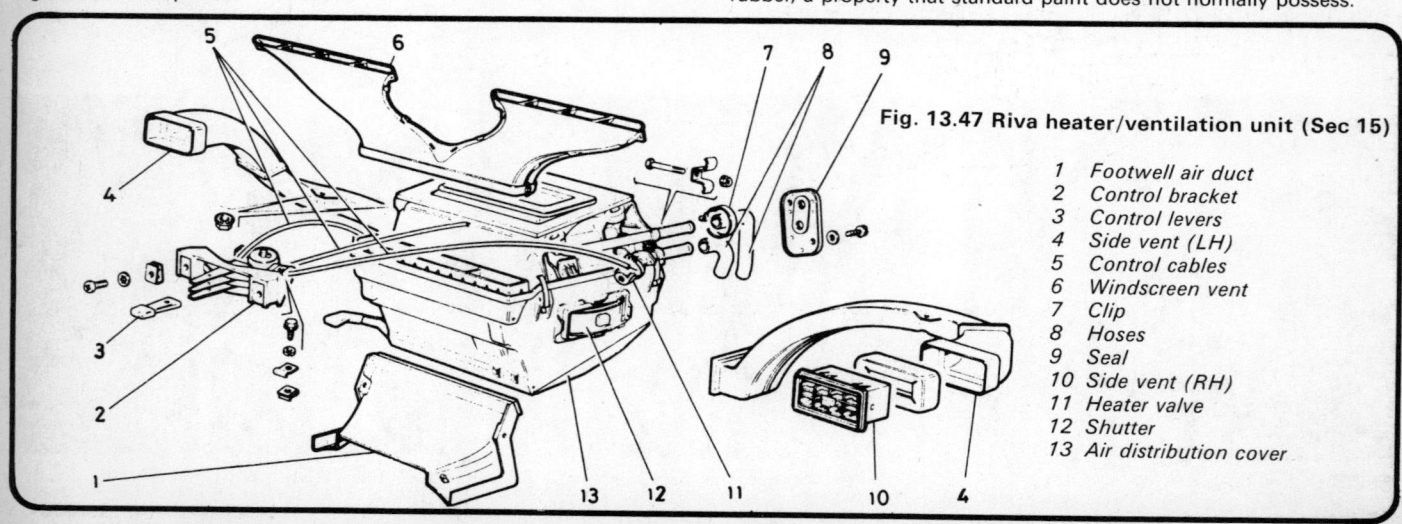

Fig. 13.47 Riva heater/ventilation unit (Sec 15)

1. Footwell air duct
2. Control bracket
3. Control levers
4. Side vent (LH)
5. Control cables
6. Windscreen vent
7. Clip
8. Hoses
9. Seal
10. Side vent (RH)
11. Heater valve
12. Shutter
13. Air distribution cover

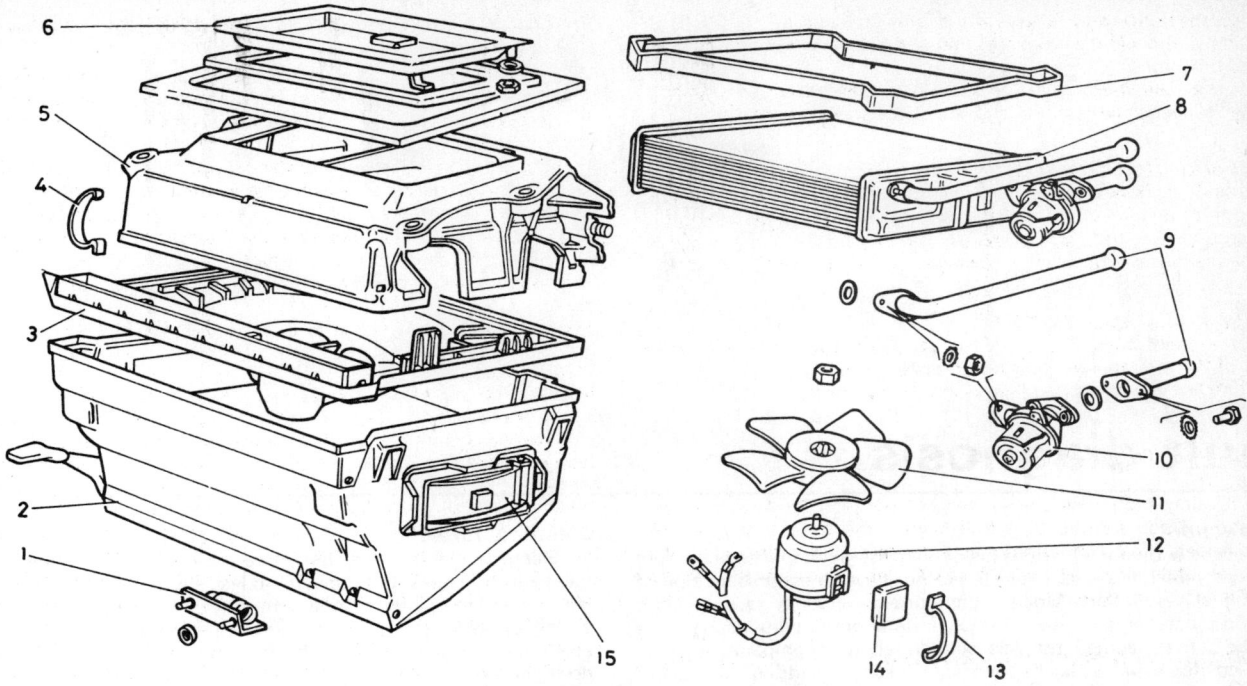

Fig. 13.48 Exploded view of Riva heater (Sec 15)

1 Series resistor	5 Matrix housing	9 Matrix pipes	13 Spring clip
2 Air distribution cover	6 Air intake cover	10 Heater valve	14 Mounting pad
3 Fan shroud	7 Gasket	11 Fan	15 Shutter
4 Spring clips	8 Matrix	12 Fan motor	

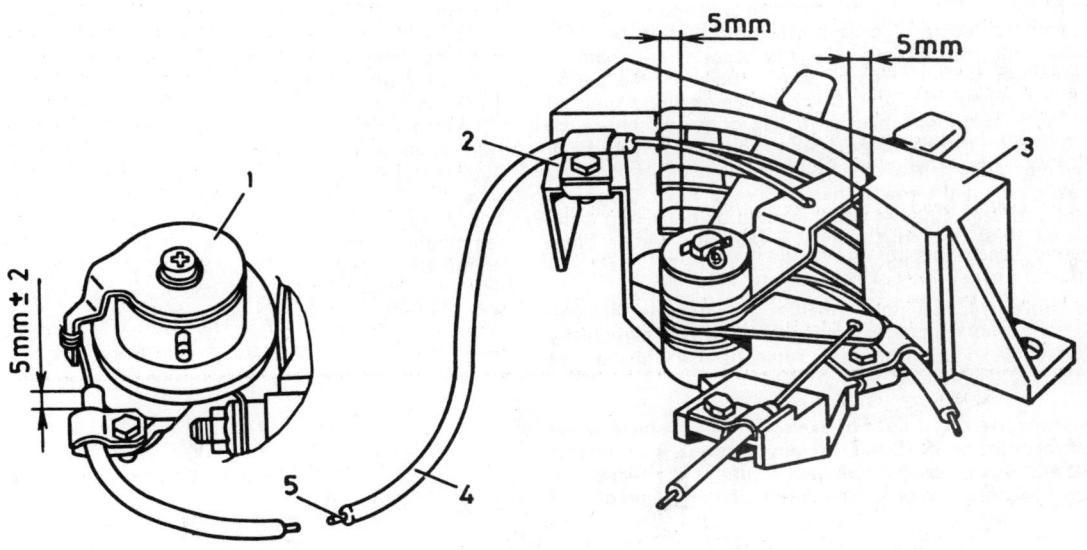

Fig. 13.49 Heater control cable adjustment diagram (Sec 15)

1 Coolant flow valve	3 Bracket	4 Outer cable (conduit)	5 Inner cable
2 Clamp			

Fault diagnosis

Introduction

The vehicle owner who does his or her own maintenance according to the recommended schedules should not have to use this section of the manual very often. Modern component reliability is such that, provided those items subject to wear or deterioration are inspected or renewed at the specified intervals, sudden failure is comparatively rare. Faults do not usually just happen as a result of sudden failure, but develop over a period of time. Major mechanical failures in particular are usually preceded by characteristic symptoms over hundreds or even thousands of miles. Those components which do occasionally fail without warning are often small and easily carried in the vehicle.

With any fault finding, the first step is to decide where to begin investigations. Sometimes this is obvious, but on other occasions a little detective work will be necessary. The owner who makes half a dozen haphazard adjustments or replacements may be successful in curing a fault (or its symptoms), but he will be none the wiser if the fault recurs and he may well have spent more time and money than was necessary. A calm and logical approach will be found to be more satisfactory in the long run. Always take into account any warning signs or abnormalities that may have been noticed in the period preceding the fault – power loss, high or low gauge readings, unusual noises or smells, etc – and remember that failure of components such as fuses or spark plugs may only be pointers to some underlying fault.

The pages which follow here are intended to help in cases of failure to start or breakdown on the road. There is also a Fault Diagnosis Section at the end of each Chapter which should be consulted if the preliminary checks prove unfruitful. Whatever the fault, certain basic principles apply. These are as follows:

Verify the fault. This is simply a matter of being sure that you know what the symptoms are before starting work. This is particularly important if you are investigating a fault for someone else who may not have described it very accurately.

Don't overlook the obvious. For example, if the vehicle won't start, is there petrol in the tank? (Don't take anyone else's word on this particular point, and don't trust the fuel gauge either!) If an electrical fault is indicated, look for loose or broken wires before digging out the test gear.

Cure the disease, not the symptom. Substituting a flat battery with a fully charged one will get you off the hard shoulder, but if the underlying cause is not attended to, the new battery will go the same way. Similarly, changing oil-fouled spark plugs for a new set will get you moving again, but remember that the reason for the fouling (if it wasn't simply an incorrect grade of plug) will have to be established and corrected.

Don't take anything for granted. Particularly, don't forget that a 'new' component may itself be defective (especially if it's been rattling round in the boot for months), and don't leave components out of a fault diagnosis sequence just because they are new or recently fitted. When you do finally diagnose a difficult fault, you'll probably realise that all the evidence was there from the start.

Electrical faults

Electrical faults can be more puzzling than straightforward mechanical failures, but they are no less susceptible to logical analysis if the basic principles of operation are understood. Vehicle electrical wiring exists in extremely unfavourable conditions – heat, vibration and chemical attack – and the first things to look for are loose or corroded connections and broken or chafed wires, especially where the wires pass through holes in the bodywork or are subject to vibration.

All metal-bodied vehicles in current production have one pole of the battery 'earthed', ie connected to the vehicle bodywork, and in nearly all modern vehicles it is the negative (–) terminal. The various electrical components – motors, bulb holders etc – are also connected to earth, either by means of a lead or directly by their mountings. Electric current flows through the component and then back to the battery via the bodywork. If the component mounting is loose or corroded, or if a good path back to the battery is not available, the circuit will be incomplete and malfunction will result. The engine and/or gearbox are also earthed by means of flexible metal straps to the body or subframe; if these straps are loose or missing, starter motor, generator and ignition trouble may result.

Assuming the earth return to be satisfactory, electrical faults will be due either to component malfunction or to defects in the current supply. Individual components are dealt with in Chapter 10. If supply wires are broken or cracked internally this results in an open-circuit, and the easiest way to check for this is to bypass the suspect wire temporarily with a length of wire having a crocodile clip or suitable connector at each end. Alternatively, a 12V test lamp can be used to verify the presence of supply voltage at various points along the wire and the break can be thus isolated.

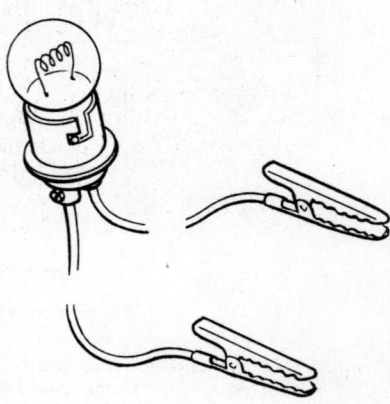

Simple test lamp is useful for tracing electrical faults

Fault diagnosis

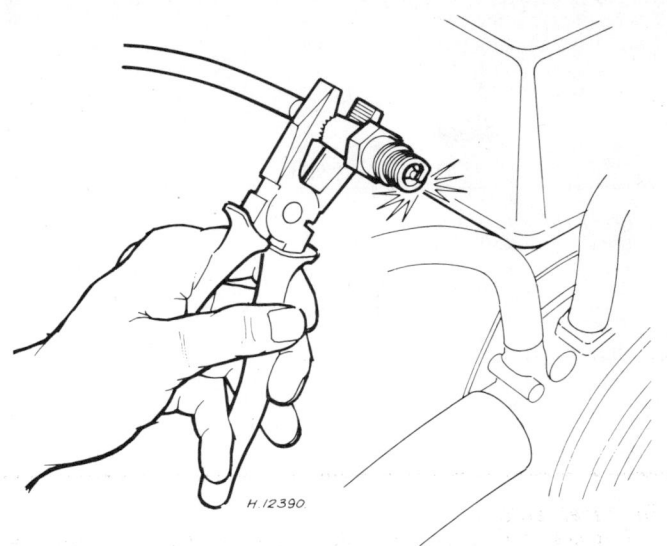

Crank engine and check for spark. Note use of insulated tool to hold plug lead

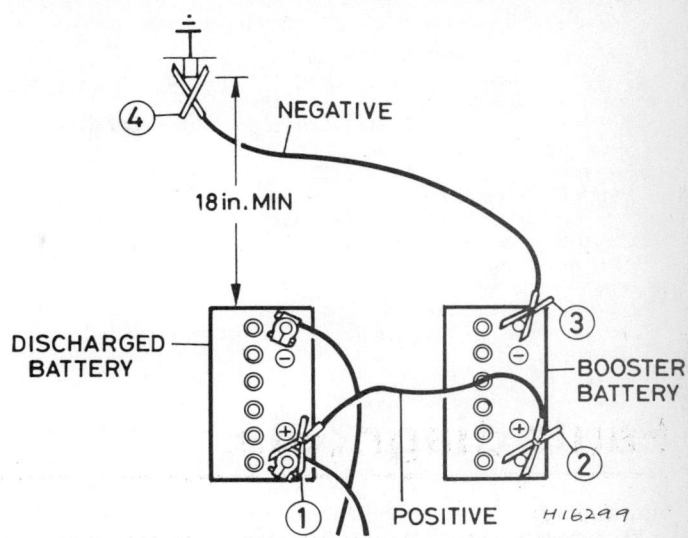

Jump start lead connections for negative earth vehicles – connect leads in order shown

Carrying a few spares can save you a long walk

If a bare portion of a live wire touches the bodywork or other earthed metal part, the electricity will take the low-resistance path thus formed back to the battery: this is known as a short-circuit. Hopefully a short-circuit will blow a fuse, but otherwise it may cause burning of the insulation (and possibly further short-circuits) or even a fire. This is why it is inadvisable to bypass persistently blowing fuses with silver foil or wire.

Spares and tool kit

Most vehicles are supplied only with sufficient tools for wheel changing; the *Maintenance and minor repair* tool kit detailed in *Tools and working facilities*, with the addition of a hammer, is probably sufficient for those repairs that most motorists would consider attempting at the roadside. In addition a few items which can be fitted without too much trouble in the event of a breakdown should be carried. Experience and available space will modify the list below, but the following may save having to call on professional assistance:

Spark plugs, clean and correctly gapped
HT lead and plug cap – long enough to reach the plug furthest from the distributor
Distributor rotor, condenser and contact breaker points
Drivebelt(s) – emergency type may suffice
Spare fuses
Set of principal light bulbs
Tin of radiator sealer and hose bandage
Exhaust bandage
Roll of insulating tape
Length of soft iron wire
Length of electrical flex
Torch or inspection lamp (can double as test lamp)
Battery jump leads
Tow-rope
Ignition water dispersant aerosol
Litre of engine oil
Sealed can of hydraulic fluid
Worm drive clips

If spare fuel is carried, a can designed for the purpose should be used to minimise risks of leakage and collision damage. A first aid kit and a warning triangle, whilst not at present compulsory in the UK, are obviously sensible items to carry in addition to the above.

When touring abroad it may be advisable to carry additional spares which, even if you cannot fit them yourself, could save having to wait while parts are obtained. The items below may be worth considering:

Choke and throttle cables
Cylinder head gasket
Alternator brushes
Fuel pump repair kit
Tyre valve core

One of the motoring organisations will be able to advise on availability of fuel etc in foreign countries.

Engine will not start

Engine fails to turn when starter operated
Flat battery (recharge, use jump leads, or push start)
Battery terminals loose or corroded
Battery earth to body defective
Engine earth strap loose or broken
Starter motor (or solenoid) wiring loose or broken
Ignition/starter switch faulty
Major mechanical failure (seizure)
Starter or solenoid internal fault (see Chapter 10)

Starter motor turns engine slowly
Partially discharged battery (recharge, use jump leads, or push start)
Battery terminals loose or corroded
Battery earth to body defective
Engine earth strap loose
Starter motor (or solenoid) wiring loose
Starter motor internal fault (see Chapter 10)

Engine turns normally but fails to start
Damp or dirty HT leads and distributor cap (crank engine and check for spark)
Dirty or incorrectly gapped distributor points (if applicable)
No fuel in tank (check for delivery at carburettor)
Excessive choke (hot engine) or insufficient choke (cold engine)
Fouled or incorrectly gapped spark plugs (remove, clean and regap)
Other ignition system fault (see Chapter 4)
Other fuel system fault (see Chapter 3)
Poor compression (see Chapter 1)
Major mechanical failure (eg camshaft drive)

Engine fires but will not run
Insufficient choke (cold engine)
Air leaks at carburettor or inlet manifold
Fuel starvation (see Chapter 3)
Ballast resistor defective, or other ignition fault (see Chapter 4)

Engine cuts out and will not restart

Engine cuts out suddenly – ignition fault
Loose or disconnected LT wires
Wet HT leads or distributor cap (after traversing water splash)
Coil or condenser failure (check for spark)
Other ignition fault (see Chapter 4)

Engine misfires before cutting out – fuel fault
Fuel tank empty
Fuel pump defective or filter blocked (check for delivery)
Fuel tank filler vent blocked (suction will be evident on releasing cap)
Carburettor needle valve sticking
Carburettor jets blocked (fuel contaminated)
Other fuel system fault (see Chapter 3)

Engine cuts out – other causes
Serious overheating
Major mechanical failure (eg camshaft drive)

Engine overheats

Ignition (no-charge) warning light illuminated
Slack or broken drivebelt – retension or renew (Chapter 2)

Ignition warning light not illuminated
Coolant loss due to internal or external leakage (see Chapter 2)
Thermostat defective
Low oil level
Brakes binding
Radiator clogged externally or internally
Electric cooling fan not operating correctly
Engine waterways clogged
Ignition timing incorrect or automatic advance malfunctioning
Mixture too weak

Note: *Do not add cold water to an overheated engine or damage may result*

Low engine oil pressure

Gauge reads low or warning light illuminated with engine running
Oil level low or incorrect grade
Defective gauge or sender unit
Wire to sender unit earthed
Engine overheating
Oil filter clogged or bypass valve defective
Oil pressure relief valve defective
Oil pick-up strainer clogged
Oil pump worn or mountings loose
Worn main or big-end bearings

Note: *Low oil pressure in a high-mileage engine at tickover is not necessarily a cause for concern. Sudden pressure loss at speed is far more significant. In any event, check the gauge or warning light sender before condemning the engine.*

Engine noises

Pre-ignition (pinking) on acceleration
Incorrect grade of fuel
Ignition timing incorrect

Distributor faulty or worn
Worn or maladjusted carburettor
Excessive carbon build-up in engine

Whistling or wheezing noises
Leaking vacuum hose
Leaking carburettor or manifold gasket
Blowing head gasket

Tapping or rattling
Incorrect valve clearances
Worn valve gear
Worn timing chain or belt
Broken piston ring (ticking noise)

Knocking or thumping
Unintentional mechanical contact (eg fan blades)
Worn drivebelt
Peripheral component fault (generator, water pump etc)
Worn big-end bearings (regular heavy knocking, perhaps less under load)
Worn main bearings (rumbling and knocking, perhaps worsening under load)
Piston slap (most noticeable when cold)

General dimensions, weights and capacities

Dimensions
1200/1300 saloon:
    Overall length .... 13 ft 4½ in (4073 mm)
    Overall width .... 5 ft 3⅜ in (1611 mm)
    Overall height (unladen) .... 4 ft 9 in (1450 mm)
    Wheelbase .... 7 ft 11 in (2424 mm)
    Rear track .... 4 ft 3 in (1305 mm)
    Front track .... 4 ft 5 in (1349 mm)
1500 saloon:
    Overall length .... 13 ft 6 in (4116 mm)
    Overall width .... 5 ft 3⅜ in (1611 mm)
    Overall height (unladen) .... 4 ft 6½ in (1446 mm)
    Wheelbase .... 7 ft 11 in (2424 mm)
    Rear track .... 4 ft 4 in (1321 mm)
    Front track .... 4 ft 6 in (1365 mm)
1600 saloon:
    Overall length .... 13 ft 8 in (4166 mm)
    Overall width .... 5 ft 3 in (1611 mm)
    Overall height (unladen) .... 4 ft 9 in (1450 mm)
    Wheelbase .... 7 ft 11 in (2424 mm)
    Rear track .... 4 ft 4 in (1321 mm)
    Front track .... 4 ft 6 in (1365 mm)
Estate versions:
Dimensions as respective saloon models except for:
    Overall height (unladen) .... 4 ft 7 in (1400 mm)
    Overall length .... 13 ft 4 in (4059 mm)
Van versions:
Dimensions as 1500 saloon model except for:
    Overall length .... 13 ft 6 in (4116 mm)
    Overall height .... 4 ft 9 in (1450 mm)

Kerb weights
1200/1300 saloon .... 2105 lb (955 kg)
1200/1300 estate .... 2222 lb (1010 kg)
1500 saloon .... 2266 lb (1030 kg)
1500 estate .... 2398 lb (1090 kg)
1600 saloon .... 2299 lb (1045 kg)
Van .... 2249 lb (1021 kg)

Towing capacity (maximum)
1200 .... 2867 lb (1302 kg)
1300 saloon .... 2867 lb (1302 kg)
1300 estate .... 3024 lb (1373 kg)
1500 estate .... 3136 lb (1424 kg)
1600 saloon .... 3080 lb (1398 kg)

Figures for other models not available at time of writing

Capacities
Fuel:
    Saloon .... 8.5 gals (38.6 litres)
    Estate and van .... 10.0 gals (45.5 litres)
Cooling system:
    Models up to 1976 .... 16.8 pints (9.6 litres)
    Models 1976 on .... 17.3 pints (9.85 litres)
Engine oil (including filter change) .... 6.6 pints (3.75 litres)
Gearbox .... 2.36 pints (1.35 litres)
Rear axle .... 2.3 pints (1.3 litres)
Steering box .... 0.38 pint (0.215 litre)

General repair procedures

Whenever servicing, repair or overhaul work is carried out on the car or its components, it is necessary to observe the following procedures and instructions. This will assist in carrying out the operation efficiently and to a professional standard of workmanship.

Joint mating faces and gaskets

Where a gasket is used between the mating faces of two components, ensure that it is renewed on reassembly, and fit it dry unless otherwise stated in the repair procedure. Make sure that the mating faces are clean and dry with all traces of old gasket removed. When cleaning a joint face, use a tool which is not likely to score or damage the face, and remove any burrs or nicks with an oilstone or fine file.

Make sure that tapped holes are cleaned with a pipe cleaner, and keep them free of jointing compound if this is being used unless specifically instructed otherwise.

Ensure that all orifices, channels or pipes are clear and blow through them, preferably using compressed air.

Oil seals

Whenever an oil seal is removed from its working location, either individually or as part of an assembly, it should be renewed.

The very fine sealing lip of the seal is easily damaged and will not seal if the surface it contacts is not completely clean and free from scratches, nicks or grooves. If the original sealing surface of the component cannot be restored, the component should be renewed.

Protect the lips of the seal from any surface which may damage them in the course of fitting. Use tape or a conical sleeve where possible. Lubricate the seal lips with oil before fitting and, on dual lipped seals, fill the space between the lips with grease.

Unless otherwise stated, oil seals must be fitted with their sealing lips toward the lubricant to be sealed.

Use a tubular drift or block of wood of the appropriate size to install the seal and, if the seal housing is shouldered, drive the seal down to the shoulder. If the seal housing is unshouldered, the seal should be fitted with its face flush with the housing top face.

Screw threads and fastenings

Always ensure that a blind tapped hole is completely free from oil, grease, water or other fluid before installing the bolt or stud. Failure to do this could cause the housing to crack due to the hydraulic action of the bolt or stud as it is screwed in.

When tightening a castellated nut to accept a split pin, tighten the nut to the specified torque, where applicable, and then tighten further to the next split pin hole. Never slacken the nut to align a split pin hole unless stated in the repair procedure.

When checking or retightening a nut or bolt to a specified torque setting, slacken the nut or bolt by a quarter of a turn, and then retighten to the specified setting.

Locknuts, locktabs and washers

Any fastening which will rotate against a component or housing in the course of tightening should always have a washer between it and the relevant component or housing.

Spring or split washers should always be renewed when they are used to lock a critical component such as a big-end bearing retaining nut or bolt.

Locktabs which are folded over to retain a nut or bolt should always be renewed.

Self-locking nuts can be reused in non-critical areas, providing resistance can be felt when the locking portion passes over the bolt or stud thread.

Split pins must always be replaced with new ones of the correct size for the hole.

Special tools

Some repair procedures in this manual entail the use of special tools such as a press, two or three-legged pullers, spring compressors etc. Wherever possible, suitable readily available alternatives to the manufacturer's special tools are described, and are shown in use. In some instances, where no alternative is possible, it has been necessary to resort to the use of a manufacturer's tool and this has been done for reasons of safety as well as the efficient completion of the repair operation. Unless you are highly skilled and have a thorough understanding of the procedure described, never attempt to bypass the use of any special tool when the procedure described specifies its use. Not only is there a very great risk of personal injury, but expensive damage could be caused to the components involved.

Safety first!

Professional motor mechanics are trained in safe working procedures. However enthusiastic you may be about getting on with the job in hand, do take the time to ensure that your safety is not put at risk. A moment's lack of attention can result in an accident, as can failure to observe certain elementary precautions.

There will always be new ways of having accidents, and the following points do not pretend to be a comprehensive list of all dangers; they are intended rather to make you aware of the risks and to encourage a safety-conscious approach to all work you carry out on your vehicle.

Essential DOs and DON'Ts

DON'T rely on a single jack when working underneath the vehicle. Always use reliable additional means of support, such as axle stands, securely placed under a part of the vehicle that you know will not give way.

DON'T attempt to loosen or tighten high-torque nuts (e.g. wheel hub nuts) while the vehicle is on a jack; it may be pulled off.

DON'T start the engine without first ascertaining that the transmission is in neutral (or 'Park' where applicable) and the parking brake applied.

DON'T suddenly remove the filler cap from a hot cooling system – cover it with a cloth and release the pressure gradually first, or you may get scalded by escaping coolant.

DON'T attempt to drain oil until you are sure it has cooled sufficiently to avoid scalding you.

DON'T grasp any part of the engine, exhaust or catalytic converter without first ascertaining that it is sufficiently cool to avoid burning you.

DON'T allow brake fluid or antifreeze to contact vehicle paintwork.

DON'T syphon toxic liquids such as fuel, brake fluid or antifreeze by mouth, or allow them to remain on your skin.

DON'T inhale dust – it may be injurious to health (see *Asbestos* below).

DON'T allow any spilt oil or grease to remain on the floor – wipe it up straight away, before someone slips on it.

DON'T use ill-fitting spanners or other tools which may slip and cause injury.

DON'T attempt to lift a heavy component which may be beyond your capability – get assistance.

DON'T rush to finish a job, or take unverified short cuts.

DON'T allow children or animals in or around an unattended vehicle.

DO wear eye protection when using power tools such as drill, sander, bench grinder etc, and when working under the vehicle.

DO use a barrier cream on your hands prior to undertaking dirty jobs – it will protect your skin from infection as well as making the dirt easier to remove afterwards; but make sure your hands aren't left slippery. Note that long-term contact with used engine oil can be a health hazard.

DO keep loose clothing (cuffs, tie etc) and long hair well out of the way of moving mechanical parts.

DO remove rings, wristwatch etc, before working on the vehicle – especially the electrical system.

DO ensure that any lifting tackle used has a safe working load rating adequate for the job.

DO keep your work area tidy – it is only too easy to fall over articles left lying around.

DO get someone to check periodically that all is well, when working alone on the vehicle.

DO carry out work in a logical sequence and check that everything is correctly assembled and tightened afterwards.

DO remember that your vehicle's safety affects that of yourself and others. If in doubt on any point, get specialist advice.

IF, in spite of following these precautions, you are unfortunate enough to injure yourself, seek medical attention as soon as possible.

Asbestos

Certain friction, insulating, sealing, and other products – such as brake linings, brake bands, clutch linings, torque converters, gaskets, etc – contain asbestos. *Extreme care must be taken to avoid inhalation of dust from such products since it is hazardous to health.* If in doubt, assume that they *do* contain asbestos.

Fire

Remember at all times that petrol (gasoline) is highly flammable. Never smoke, or have any kind of naked flame around, when working on the vehicle. But the risk does not end there – a spark caused by an electrical short-circuit, by two metal surfaces contacting each other, by careless use of tools, or even by static electricity built up in your body under certain conditions, can ignite petrol vapour, which in a confined space is highly explosive.

Always disconnect the battery earth (ground) terminal before working on any part of the fuel or electrical system, and never risk spilling fuel on to a hot engine or exhaust.

It is recommended that a fire extinguisher of a type suitable for fuel and electrical fires is kept handy in the garage or workplace at all times. Never try to extinguish a fuel or electrical fire with water.

Note: *Any reference to a 'torch' appearing in this manual should always be taken to mean a hand-held battery-operated electric lamp or flashlight. It does NOT mean a welding/gas torch or blowlamp.*

Fumes

Certain fumes are highly toxic and can quickly cause unconsciousness and even death if inhaled to any extent. Petrol (gasoline) vapour comes into this category, as do the vapours from certain solvents such as trichloroethylene. Any draining or pouring of such volatile fluids should be done in a well ventilated area.

When using cleaning fluids and solvents, read the instructions carefully. Never use materials from unmarked containers – they may give off poisonous vapours.

Never run the engine of a motor vehicle in an enclosed space such as a garage. Exhaust fumes contain carbon monoxide which is extremely poisonous; if you need to run the engine, always do so in the open air or at least have the rear of the vehicle outside the workplace.

If you are fortunate enough to have the use of an inspection pit, never drain or pour petrol, and never run the engine, while the vehicle is standing over it; the fumes, being heavier than air, will concentrate in the pit with possibly lethal results.

The battery

Never cause a spark, or allow a naked light, near the vehicle's battery. It will normally be giving off a certain amount of hydrogen gas, which is highly explosive.

Always disconnect the battery earth (ground) terminal before working on the fuel or electrical systems.

If possible, loosen the filler plugs or cover when charging the battery from an external source. Do not charge at an excessive rate or the battery may burst.

Take care when topping up and when carrying the battery. The acid electrolyte, even when diluted, is very corrosive and should not be allowed to contact the eyes or skin.

If you ever need to prepare electrolyte yourself, always add the acid slowly to the water, and never the other way round. Protect against splashes by wearing rubber gloves and goggles.

When jump starting a car using a booster battery, for negative earth (ground) vehicles, connect the jump leads in the following sequence: First connect one jump lead between the positive (+) terminals of the two batteries. Then connect the other jump lead first to the negative (–) terminal of the booster battery, and then to a good earthing (ground) point on the vehicle to be started, at least 18 in (45 cm) from the battery if possible. Ensure that hands and jump leads are clear of any moving parts, and that the two vehicles do not touch. Disconnect the leads in the reverse order.

Mains electricity and electrical equipment

When using an electric power tool, inspection light etc, always ensure that the appliance is correctly connected to its plug and that, where necessary, it is properly earthed (grounded). Do not use such appliances in damp conditions and, again, beware of creating a spark or applying excessive heat in the vicinity of fuel or fuel vapour. Also ensure that the appliances meet the relevant national safety standards.

Ignition HT voltage

A severe electric shock can result from touching certain parts of the ignition system, such as the HT leads, when the engine is running or being cranked, particularly if components are damp or the insulation is defective. Where an electronic ignition system is fitted, the HT voltage is much higher and could prove fatal.

Conversion factors

Length (distance)
Inches (in)	X 25.4	= Millimetres (mm)	X 0.0394	= Inches (in)
Feet (ft)	X 0.305	= Metres (m)	X 3.281	= Feet (ft)
Miles	X 1.609	= Kilometres (km)	X 0.621	= Miles

Volume (capacity)
Cubic inches (cu in; in³)	X 16.387	= Cubic centimetres (cc; cm³)	X 0.061	= Cubic inches (cu in; in³)
Imperial pints (Imp pt)	X 0.568	= Litres (l)	X 1.76	= Imperial pints (Imp pt)
Imperial quarts (Imp qt)	X 1.137	= Litres (l)	X 0.88	= Imperial quarts (Imp qt)
Imperial quarts (Imp qt)	X 1.201	= US quarts (US qt)	X 0.833	= Imperial quarts (Imp qt)
US quarts (US qt)	X 0.946	= Litres (l)	X 1.057	= US quarts (US qt)
Imperial gallons (Imp gal)	X 4.546	= Litres (l)	X 0.22	= Imperial gallons (Imp gal)
Imperial gallons (Imp gal)	X 1.201	= US gallons (US gal)	X 0.833	= Imperial gallons (Imp gal)
US gallons (US gal)	X 3.785	= Litres (l)	X 0.264	= US gallons (US gal)

Mass (weight)
Ounces (oz)	X 28.35	= Grams (g)	X 0.035	= Ounces (oz)
Pounds (lb)	X 0.454	= Kilograms (kg)	X 2.205	= Pounds (lb)

Force
Ounces-force (ozf; oz)	X 0.278	= Newtons (N)	X 3.6	= Ounces-force (ozf; oz)
Pounds-force (lbf; lb)	X 4.448	= Newtons (N)	X 0.225	= Pounds-force (lbf; lb)
Newtons (N)	X 0.1	= Kilograms-force (kgf; kg)	X 9.81	= Newtons (N)

Pressure
Pounds-force per square inch (psi; lbf/in²; lb/in²)	X 0.070	= Kilograms-force per square centimetre (kgf/cm²; kg/cm²)	X 14.223	= Pounds-force per square inch (psi; lbf/in²; lb/in²)
Pounds-force per square inch (psi; lbf/in²; lb/in²)	X 0.068	= Atmospheres (atm)	X 14.696	= Pounds-force per square inch (psi; lbf/in²; lb/in²)
Pounds-force per square inch (psi; lbf/in²; lb/in²)	X 0.069	= Bars	X 14.5	= Pounds-force per square inch (psi; lbf/in²; lb/in²)
Pounds-force per square inch (psi; lbf/in²; lb/in²)	X 6.895	= Kilopascals (kPa)	X 0.145	= Pounds-force per square inch (psi; lbf/in²; lb/in²)
Kilopascals (kPa)	X 0.01	= Kilograms-force per square centimetre (kgf/cm²; kg/cm²)	X 98.1	= Kilopascals (kPa)
Millibar (mbar)	X 100	= Pascals (Pa)	X 0.01	= Millibar (mbar)
Millibar (mbar)	X 0.0145	= Pounds-force per square inch (psi; lbf/in²; lb/in²)	X 68.947	= Millibar (mbar)
Millibar (mbar)	X 0.75	= Millimetres of mercury (mmHg)	X 1.333	= Millibar (mbar)
Millibar (mbar)	X 0.401	= Inches of water (inH₂O)	X 2.491	= Millibar (mbar)
Millimetres of mercury (mmHg)	X 0.535	= Inches of water (inH₂O)	X 1.868	= Millimetres of mercury (mmHg)
Inches of water (inH₂O)	X 0.036	= Pounds-force per square inch (psi; lbf/in²; lb/in²)	X 27.68	= Inches of water (inH₂O)

Torque (moment of force)
Pounds-force inches (lbf in; lb in)	X 1.152	= Kilograms-force centimetre (kgf cm; kg cm)	X 0.868	= Pounds-force inches (lbf in; lb in)
Pounds-force inches (lbf in; lb in)	X 0.113	= Newton metres (Nm)	X 8.85	= Pounds-force inches (lbf in; lb in)
Pounds-force inches (lbf in; lb in)	X 0.083	= Pounds-force feet (lbf ft; lb ft)	X 12	= Pounds-force inches (lbf in; lb in)
Pounds-force feet (lbf ft; lb ft)	X 0.138	= Kilograms-force metres (kgf m; kg m)	X 7.233	= Pounds-force feet (lbf ft; lb ft)
Pounds-force feet (lbf ft; lb ft)	X 1.356	= Newton metres (Nm)	X 0.738	= Pounds-force feet (lbf ft; lb ft)
Newton metres (Nm)	X 0.102	= Kilograms-force metres (kgf m; kg m)	X 9.804	= Newton metres (Nm)

Power
Horsepower (hp)	X 745.7	= Watts (W)	X 0.0013	= Horsepower (hp)

Velocity (speed)
Miles per hour (miles/hr; mph)	X 1.609	= Kilometres per hour (km/hr; kph)	X 0.621	= Miles per hour (miles/hr; mph)

*Fuel consumption**
Miles per gallon, Imperial (mpg)	X 0.354	= Kilometres per litre (km/l)	X 2.825	= Miles per gallon, Imperial (mpg)
Miles per gallon, US (mpg)	X 0.425	= Kilometres per litre (km/l)	X 2.352	= Miles per gallon, US (mpg)

Temperature

Degrees Fahrenheit = (°C x 1.8) + 32   Degrees Celsius (Degrees Centigrade; °C) = (°F - 32) x 0.56

*It is common practice to convert from miles per gallon (mpg) to litres/100 kilometres (l/100km), where mpg (Imperial) x l/100 km = 282 and mpg (US) x l/100 km = 235

Index

A

About this manual – 5
Air cleaner
    adjustment (except Riva models) – 170
    description (Riva models) – 170
Air cleaner and element
    removal and refitting – 45
Alternator
    brushes renewal – 109
    description – 107, 183
    fault diagnosis and repair – 109
    maintenance, removal, refitting and special procedures – 107
Antifreeze and corrosion inhibitors – 43
Anti-roll bar (front suspension)
    removal and refitting – 140
Auxiliary driveshaft
    removal and refitting – 23, 169

B

Balljoints (front suspension) – 139
Balljoints (steering) – 146
Battery
    charging – 107
    electrolyte replenishment and testing – 107
    maintenance and inspection – 106, 186
    removal and refitting – 106
Big-end bearings
    examination and renovation – 26
    removal – 23
Bodywork and underframe – 149 *et seq*, 204 *et seq*
Bodywork and underframe
    centre console – 209
    damage repair – 149, 150, 151
    description – 149
    doors – 152 to 154, 205
    facia panel – 151, 205
    grab rails – 210
    maintenance
        bodywork and underframe – 149
        interior – 149
        locks and hinges – 151
        vinyl roof covering – 149
        plastic components – 212
        radiator grille – 151, 210
        rear window and fixed side windows (Estate) – 151
    repair sequence (colour) – *see colour pages between pages 32 and 33*
    tailgate – 155, 212
    windscreen – 151
Bonnet
    adjustment and removal – 155
    strut (later models) – 212
Boot lid adjustment and removal – 155
Braking system – 93 *et seq*, 180 *et seq*
Braking system
    bleeding the hydraulic system – 101, 182
    description – 93
    fault diagnosis – 104
    flexible brake hoses inspection, removal and refitting – 96
    fluid warning switch checking – 180
    front disc brakes – 96, 97, 104
    handbrake – 101, 102, 182
    maintenance – 93
    master cylinder – 98
    pedal – 102, 182
    rear drum brakes – 94, 96, 99, 104, 180
    rigid brake lines inspection, removal and refitting – 96
    servo unit – 103, 182
    specifications – 93, 162
    torque wrench settings – 93, 162, 164
Bulbs *also see* **Light units**
    specifications – 106, 162
Bumpers
    removal and refitting – 151, 204

C

Camshaft
    examination and renovation – 27
    refitting – 32
    removal – 20
Camshaft front oil seal renewal (Riva models) – 169
Capacities, general – 218
Carburettor
    adjustment – 51
    description – 47
    dismantling, inspection and reassembly – 50
    removal and refitting – 49
    specifications – 44, 45
Carburettor (Ozone)
    description – 170
    idle speed and mixture adjustment – 170
    overhaul and adjustments – 172
    specifications – 160, 161
Choke
    adjustment – 50
Clutch – 61 *et seq*, 175
Clutch
    adjustment – 62
    bleeding the hydraulic system – 63
    description – 61
    fault diagnosis – 66
    inspection and renovation – 65, 175
    master cylinder – 63
    pedal – 63
    refitting – 65
    release bearing removal and refitting – 65
    release lever removal and refitting – 175
    removal – 64
    slave (operating) cylinder – 64
    specifications – 61, 161
    torque wrench setting – 61, 161
Coil spring
    removal and refitting
        front suspension – 136
        rear suspension – 140

Index

Condenser
   removal, testing and refitting – 57
Connecting rods
   examination and renovation – 27
   reassembly – 30
   refitting – 31
   removal and dismantling – 24
Connecting rods/big-end bearings
   refitting to crankshaft – 31
Contact breaker points
   adjustment – 56
   removal and refitting – 56
Control arm (front suspension), lower and upper
   overhaul, removal and refitting – 139
Conversion factors – 221
Cooling fan, electric
   removal and refitting – 43, 170
Cooling system – 38 *et seq*, 170
Cooling system
   antifreeze and corrosion inhibitors – 43
   description – 38
   draining, flushing and filling – 39, 170
   drivebelt – 42
   expansion tank – 42
   fault diagnosis – 43
   radiator – 39
   specifications – 38, 160
   thermostat – 40
   water pump – 41
   water temperature gauge – 42
Crankcase ventilation system, closed circuit – 54
Crankshaft
   examination and renovation – 26
   rear oil seal removal – 23
   refitting – 29
   removal – 23
Crossmember (front suspension)
   removal and refitting – 140
Cylinder block
   dismantling – 23
Cylinder bores
   examination and renovation – 26
Cylinder head
   decarbonisation – 28
   dismantling – 20
   refitting – 32
   removal
      engine in car – 20
      engine out of car – 20
   removal and reassembly (Riva models) – 169

D

Decarbonisation – 28
Differential carrier – 88, 89
Dimensions, vehicle – 218
Direction indicator
   fault diagnosis – 115
   switch removal and refitting – 115
Distributor
   description – 173
   inspection and overhaul – 57
   refitting – 35, 57
   removal – 57
Distributor, oil pump and fuel pump drive
   examination and renovation – 28
Doors
   checking and adjustment – 152
   dismantling and reassembly (Riva models) – 205
   removal and dismantling – 153
Drivebelt
   adjustment – 42
   removal and refitting – 42

E

Electrical system – 105 *et seq*, 183 *et seq*
Electrical system
   alternator – 107, 109, 183
   battery – 106, 107, 186
   description – 106
   direction indicators – 115
   fault diagnosis – 125
   fuses – 105, 112, 162, 184
   headlights – 115, 116, 184 to 187
   horns – 115
   lamp bulbs – 116 to 121, 185, 187
   instrument panel (Riva models) – 187
   radios and tape players – 122 to 125
   relays – 122, 184
   specifications – 105, 106, 162
   speedometer cable – 122
   starter motor – 109 to 112
   switches – 115, 122, 174
   voltage regulator – 112
   washer pump – 184
   windscreen wipers – 113, 115
   wiring diagrams – 128 to 133, 189 to 199
Electronically-controlled idle circuit checking – 171
Engine – 13 *et seq*, 166 *et seq*
Engine
   auxiliary driveshaft – 23
   big-end bearings – 23, 26, 31
   camshaft and housing – 20, 27, 32
   connecting rods – 24, 27, 30, 31
   crankshaft – 23, 26, 29
   cylinder block – 23
   cylinder bores – 26
   cylinder head – 20, 28, 32
   decarbonisation – 28
   description – 16
   dismantling – 19
   distributor, oil pump and fuel pump drive – 28
   examination and renovation – 26
   fault diagnosis – 37
   final assembly – 36
   firing order – 13
   flywheel – 23, 28, 32
   lubrication system – 24
   main bearing – 23, 26
   oil filter cartridge – 25
   oil pump – 24, 32
   operations possible with engine in car – 16
   operations requiring engine removal – 16
   piston rings – 24, 27, 31
   pistons – 24, 27, 30, 31
   reassembly – 29
   refitting – 36
   removal – 16
   removal ancillary components – 20
   rockers – 20
   specifications – 13 to 16, 160
   starter ring gear – 28
   start-up after overhaul – 37
   sump – 32
   timing chain – 21, 32
   torque wrench setting – 16, 160
   valve guides – 28
   valve seats – 27
   valves – 20, 27, 32, 34
Engine (1300 cc Riva)
   auxiliary driveshaft – 169
   camshaft front oil seal – 169
   cylinder head – 169
   description – 166
   specifications – 160
   timing belt – 166
   timing cover oil seals – 167

Index

Exhaust system – 52
Expansion tank – 42

F

Facia panel removal and refitting – 151, 205
Fault diagnosis – 214 et seq
Fault diagnosis
    braking system – 104
    clutch – 66
    cooling system – 43
    electrical system – 125, 214
    engine – 37, 216
    fuel system – 54
    gearbox – 80
    ignition system – 59
    propeller shaft – 84
    rear axle – 92
    suspension and steering – 148
Firing order – 13
Flywheel
    examination and renovation – 28
    refitting – 32
    removal – 23
Front disc brakes
    caliper unit
        dismantling and reassembly – 97
        removal and refitting – 97
    disc examination and renovation – 104
    pads renewal – 96
Fuel and exhaust systems – 44 et seq, 170 et seq
Fuel gauge sender unit
    fault finding – 51
Fuel pipes and lines inspection – 51
Fuel pump
    dismantling, inspection and reassembly – 46
    removal and refitting – 46
    testing – 46
Fuel system
    accelerator linkage – 173
    air cleaner and element – 45, 170
    carburettor – 47 to 51, 170, 172
    choke – 50
    closed circuit crankcase ventilation system – 54
    clutch (cold start) control cable renewal – 173
    description – 45
    electronically-controlled idle circuit – 172
    fault diagnosis – 54
    specification – 44 to 45, 160, 161
Fuel tank
    cleaning and repair – 51
    removal and refitting – 51
Fuses
    general – 112
    Riva models – 162, 184
    specifications – 105, 163

G

Gearbox – 67 et seq, 176
Gearbox
    components inspection and renewal – 73
    description – 67
    dismantling – 69
    fault diagnosis – 80
    mainshaft
        dismantling – 71
        reassembly – 73
    reassembly – 73
    removal and refitting – 68
    specifications – 67, 161
    speedometer drive gear – 78
    torque wrench settings – 67, 161

Gearbox (five-speed)
    description – 176
    inspection and renewal of components – 176
    mainshaft dismantling and reassembly – 176
    overhaul – 176
    reassembly – 176
    removal and refitting – 176

H

Handbrake
    adjustment – 101
    cable renewal – 101
    lever assembly removal and refitting – 102
    warning switch removal and refitting – 182
Headlights
    adjustment, removal and refitting – 116
    bulb renewal – 116, 185
    dim-dip system – 187
    hydraulic adjuster (Riva models) – 186
    removal and refitting (Riva models) – 187
    switch – 115
    wiper motor description, removal and refitting – 184
Heating and ventilation system
    description – 155, 212
    removal, overhaul and refitting – 155
    unit dismantling and reassembly (Riva models) – 212
    unit removal and refitting (Riva models) – 212
High tension leads – 58
Horns and horn switch – 115
Hubs, front
    bearings adjustment and lubrication – 138
    overhaul – 138
Hydraulic system, bleeding
    brakes – 101, 182
    clutch – 63

I

Ignition switch (with steering lock) – 122, 174
Ignition system – 55 et seq, 173 et seq
Ignition system
    condenser – 57
    contact breaker points – 56
    description – 55
    distributor – 57, 173
    fault diagnosis – 60
    spark plugs and high tension leads – 58, 60
    specifications – 55, 161
    timing – 58, 174
    torque wrench setting – 55
Ignition warning lamp – 184
Instrument panel (Riva) removal and refitting – 186
Introduction to the Lada – 5

J

Jacking – 9

L

Light units
    removal, refitting and bulb renewal
        boot – 120
        engine compartment – 120
        engine compartment – 120
        flasher side repeater – 120
        front direction indicator (Riva models) – 185
        front parking (Riva models) – 185
        front/side and flasher light assemblies – 118
        glove compartment – 120
        headlight – 116, 185, 187

Index

instrument cluster – 120
interior – 120
number plate (Riva models) – 185
passenger compartment – 120
rear lamp cluster (Riva models) – 185
rear light assemblies – 118
Lubricants and fluids, recommended – 10
Lubrication chart – 10
Lubrication system, engine – 24

M

Main bearings
examination and renovation – 26
removal – 23
Maintenance, routine – 11 *et seq*, 163, 166
Maintenance
bodywork and underframe – 149, 151
braking system – 93
schedules – 11, 163, 165
suspension and steering – 133
Manifolds – 52
Master cylinder
braking system
dismantling and reassembly – 98
removal and refitting – 98
clutch: removal, dismantling and reassembly – 63
Mirror, remote control exterior
except Riva models – 210
Riva models – 210

O

Oil filter cartridge
renewal – 25
Oil pump
refitting – 32
removal, dismantling and inspection – 24

P

Pedal
brake
adjustment – 103, 182
removal, renovation and refitting – 102
clutch
adjustment – 63
removal and refitting – 63, 101
Piston/connecting rod assemblies
refitting – 31
removal and dismantling – 24
Piston rings
examination and renovation – 27
refitting – 31
removal and dismantling – 24
Pistons
examination and renovation – 27
reassembly – 30
Propeller shaft – 81 *et seq*, 179
Propeller shaft
balancing – 83
centre bearing removal and refitting – 83
description – 81
fault diagnosis – 84
flexible coupling removal and refitting – 83
removal and refitting – 81
specifications – 81, 161
torque wrench settings – 81
universal joints – 82, 179

R

Radiator
removal, inspection, cleaning and refitting – 39
Radiator grille
removal and refitting – 152, 210
Radios and tape players
fitting – 122
suppression of interference – 123
Rear axle – 85 *et seq*, 179
Rear axle
assembly, removal and refitting – 88
axle shaft bearings
checking – 179
renewal – 85
axle shafts and oil seals removal and refitting – 85
description – 85
differential carrier
overhaul – 89
removal and refitting – 88
fault diagnosis – 92
locating arms removal, overhaul and refitting – 143
pinion oil seal renewal – 85, 180
specifications – 85, 162
torque wrench settings – 85
Rear drum brakes
adjustment – 94
drum
examination and renovation – 104
removal – 175
pressure compensator valve – 99, 182
shoes renewal – 96, 180
wheel cylinder removal, overhaul and refitting – 96
Rear drum brakes (Riva and later models)
description – 180
shoes renewal – 180
wheel cylinder removal, refitting and overhaul – 181
Rear window and fixed side windows (Estate)
removal and refitting – 151
Relays – 122, 184
Repair procedures, general – 219
Rockers
removal – 20
Routine maintenance *see* **Maintenance, routine**

S

Safety precautions – 220
Seat belts – 210
Seats removal and refitting – 209
Servo unit (braking system)
adjustment and checking – 182
description – 103
removal and refitting – 104
Shock absorber
front: removal, testing and refitting – 135
rear: removal and refitting – 140
Slave (operating) cylinder, clutch
removal, dismantling and reassembly – 64
Spare parts
buying – 6
to carry in vehicle – 215
Spark plugs
conditions (colour chart) – *see colour pages between pages 32 and 33*
general – 58
Speedometer
cable – 122
drive gear – 78
Starter motor
circuit testing – 109
description – 109

Index

dismantling, repair and reassembly – 110
drive pinion inspection and repair – 112
removal and refitting – 110
Starter ring gear
    examination and renovation – 28
Steering
    box
        dismantling, examination and reassembly – 145
        overhaul (Riva models) – 203
        removal and refitting – 110
        removal and refitting (Riva models) – 200
    column and shaft
        description (Riva models) – 200
        overhaul (Riva models) – 200
        removal and refitting – 143, 200
    front wheel alignment – 146
    gear adjustment – 146
    idler arm
        removal and refitting – 146
    rods and balljoints – 146
    specifications – 134, 162
    torque wrench settings – 135, 163
    wheel removal and refitting – 143
Sump
    refitting – 32
Supplement: revisions/information on later models – 159 *et seq*
Suspension and steering – 134 *et seq*, 200 *et seq*
Suspension and steering
    description and maintenance – 135
    fault diagnosis – 148
Suspension (front)
    anti-roll bar – 140
    balljoints – 139
    coil spring – 136
    crossmember – 140
    hubs – 138
    lower control arm – 139
    shock absorber – 135
    specifications – 134
    torque wrench settings – 135, 163
    upper control arm – 139
Suspension (rear)
    axle locating arms – 143
    coil spring – 140
    shock absorber – 140
    specifications – 134, 162
    torque wrench settings – 135
Switches
    removal and refitting
        direction indicator – 115
        headlights – 115
        horn – 115
        ignition (with steering lock) – 122, 174
        windscreen wiper – 115

T

Tailgate
    removal and refitting – 155, 212
Tape players – 122, 123
Thermostat
    removal, testing and refitting – 40
Timing belt (Riva models)
    renewal – 166
Timing chain and sprockets
    refitting – 21, 32
    removal – 21
Timing cover oil seals renewal (Riva models) – 167

Timing (ignition) – 58, 174
Tools
    general – 7
    to carry in vehicles – 215
Towing – 9
Tyres
    general – 148
    maintenance – 203
    pressures – 135
    size – 134, 163

U

Universal joints
    dismantling, inspection, repair and reassembly – 82, 179
Unleaded fuel – 173

V

Valve guides
    examination and renovation – 28
Valve seats
    examination and renovation – 27
Valves
    clearances adjustment – 34
    examination and renovation – 27
    refitting to cylinder head – 32
    removal – 20
Vehicle identification numbers – 6
Voltage regulator
    description – 112
    maintenance and renewal – 112

W

Washer pump (electric)
    description, removal and refitting – 184
Washer pump (foot-operated)
    description – 184
Water pump
    dismantling and overhaul – 41
    removal and refitting – 41
Water temperature gauge
    fault finding – 42
Weights, Kerb – 218
Wheels
    alignment (front wheel) – 146
    general – 148
    maintenance – 203
    size – 134, 162
Windscreen
    removal and refitting – 151
Windscreen wipers
    fault diagnosis – 113
    motor and linkage
        dismantling and reassembly – 113
        removal and refitting – 113
    switches removal and refitting – 115
Wiring diagrams – 128 to 133, 189 to 199
Working facilities – 8

Printed by
J H Haynes & Co Ltd
Sparkford Nr Yeovil
Somerset BA22 7JJ England